Reinforced Concrete with FRP Bars

Mechanics and Design

Reinforced Concrete with FRP Bars

Mechanics and Design

Antonio Nanni
Antonio De Luca
Hany Jawaheri Zadeh

CRC Press
Taylor & Francis Group
Boca Raton London New York

CRC Press is an imprint of the
Taylor & Francis Group, an **informa** business

A SPON PRESS BOOK

CRC Press
Taylor & Francis Group
6000 Broken Sound Parkway NW, Suite 300
Boca Raton, FL 33487-2742

First issued in paperback 2019

ISBN-13: 978-0-415-77882-4 (hbk)
ISBN-13: 978-0-367-86499-6 (pbk)

Library of Congress Cataloging-in-Publication Data

Nanni, Antonio.
    Reinforced concrete with FRP bars : mechanics and design / Antonio Nanni, Antonio De Luca, and Hany Jawaheri Zadeh.
        pages cm
    "A CRC title."
    Includes bibliographical references and index.
    ISBN 978-0-415-77882-4 (alk. paper)
    1. Reinforced concrete. 2. Reinforcing bars. 3. Fiber-reinforced concrete. 4. Polymer-impregnated concrete. I. De Luca, Antonio (Civil engineer) II. Jawaheri Zadeh, Hany. III. Title. IV. Title: Reinforced concrete with fiber-reinforced polymer bars.

TA444.N28 2014
620.1'37--dc23   2013039475

Visit the Taylor & Francis Web site at
http://www.taylorandfrancis.com

and the CRC Press Web site at
http://www.crcpress.com

To Our Families—

Near and Afar

Contents

8 Design of a two-way slab   303

Preface

After 22 years since the formation of American Concrete Institute (ACI) Committee 440 and almost half a century of research endeavors, fiber-reinforced polymer (FRP) reinforcement for concrete members is about to see full market acceptance and implementation. ACI Committee 440 has recently started the effort to create a mandatory-language design code that, in addition to other ACI reports, guides, and specifications, and ASTM test methods and material specifications, will be the instrument for this takeoff not just in North America but all over the world. For practitioners and owners, the primary motivation for the use of FRP is the need to improve the durability of concrete structures.

This book is mainly intended for practitioners and focuses on ACI technical literature covering the fundamentals of performance and design of concrete members with FRP reinforcement and reinforcement detailing. Graduate students and researchers can use it as a valuable resource to guide their studies and creative work. The book covers only internal, nonprestressed FRP reinforcement and excludes prestressing and near-surface-mounted reinforcement applications. It is assumed that the reader already has familiarity with concrete as a material and reinforced concrete as a construction technology (i.e., fabrication, analysis, and design). The book is divided into parts that follow the typical approach to design of conventional reinforced concrete.

PART I—MATERIALS AND TEST METHODS

Chapter 1 deals with the historical background and the state of the art in research worldwide. Reference is made to existing design guides and significant institutional-type literature. Some considerations are provided on limitations in use that are primarily due to a lack of experience rather than engineering. The chapter closes with an illustration of relevant completed projects.

Chapter 2 informs the reader about the characteristics and peculiarities of FRP constituents. Following the spirit of the book, the chapter is limited to the items of primary interest to a designer/practitioner and reference is made to more exhaustive literature on the subject. Attention is devoted to issues regarding testing and quality control as needed for the execution of field projects. Different forms of internal FRP reinforcement are mentioned.

Chapter 3 describes available test methods necessary for the determination of the mechanical and physical properties of FRP bars with reference made to more exhaustive literature and available American Society for Testing and Materials (ASTM International) standards. Attention is devoted to issues regarding testing and quality control as needed for the execution of field projects.

PART 2—ANALYSIS AND DESIGN

Chapter 4 covers flexural members and provides a detailed explanation of flexural and shear behavior. Types of members covered are slabs (one-way and two-way), footings, and beams. Emphasis is placed on structural reliability and the derivation of the strength-reduction factors. The examples shown in this chapter are only provided for clarification, while more exhaustive design examples are given in Part 3. A section on torsion completes the chapter.

Chapter 5 covers members subject to combined axial force and bending moment. This chapter lays the foundation for the acceptance of FRP reinforcement in column-type members, a topic presently ignored by existing design guides. Similarly to Chapter 4, the reader is referred to Part 3 for an exhaustive design example. The chapter covers rectangular and circular cross-section columns and shear walls.

PART 3—DESIGN EXAMPLES

Taking a two-story medical facility building as the case study, Part 3 deals with the design of slabs on the second floor (i.e., Chapter 6 for one-way and Chapter 8 for two-way), internal beams (i.e., Chapter 7), column of the first story (i.e., Chapter 9), and isolated column footing (i.e., Chapter 10). It was decided to show the practical implications of design on the key members of a building through the use of Mathcad©. With this powerful computational software, mathematical expressions are created and manipulated in the same graphical format as they are presented so that the reader can easily comprehend the design flow and use the solved examples as a template for real projects.

The idea of this book started many years ago with university students and industry colleagues with the goal of facilitating the implementation of FRP reinforcement in construction and disseminating the experience gathered in the laboratory and numerous field applications. Among the many individuals who directly and indirectly contributed, we must single out the following for a special thank you: Doug Gremel, Fabio Matta, and Renato Parretti.

About the authors

Antonio Nanni, PhD, PE, FACI, FASCE, FIIFC, is a structural engineer interested in construction materials, their structural performance, and field application. His specific interests are civil infrastructure sustainability and renewal. In the past 28 years, he has obtained experience in concrete- and advanced composite-based systems as the principal investigator of projects sponsored by federal and state agencies, and private industry. Over the course of this time, his constant efforts in materials research have impacted the work of several ACI committees such as 325, 437, 440, 544, 549, and 562. Dr. Nanni has served on the Executive Committee of ASCE Materials Division, is the editor-in-chief of the *ASCE Journal of Materials in Civil Engineering,* and serves on the editorial board of other technical journals. He has advised over 60 graduate students pursuing MSc and PhD degrees and published over 175 and 300 papers in refereed journals and conference proceedings, respectively. Dr. Nanni has maintained a balance between academic and practical experience and has received several awards, including the ASCE 2012 Henry L. Michel Award for Industry Advancement of Research and the Engineering News-Record Award of Excellence for 1997 (Top 25 Newsmakers in Construction). He is a registered PE in Italy, and in the United States in Florida, Pennsylvania, Missouri, and Oklahoma.

Antonio De Luca received his PhD degree in structural engineering from the University of Miami, Coral Gables, Florida. He also earned a BS in civil engineering and an MSc in structural and geotechnical engineering, both from the University of Naples, Federico II, Italy. After completing his PhD, Dr. De Luca had a brief experience in academia working as a postdoctoral associate at the University of Miami. Dr. De Luca's research interests are focused on sustainable material systems and technologies for new construction and rehabilitation. Before joining Simpson Gumpertz & Heger, Dr. De Luca was a graduate engineer for the diagnostics group of Walter P Moore, Inc., Dallas, Texas. In this role, he gained experience with repair and rehabilitation design, structural and architectural assessment, and nondestructive evaluation of concrete structures.

Hany Jawaheri Zadeh obtained his PhD degree in structural engineering from the Department of Civil, Architectural, and Environmental Engineering at the University of Miami, Coral Gables, Florida. He received his BS from Tehran University, Iran, and his MSc from Sharif Institute of Technology, Tehran, Iran. His research interests include the use of composite material systems as internal and external reinforcement.

Part I

Materials and test methods

Chapter 1

Introduction

1.1 BACKGROUND

Plain concrete is strong in compression, but weak in tension. For this reason, it was originally used for simple, massive structures, such as foundations, bridge piers, and heavy walls. Over the second half of the nineteenth century, designers and builders developed the technique of embedding steel bars into concrete members in order to provide additional capacity to resist tensile stresses. This pioneering effort has resulted in what we now call reinforced concrete (RC).

Until a few decades ago, steel bars were practically the only option for reinforcement of concrete structures. The combination of steel bars and concrete is mutually beneficial. Steel bars provide the capacity to resist tensile stresses. Concrete resists compression well and provides a high degree of protection to the reinforcing steel against corrosion as a result of its alkalinity.

Combinations of chlorides (depassivation of steel) and CO_2 (carbonation of concrete) in presence of moisture produce corrosion of the steel reinforcement. This phenomenon causes the deterioration of the concrete and, ultimately, the loss of the usability of the structure [1]. Over the second half of the 1900s, the deterioration of several RC structures due to the chloride-ion induced corrosion of the internal steel reinforcement became a major concern. Various solutions were investigated for applications in aggressive corrosion environments [2]. These included galvanized coatings, electrostatic spray fusion-bonded (powder resin) coatings, and polymer-impregnated concrete epoxy coatings. Eventually, fiber-reinforced polymer (FRP) reinforcing bars were considered as an alternative to steel bars [3,4].

The FRP reinforcing bar became a commercially available viable solution as internal reinforcement for concrete structures in the late 1980s, when the market demand for electromagnetic-transparent (therefore nonferrous) reinforcing bars increased.

1.2 FRP REINFORCEMENT

FRP composites are the latest version of the very old concept of making a better material by combining two different ones. They consist of a reinforcing phase (fibers) embedded into a matrix (polymer). The individual components, fibers and polymer, do not serve the function of structural materials by themselves but do when put together. The fibers provide strength and stiffness to the composite and carry most of the applied load [3] while the resin encapsulates them, thus transferring stresses and providing protection.

After World War II, the need to satisfy aerospace industry demand not met by traditional materials induced researchers and scientists to look for new solutions. The answer was found in developing new material systems by embedding strong fibers into a polymeric matrix. The so-called FRP composite materials offered several advantages with respect to traditional metallic materials. Their innovative properties, such as high tensile strength and modulus, lightness, corrosion resistance, electromagnetic transparency, and the possibility to "engineer" their mechanical properties by changing constituent composition and fiber type and orientation, made FRP composites suitable for a number of applications in different industries [3,5]. The aerospace industry began to use FRP composites as lightweight material with superior strength and stiffness, which reduced the weight of aircraft structures. Later, other industries like naval, defense, and sporting goods started using FRP composites on an extensive basis [5].

FRP reinforcement for concrete structures has been under development since the 1960s in the United States [6] and the 1970s in Europe [7] and Japan [8]. However, it was in the 1980s that the overall level of research, field demonstration, and commercialization became remarkable [9].

FRP reinforcing bars (rebars) are anisotropic. Strength and stiffness of the FRP rebar in the direction of the fibers are significantly affected by the types of fibers and the ratio of the volume of fiber to the overall volume of the FRP. The type of resin affects the failure mechanism and the fracture toughness of the composite. Other factors influencing the properties of FRP rebars are fiber orientation, rate of resin curing, and manufacturing process and its quality control [3,9–11].

Fibers commonly used to make FRP bars are glass, carbon, and aramid. Recently, continuous basalt fibers have become commercially available as an alternative to glass fibers. The matrix bonds and protects the fibers and allows the transfer of stresses from fiber to fiber through shear stresses [3]. Matrices are typically thermosetting resins such as epoxies, polyesters, and vinyl esters. Epoxy is the most common type of matrix material used with carbon fibers. Vinyl ester resins are generally coupled with glass fibers.

The techniques used to manufacture FRP rebars are pultrusion, braiding, and weaving. The typical cross-sectional shape is solid and round, but hollow and other shapes are available. Bars cannot be bent after resin curing: Bends must be incorporated during manufacturing. A bar surface deformation or texture, such as wound fibers, sand coatings, and separately formed deformations, is induced so that mechanical bonding is developed between FRP rebars and concrete. The longitudinal tensile strength of FRP rebars is bar size dependent [11] due to a phenomenon known as "shear-lag." The lower cost-to-performance advantage of glass over carbon fibers makes glass FRP (GFRP) rebars preferable in conventional concrete members. However, for special requirements, carbon FRP (CFRP) rebars may be the ideal choice.

Internal FRP reinforcements are also available in multidimensional shapes [10] with the most common being prefabricated, orthogonal, two-dimensional grids. Multidimensional FRP reinforcements can also be fabricated on-site by hand placement and tying of one-dimensional shapes [9].

To minimize uncertainty in their performance and specification, several standards development organizations have developed consensus-based test methods for the characterization of the short- and long-term mechanical, thermomechanical, and durability properties of FRP reinforcements. The recommended test methods are based on the knowledge gained from research results and literature worldwide. The first document that introduced test methods for FRP rebars was "Recommendation for Design and Construction of Concrete Structures Using Continuous Fiber Reinforcing Materials," which was published in 1997 by the Japan Society for Civil Engineering (JSCE) [12]. ASTM International and the Organization for Standards (ISO) offer standardized test methods related to the use of FRP composites in structural engineering. Model test methods for FRP bars are recommended by the American Concrete Institute (ACI) in document 440.3R, "Guide Test Methods for Fiber-Reinforced Polymers (FRPs) for Reinforcing or Strengthening Concrete and Masonry Structures" [13], effective since 2004. Testing procedures have also been developed by the Canadian Standards Association (CSA).

1.3 FRP REINFORCED CONCRETE

Over the past two decades, laboratory tests have demonstrated that FRP bars can be used successfully and practically as internal reinforcement in concrete structures. The role of industry/university cooperative research became key in transferring the use of internal FRP reinforcement for concrete from the laboratory to the field. To date, reinforcing bars made of FRP have gained acceptance as internal reinforcement in concrete structures.

The mechanical behavior of FRP rebars differs from the behavior of conventional steel rebars. FRP composites are anisotropic, linear, and elastic until failure and are characterized by high tensile strength only in the direction of the fibers [14–17], with no yielding. A flexural concrete member reinforced with FRP rebars generally experiences extensive cracking and large deflections prior to failure, which is, typically, sudden and catastrophic. The shear strength and dowel action of FRP rebars as well as the bond performance are affected by the anisotropic behavior of the bars [3]. Furthermore, the behavior of FRP bars in compression is not as good as the one in tension. Due to the FRP anisotropic and nonhomogeneous nature, the compressive modulus is lower than the tensile one [15,18]. There is still little experience in the use of FRP reinforcement in compression members (columns) and for moment frames or zones where moment redistribution is required [19].

Several global activities have taken place to implement FRP rebars into design codes and guidelines since the 1980s. In the United States, the initiatives and vision of the National Science Foundation and the Federal Highway Administration promoted the development of this technology supporting research at different universities and research institutions [9]. In 1991, the ACI established Committee 440, "FRP Reinforcement." The objective of the committee was to provide the construction industry with science-based design guidelines, construction specifications, and inspection and quality control recommendations related to the use of FRP rebars for concrete structures. In 2001, Committee 440 published the first version of the document "Guide for the Design and Construction of Structural Concrete Reinforced with FRP Bars" [20]. The availability of this document further expedited the adoption of FRP rebars.

While the use of FRP reinforcement in buildings in the United States is within the jurisdiction of ACI, new bridges financed with federal funds have to be designed following the American Association of State Highway and Transportation Officials (AASHTO) load and resistance factor design (LRFD) bridge design specification. The lack of AASHTO limit-state-based specifications covering the design of FRP reinforced concrete bridge deck systems was the last barrier to sanction the acceptance of this innovative and already competitive technology. In 2007, a task force led by researchers, consultants, and representatives from State Departments of Transportation and the US Federal Highway Administration developed LRFD design specifications written in mandatory language. While maintaining the AASHTO provisions for the definition of loads, load factors, and limit states, the document covered specific material properties and detailing of FRP reinforcement, and defined applicable design algorithms and resistance factors. The proposed guide, "AASHTO LRFD Bridge Design Guide Specifications for GFRP-Reinforced Concrete Bridge Decks and Traffic Railings," was approved by the Subcommittee on Bridges and Structures in May 2008 and published in December 2009 [21].

In addition to FIB (Fédération Internationale du Béton) bulletin 40, "FRP Reinforcement in RC Structures," published by the International Federation for Structural Concrete [22], some historical and well-known guidelines specific to FRP-RC available around the world follow [12,13,20,21,23–25]:

Asia
- Japan
  - "Recommendation for Design and Construction of Concrete Structures Using Continuous Fiber Reinforced Materials" (2007), published by JSCE

Europe
- Italy
  - CNR-DT 203/2006 (2006), "Guide for the Design and Construction of Concrete Structures Reinforced with Fiber-Reinforced Polymer Bars," published by the Italian National Research Council (CNR)
- Norway
  - SINTEF Report STF22 A98741, "Modifications to NS3473 When Using Fiber-Reinforced Plastic Reinforcement 2.24" (2002), published by the Norwegian Council for Building Standardization (NBR)
- United Kingdom
  - "Interim Guidance on the Design of Reinforced Concrete Structures Using Fiber Composite Reinforcement" (1999), published by the Institution of Structural Engineers

North America
- Canada
  - CAN/CSA-S806-12 (2002 and 2012), "Design and Construction of Building Structures with Fiber-Reinforced Polymers," published by CSA
  - CAN/CSA-S807-10 (2010), "Specification for Fiber-Reinforced Polymers," published by CSA
  - CAN/CSA-S6-06 (2006) plus CAN/CSA S6S1-10 (2010 Supplement), "Canadian Highway Bridge Design Code," published by CSA
- United States
  - ACI 440.1R (2001 and 2006), "Guide for the Design and Construction of Structural Concrete Reinforced with FRP Bars," published by ACI
  - ACI 440.3R-04 (2004 and 2012), "Guide Test Methods for Fiber-Reinforced Polymers (FRPs) for Reinforcing or Strengthening Concrete Structures," published by ACI
  - "AASHTO LRFD Bridge Design Guide Specifications for GFRP Reinforced Concrete Bridge Decks and Traffic Railings" (2009), published by the American Association of State Highway and Transportation Officials (AASHTO)

1.4 ACCEPTANCE BY BUILDING OFFICIALS

1.4.1 Premise on code adoption

Standardization is the most rigorous consensus process used by public and professional agencies worldwide. It provides the widest input and highest overall quality assurance for a document. In the United States, the standardization process is approved by the American National Standards Institute (ANSI). Documents that go through this process are identified as standards. Standards are written in mandatory language and can be referenced by model codes, authorities having jurisdiction over local building codes, persons or agencies that provide specifications, or in legal documents such as project specifications. There are different types of standards:

- Design standards that are directed to the design professional, not the construction team
- Design specifications that are available for reference in legal documents other than building codes, such as federal government contracts
- Construction standards that are written to direct the producers, testing agencies, and construction teams rather than the design professional
- Construction specifications that are reference documents to be included as part of a contract between an owner and a contractor
- Material specifications that are reference documents to prescribe requirements for materials used in projects are written to the producer, are incorporated by reference in contract documents, and may be incorporated by reference into construction specifications or into contract documents
- Test methods that prescribe means of testing for compliance of materials or construction methods that are proposed for or used in projects—written to the testing agency and may be incorporated by reference in material specifications, construction specifications, or contract documents
- Inspection services specifications that are reference documents written as part of a contract between an owner and an inspection agency
- Testing services specifications that are reference documents written as part of a contract between an owner and a testing agency or between a contractor and a testing agency

For a design standard to become law it must be adopted by a model building code or by a regulatory agency. In the United States (and other parts of the world including the United Nations), the International Building Code (IBC) [26] part of the family of International Codes (I-Codes) is the predominant "model code" (adopted by all 50 states, Puerto Rico, and the US Virgin Islands) and covers the design and construction of new buildings. For current and well-established materials systems and technologies (for example, reinforced concrete), IBC references other standard documents (in the case of the example and for design: ACI 318-11 [27]) de facto, making them part of the model code itself. Once IBC is adopted by a state or other legal jurisdiction, it becomes law and, with it, its referenced standards.

Based on the preceding (except the case when a standard is directly adopted by a jurisdiction), in order for any standard document to have legal status and thus be enforceable by a building official, it must be referenced directly by IBC or any of the other I-Codes. As of today, notwithstanding the availability of guides, test methods, and construction and materials specifications, neither IBC nor any of the I-Codes references FRP reinforcement for concrete, thus making it impossible for a building official to approve the use of FRP without special consideration.

1.4.2 The role of acceptance criteria from ICC-ES

Section 104.11 of IBC [26] (and equivalent ones in the other I-Codes) allows alternative materials by stating that "the provisions of this code are not intended to prevent the installation of any materials or to prohibit any design or method of construction not specifically prescribed by this code, provided that any such alternative has been approved...".

More specifically, Section 104.11.1 of IBC states that a "research report" is the source of information on and the means for building officials' approval for alternative materials: "Supporting data, where necessary to assist in the approval of materials or assemblies not specifically provided for use in this code, shall consist of valid Research Reports from approved sources."

The existence of a set of protocols and provisions is therefore necessary in order to conduct the tests, the analysis of the results, the design, and the installation of the product on which to base the "research report." To this end, ICC Evaluation Services (ICC-ES) develops in partnership with the proposers of new technology-specific documents called "acceptance criteria (AC)" for the purpose of issuing "evaluation (research) reports." Once it is demonstrated that the product is manufactured under an approved quality control program, the research program outlined in the AC is conducted by a certified independent laboratory, its outcomes are evaluated by ICC-ES, and, assuming compliance, a research report is issued. Thus, the alternative material/technology now has official recognition.

Recently, ICC-ES has developed a new document: "AC454-Proposed Acceptance Criteria for Glass Fiber-Reinforced Polymer (GFRP) Bars for Internal Reinforcement of Concrete and Masonry Members" [28]. The purpose of this AC is to establish requirements for GFRP bars to be recognized in an ICC-ES research report IBC and other I-Codes. Basis of recognition is

IBC Section 104.11. The reason for the development of this AC is to provide guidelines for the evaluation of an alternative reinforcement for concrete and masonry structures, where the codes do not provide requirements for testing and determination of physical and mechanical properties of this type of reinforcement product. AC454 applies to deformed GFRP bars used to reinforce concrete and masonry structural elements in cut lengths and bent shapes. Properties evaluated include performance under accelerated environmental exposures, performance under exposure to fire conditions, and structural design procedures.

A summary of tests required by AC454 [28] and their frequency is shown in Table 1.1. The test methods referenced are listed in Table 1.2 and a more detailed discussion of physical, mechanical, and durability properties that they intend to capture is offered in Chapters 2 and 3. In addition, AC454 establishes minimum requirements for some of these properties. It should also be noted that AC454 adopts a clear distinction between nominal cross-sectional area (and diameter) and the measured or "real" values. In fact, the nominal cross-sectional area and the nominal diameter of an equivalent FRP round solid bar are to be used for the purpose of classification and initial design (as the designer does not select a specific product). These two nominal values are to allow the designer to establish a relationship with steel reinforcing bars, thus facilitating initial design and dimensioning.

1.5 APPLICATIONS

FRP rebars are suitable alternatives to steel, epoxy-coated steel, and stainless steel bars in reinforced concrete applications if durability, electromagnetic transparency, or ease of demolition in temporary applications is sought.

The majority of applications (Figures 1.1 through 1.8) utilize FRP rebars to mitigate the risk of corrosion in concrete structures that operate in aggressive marine environments or are exposed to deicing salts. The service life of these types of structures is strictly contingent with the durability of the internal reinforcement. Although their initial cost (raw material and manufacturing costs) and environmental impact (CO_2 emission during the manufacturing process) may be slightly higher than that of conventional steel, the use of FRP rebars in concrete structures subjected to harsh environments generates a significant potential for extending the service life of these structures and lowering their overall life cycle cost [2,24]. Applications of this type include:

- Bridges at sea, retaining/sea walls, ports infrastructure, and dry docks (Figures 1.1 through 1.4)
- Bridge decks and railings where deicing salts are used (Figures 1.5 to 1.7)
- Locks and dam weirs (Figure 1.8)

Table 1.1 Summary of tests and repetitions proposed by AC454

Property	Test or calculation method	No. of repetitions
Physical		
Fiber content	ASTM D2584	For each bar size: total 15 (five from three separate lots)
Glass transition temperature	ASTM E1640	For four available bar sizes: total 10 (five from smallest and largest bar size each) For five or more available bar sizes: total 15 (five from smallest, median, and largest bar size each)
Actual cross-sectional area	ASTM D7205/D7205M	For each bar size: total 15 (five from three separate lots)
Nominal area and diameter	Equivalency with round solid bar sizes no. 2 to 13	N/A
Maximum and minimum cross-sectional dimensions	ISO 17025 calibrated micrometer (reading accuracy to within 1% of the intended measurement)	For each bar size: total 15 (five from three separate lots)
Mechanical		
Tensile strength	ASTM D7205/D7205M	For each bar size: total 15 (five from three separate lots)
Tensile modulus of elasticity	ASTM D7205/D7205M	
Shear strength (perpendicular to the bar)	ASTM D7617	
Ultimate tensile strain	Tensile strength to modulus of elasticity ratio	
Bond strength	ACI 440.3R (B.3)	For four available bar sizes: total 10 (five from smallest and largest bar size each) For five or more available bar sizes: total 15 (five from smallest, median, and largest bar size each)
Durability		
Moisture absorption	ASTM D570 or ASTM D5229/D5229M	For four available bar sizes: total 10 (five from smallest and largest bar size each) For five or more available bar sizes: total 15 (five from smallest, median, and largest bar size each)
Resistance to alkaline environment	ACI 440.3R (B.6) Exposure for 3000 h	
Void content or longitudinal wicking	ASTM D5117	

Continued

Table 1.1 (Continued) Summary of tests and repetitions proposed by AC454

Property	Test or calculation method	No. of repetitions
Bends		
Strength of bend	ACI 440.3R (B.5)	For four available bar sizes: total 10 (five from smallest and largest bar size each) For five or more available bar sizes: total 15 (five from smallest, median, and largest bar size each)

Table 1.2 Test methods cited by AC454

ASTM test methods

ASTM A615/A615M-09 (2012 IBC), -04a (2009 IBC): Standard Specification for Deformed and Plain Carbon Steel, ASTM International

ASTM C904-01 (2006): Standard Terminology Relating to Chemical-Resistant Nonmetallic Materials, ASTM International

ASTM D570-98 (1010): Standard Test Method for Water Absorption of Plastics, ASTM International

ASTM D792-08: Standard Test Methods for Density and Specific Gravity (Relative Density) of Plastics by Displacement

ASTM D2584-11: Test Method for Ignition Loss of Cured Reinforced Resins, ASTM International

ASTM D5117-09: Standard Test Method for Dye Penetration of Solid Fiberglass Reinforced Pultruded Stock, ASTM International

ASTM D5229/D5229M-92(2010): Standard Test Method for Moisture Absorption Properties and Equilibrium Conditioning of Polymer Matrix Composite Materials, ASTM International

ASTM D7205/D7205M-06: Standard Test Method for Tensile Properties of Fiber-Reinforced Polymer Matrix Composite Bars, ASTM International

ASTM D7617/D7617M-11: Standard Test Method for Transverse Shear Strength of Fiber-Reinforced Polymer Matrix Composite Bars, ASTM International

ASTM E1356-08: Standard Test Method for Assignment of the Glass Transition Temperatures by Differential Scanning Calorimetry, ASTM International

ASTM E1640-09: Standard Test Method for Assignment of the Glass Transition Temperature by Dynamic Mechanical Analysis, ASTM International

ACI guide

ACI 440.3R-12: Guide Test Methods for Fiber-Reinforced Polymers (FRPs) for Reinforcing or Strengthening Concrete Structures, American Concrete Institute

B.3: Test method for bond strength of FRP bars by pullout testing

B.5: Test method for strength of FRP bent bars and stirrups at bend locations

B.6: Accelerated test method for alkali resistance of FRP bars (Note: While this document suggests various exposure periods, for the purpose of this document and consistently with AC125, the exposure periods to be considered are 1000 and 3000 hours)

The use of FRP rebars is also particularly attractive for buildings that host equipment sensitive to electromagnetic fields, such as magnetic resonance imaging (MRI) units; for bases of large motors; or for railway systems (Figures 1.9 through 1.11).

Furthermore, FRP reinforcement is the ideal material to reinforce concrete structures temporarily, such as "soft-eyes" that have to be demolished partially by tunnel boring machines (TBMs). The "soft-eye" consists of a reinforcing cage using GFRP bars, which can be easily cut by the TBM (Figures 1.12 through 1.15).

(a)

(b)

Figure 1.1 CFRP grid-reinforced concrete bridge (a) view of completed bridge (insert shows pier reinforcement cage); (b) reinforcement cage for deck (Fukushima Prefecture, Japan).

Figure 1.2 Honopapiilani highway retaining sea wall south (Lahaina, Maui Hawaii).

Figure 1.3 Dowel bars in concrete pavements (Port of Rotterdam, The Netherlands).

Figure 1.4 GFRP reinforced-concrete repair for Pearl Harbor dry docks (Honolulu, Hawaii).

Figure 1.5 GFRP reinforced-concrete bridge deck (Morristown, Vermont).

Figure 1.6 GFRP reinforced-concrete bridge deck (Cookshire-Eaton, Quebec, Canada).

Figure 1.7 GFRP cages prior to casting a bridge railing (Greene county, Missouri).

Figure 1.8 GFRP reinforced-concrete for Ice Harbor lock and dam fish weir (Walla Walla, Washington).

Figure 1.9 GFRP reinforced-concrete slab for MRI rooms in hospital (York, Maine).

Figure 1.10 GFRP reinforced-concrete slab (Oran, Algeria).

Figure 1.11 GFRP reinforced-concrete rail track structure: bars in concrete rail plinths ((a) and (b)), deck bars for segmental precast elements (c), and high voltage pedestals in overhead rail guideway (d) (Miami, Florida).

Figure 1.12 GFRP reinforcement cage for soft-eye construction at a manufacturing plant (Angri, Italy).

Figure 1.13 GFRP reinforced-concrete soft-eye for Washington Dulles International Airport people mover (Dulles, Virginia).

Figure 1.14 GFRP reinforced-concrete soft-eye for Beacon Hill light rail transit (Seattle, Washington).

(a)   (b)

Figure 1.15 GFRP soft-eyes for tunnel excavation (a) and (b) view of bars when TBM emerges (London, UK).

REFERENCES

1. S. Ahmad. Reinforcement corrosion in concrete structures, its monitoring and service life prediction—A review. *Cement and Concrete Composites* 25:459–471 (2003).

2. J. Plecnik and S. H. Ahmad. Transfer of composite technology to design and construction of bridges. Final report to US DOT, contract no. DTRS 5683-C000043 (1988).

3. ACI Committee 440. Report on fiber-reinforced polymer (FRP) reinforcement for concrete structures. ACI 440R -07, American Concrete Institute, Farmington Hills, MI (2007).

4. A. Nanni. Composites: Coming on strong. Concrete Construction 44 (1): 120–124 (1999).

5. A. K. Kaw. *Mechanics of composite materials,* 2nd ed. Boca Raton, FL: Taylor and Francis (2005).

6. L. C. Bank. Composites for construction: Structural design with FRP materials. New York: John Wiley & Sons (2006).

7. L. R. Taerwe and S. Matthys. FRP for concrete construction. *Concrete International* 21 (10): 33–36 (1999).

8. H. Fukuyama. FRP composites in Japan. *Concrete International* 21 (10): 29–32 (1999).

9. C. E. Bakis, L. C. Bank, V. L. Brown, E. Cosenza, J. F. Davalos and J. J. Lesko, et al. Fiber-reinforced polymer composites for construction: State-of-the art review. *ASCE Journal of Composites for Construction* 6 (2): 73–87 (2002).

10. C. E. Bakis. FRP composites: Materials and manufacturing. In *fiber-reinforced-plastic (FRP) reinforcement for concrete structures: Properties and applications. Developments in civil engineering,* ed. A. Nanni, 42, 13–58. Amsterdam: Elsevier (1993).

11. L. C. Bank. Properties of FRP reinforcement for concrete. In fiber-reinforced-plastic (FRP) reinforcement for concrete structures: Properties and applications. Developments in civil engineering, ed. A. Nanni, 42, 59–86. Amsterdam: Elsevier (1993).

12. Japan Society of Civil Engineering (JSCE). Design guidelines of FRP reinforced concrete building structures. Concrete engineering series no. 23, Tokyo, Japan (1997).

13. ACI Committee 440. Guide test methods for fiber-reinforced polymers (FRPs) for reinforcing or strengthening concrete structures. ACI 440.3R-12. American Concrete Institute, Farmington Hills, MI (2012).

14. A. Nanni. *Fiber-reinforced-plastic for concrete structures: Properties and applications.* Amsterdam: Elsevier Science (1993).

15. W. P. Wu. Thermomechanical properties of fiber reinforced plastic (FRP) bars, PhD dissertation, West Virginia University, Morgantown, WV (1990).

16. T. Tamura. FiBRA. In *fiber-reinforced-plastic (FRP) reinforcement for concrete structures: Properties and applications. Developments in civil engineering,* ed. A. Nanni, 42, 291–303. Amsterdam: Elsevier (1993).

17. A. Nanni, S. Rizkalla, C. E. Bakis, J. O. Conrad, and A. A. Abdelrahman. Characterization of GFRP ribbed rod used for reinforced concrete construction.

Proceedings of the International Composites Exhibition (ICE-98), Nashville, TN, pp. 16A/1-6 (1998).

18. P. K. Mallick. *Fiber reinforced composites, materials, manufacturing, and design.* New York: Marcell Dekker, Inc. (1988).

19. A. Mirmiran, W. Yuan, and X. Chen. Design for slenderness in concrete columns internally reinforced with fiber-reinforced polymer bars. *ACI Structural Journal* 98 (1): 116–125 (2001).

20. ACI Committee 440. Guide for the design and construction of structural concrete reinforced with FRP bars. ACI 440.1R-06. American Concrete Institute, Farmington Hills, MI (2006).

21. AASHTO LRFD. Bridge design guide specifications for GFRP-reinforced concrete bridge decks and traffic railings. Washington, DC: American Association of State Highway and Transportation Officials (2009).

22. fib. Bulletin 40, FRP reinforcement in RC structures. International Federation for Structural Concrete (FIB—Fédération Internationale du Béton), Lausanne, Switzerland (2007).

23. CNR Advisory Committee on Technical Recommendations for Construction. Guide for the design and construction of concrete structures with fiber-reinforced polymer bars, CNR-DT 203/2006. National Research Council, Rome, Italy (2006).

24. SINTEF Report STF22 A98741. Modifications to NS3473 when using fiber reinforced plastic reinforcement 2.24. Norwegian Council for Building Standardization, Oslo, Norway (2002).

25. Interim guidance on the design of reinforced concrete structures using fiber composite reinforcement. Institution of Structural Engineers, London, UK (1999).

26. *2012 International existing building code.* International Code Council, Country Club Hills, IL, 294 pp. (2011).

27. ACI Committee 318, 2011. Building code requirements for reinforced concrete. ACI 318-11. American Concrete Institute, Farmington Hills, MI (2006).

28. AC454. Proposed acceptance criteria for glass fiber-reinforced polymer (GFRP) bars for internal reinforcement of concrete and masonry members. ICC Evaluation Service, Inc., Los Angeles, CA (2013).

Chapter 2

Material properties

2.1 INTRODUCTION

Even though other fabrication methods (e.g., braiding) and other fiber-reinforced polymer (FRP) forms (e.g., two-directional grids) are commercially available, this chapter intentionally deals with FRP bar products fabricated by pultrusion. A brief summary of the properties of the most commonly used types of fibers and polymers is provided as available in the ACI Guide 440.1R-06 [1], the *FIB Bulletin* 40 [2], and the Italian CNR Guide [3]. From a reference perspective, the most authoritative and comprehensive source of information on composites is the *Wiley Encyclopedia of Composites* [4].

2.2 FRP BAR

An FRP bar is made of continuous fibers embedded in a matrix made of a polymeric resin. The fibers have the function of carrying the load; the resin has the function of binding together the fibers, transferring the load to the fibers, and protecting the fibers. The fiber and volume fraction significantly affect strength and stiffness of the FRP, while the resin type affects the failure mechanism and the fracture toughness.

An FRP bar is anisotropic and can be manufactured using different techniques such as pultrusion, braiding, and weaving. Other factors influencing the properties of the bar are fiber orientation, rate of resin curing, manufacturing process, and quality control during manufacturing.

2.3 CONSTITUENT MATERIALS: FIBERS AND RESIN MATRICES

In this section, fibers and resin matrices that are most commonly used to manufacture FRP bars are introduced and briefly discussed. Fiber and matrix properties listed here are to be considered as generic.

2.3.1 Fibers

Fibers commonly used are glass, carbon, aramid, and basalt. Glass fibers offer an economical balance between cost and specific strength properties; this makes them preferable to carbon and aramid in most reinforced concrete (RC) applications. Basalt fibers have recently emerged as an alternative to glass fibers. Typical properties of glass, carbon, aramid, and basalt fibers are listed in Table 2.1.

2.3.1.1 Glass fiber

Glass fiber is primarily made from silica sand and is commercially available in different grades. The most common types of glass are electrical (E-glass), high-strength (S-glass), and alkali-resistance (AR-glass). E-glass presents high electrical insulating properties, low susceptibility to moisture, and high mechanical properties. S-glass has higher tensile strength and modulus, but its higher cost makes it less preferable than E-glass. AR-glass is highly resistant to alkali attack in cement-based matrices, but, at the moment, sizings compatible with thermoset resins that are commonly used to pultrude FRP bars are not available. Composites made from glass fiber exhibit good electrical and thermal insulation properties.

2.3.1.2 Carbon fiber

Carbon fiber is made from polyacrylonitrile (PAN), pitch, or rayon fiber precursors. PAN-based carbon fiber is the predominant form used in civil engineering applications. PAN-based carbon fiber presents high strength and relatively high modulus. Pitch-based carbon fiber has higher modulus but lower strength, which makes it suitable for aerospace applications. Rayon and isotropic pitch precursors are used to produce low-modulus carbon fiber. Based on its mechanical properties, carbon fiber can be classified as high modulus and low modulus. Carbon fiber has high fatigue strength, high resistance to alkali or acid attack, a low coefficient of thermal expansion (CTE), relatively low impact resistance, and high electrical conductivity; it can cause galvanic corrosion when in direct contact with metals. Moreover, it is not easily wet by resins; therefore, sizing is necessary before embedding it in the resin. Generally, carbon fiber is about 10 times more expansive than glass fiber and exhibits strength and modulus about three times higher than glass.

2.3.1.3 Aramid fiber

Aramid fiber is an aromatic polyamide organic fiber. It offers good mechanical properties at a low density, high toughness, and high impact

Table 2.1 Typical properties of fibers (single filament)

Type of fiber	Density (lb/yd³)	Tensile strength (ksi)	Tensile modulus (msi)	Ultimate tensile strain (%)	CTE ($10^{-6}/°F$)	Poisson's ratio
E-glass	4215	500	10.5	2.4	0.15	0.22
S-glass	4215	660	12.4	3.3	0.086	0.22
AR-glass	3800	260 ÷ 500	10.1 ÷ 11.0	2.0 ÷ 3.0	N/A	N/A
High-modulus carbon	3290	360 ÷ 580	50.7 ÷ 94.3	0.5	−0.036 ÷ −0.059	0.2
Low-modulus carbon	2950	507	34.8	1.1	−0.018 ÷ −0.059	0.2
Aramid (Kevlar 29)	2428	400	9.0	4.4	−0.059 long. (1.7 radial)	0.35
Aramid (Kevlar 49)	2428	525	18.0	2.2	−0.059 long. (1.7 radial)	0.35
Aramid (Kevlar 149)	2428	500	25.4	1.4	−0.059 long. (1.7 radial)	0.35
Basalt	4720	700	12.9	3.1	0.24	N/A

Notes: 1 lb/yd³ = 0.593 kg/m³; 1 ksi = 6.89 N/mm²; 1 msi = 0.645 m²; 1/°F = 1/(9/5 * °C + 32).

resistance. Aramid fiber is a good insulator of both electricity and heat, and it is resistant to organic solvents, fuels, and lubricants. It is sensitive to ultraviolet (UV) light, high temperature, and high humidity. The tensile strength of aramid fiber is higher than that of glass, and also the modulus is about 50% higher than that of glass. Kevlar is the most common type of aramid fiber and commercially available as Kevlar 29, 49, and 149. High cost limits the use of this type of fiber for manufacturing FRP bars.

2.3.1.4 Basalt fiber

Basalt fiber is slightly stronger and stiffer than E-glass, environmentally safe, nontoxic, noncorrosive, and nonmagnetic, and has high-heat stability and insulating characteristics [5–8]. Although basalt fiber is manufactured with the same technology utilized for E-glass fiber, its production process requires less energy, and the primary raw material (basalt rock) is available all around the world. Basalt fiber may offer the opportunity to engineer its mechanical properties by modification of the chemical composition resulting in a fiber having an elastic modulus higher than that of E-glass and, at the same time, high biosolubility. The latter property represents the capability of basalt fiber to dissolve in the medium- to long term when in contact with biological liquids. Biosolubility is a requirement that recent international directives have introduced in the glass fiber industry and its fulfillment is considered the principal element to guide future market demand [5]. Research is ongoing to investigate the feasibility of basalt fiber for the production of FRP bars.

2.3.2 Matrices

Matrices are commonly thermoset polymeric resins. In their initial form, thermoset resins are usually liquids or low melting-point solids and they are cured with a catalyst and heat, or a combination of the two. Unlike thermoplastic resins, once cured, solid thermoset resins cannot be converted back to their original liquid form or reshaped. The most common thermosetting resins used in the composites industry are epoxies, polyesters, and vinyl esters. Additives and fillers may be mixed with the resin to impart performance improvements, tailor the performance of the composites, and reduce costs. Typical properties of epoxy, polyester, and vinyl ester resins are listed in Table 2.2.

2.3.2.1 Epoxies

The principal advantages of epoxy resin are high mechanical properties, ease of processing, low shrinkage during cure, and good adhesion to

Table 2.2 Typical properties of resin matrices

Type of resin matrix	Density (lb/yd³)	Tensile strength (ksi)	Longitudinal modulus (ksi)	Poisson's ratio	CTE (10⁻⁶/°F)	Moisture content (%)	Glass transition temperature (°F)
Epoxy	2000 ÷ 2400	5 ÷ 15	300 ÷ 500	0.35 ÷ 0.39	1.6 ÷ 3.0	0.15 ÷ 0.60	203 ÷ 347
Polyester	2000 ÷ 2400	7 ÷ 19	400 ÷ 600	0.38 ÷ 0.40	1.3 ÷ 1.9	0.08 ÷ 0.15	158 ÷ 212
Vinyl ester	1900 ÷ 2300	10 ÷ 11	435 ÷ 500	0.36 ÷ 0.39	1.5 ÷ 2.2	0.14 ÷ 0.30	158 ÷ 329

Notes: 1 lb/yd³ = 0.593 kg/m³; 1 ksi = 6.89 N/mm²; 1/°F = 1/(9/5 * °C + 32); °F = (9/5 * °C + 32).

a wide variety of fibers. Epoxies have high corrosion resistance and are less affected by water and heat than other polymeric matrices. Their disadvantages are a high cost and long curing period (a postcuring process is generally required). Epoxy resin can also be formulated with different materials or blended with other resins to achieve specific performance features. Epoxies are primarily used for fabricating high-performance composites with superior mechanical properties, resistance to corrosive liquids and environments, superior electrical properties, and good performance at elevated temperatures. Epoxy resins are compatible with glass, carbon, aramid, and basalt fibers. However, their use in the pultrusion industry is limited.

2.3.2.2 Polyesters

The main advantage of polyester resins is a balance of good mechanical, chemical, and electrical properties; dimensional stability; cost; and ease of processing. Polyester resins are, generally, relatively inexpensive and offer good mechanical and electrical performance. Because polyesters can be chemically tailored to meet the requirements of a wide range of applications, a number of specialty polyesters, which address specific performance such as flexibility, electrical insulation, corrosion resistance, heat and UV light resistance, fire retardancy, and optical translucence, are available. Styrene is usually mixed in large quantities (more than 10% by mass of the polymer resin) to give a low viscosity liquid. Their use in the manufacturing of FRP bars is discouraged because of lower chemical resistance as compared to vinyl esters.

2.3.2.3 Vinyl esters

Vinyl esters exhibit some of the beneficial characteristics of epoxies such as chemical resistance and high strength as well as those properties of polyester such as viscosity and fast curing. Vinyl esters exhibit good alkali resistance and have good wet-out and good adhesion with glass fiber, which makes them the preferred choice to manufacture GFRP (glass FRP) composites.

2.4 MANUFACTURING BY PULTRUSION

The FRP bar is typically manufactured by pultrusion or variations of this process. Pultrusion is a continuous molding process that combines fiber reinforcement and thermosetting resin. This process is ideal for the continuous fabrication of composite parts that have a constant cross-sectional profile such as bars.

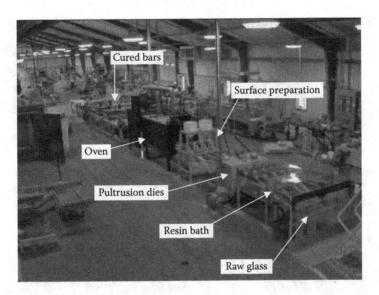

Figure 2.1 Overall view of the production line of GFRP bars.

As an example, the pultrusion process to manufacture a particular type of GFRP (glass fiber-reinforced polymer) bar is illustrated in Figure 2.1 and Figure 2.2. Glass fibers, initially packaged in rovings, are drawn through a resin bath where the material is thoroughly impregnated with a liquid thermoset resin. Before entering the resin bath, fibers are spread out to allow for even wetting. The resin-impregnated fibers are first guided through a metal die that defines the size of the final bar and, then, enter a curing oven. Before entering the oven, sand coating and helicoidal wraps are applied on the surface of the bar. Inside the oven, heat is transferred under precise temperature control to the bar. The heat activates the resin curing, changing it from a liquid to a solid. The solid bar emerges from the curing oven to the exact size of the die cavity. The bar is continuously pulled and, finally, cut to the desired length. The duration of the process varies with the size of the final bar; typically, production speed is 3 ft (0.91 m) per minute.

The surface deformation of the FRP bar is critical to develop bond to concrete. Selected types of surface deformation patterns that are currently commercially available are illustrated in Figure 2.3. GFRP bars are generally produced in sizes ranging from 3/8 to 1 3/8 in. (9 to 41 mm) in diameter (i.e., no. 3 to no. 13). CFRP bars, instead, are typically only available in diameters from 3/8 to 6/8 in. (i.e., no. 3 to no. 6) (9 to 18 mm).

When using thermoset resin, FRP bars cannot be bent after resin curing; bends must be incorporated during manufacturing. In fact, after

(a)   (b)

(c)   (d)

Figure 2.2 Phases of the manufacturing process of GFRP bars: (a) raw glass fibers in rovings; (b) glass fibers drawn through a resin bath for impregnation and a metal die; (c) solid bar emerging from the curing oven; (d) cured bar ready to be cut at the desired length.

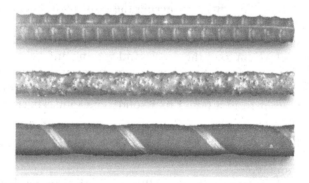

Figure 2.3 Surface deformation patterns for commercially available FRP bars: ribbed (top); sand-coated (middle); and wrapped and sand-coated (bottom).

the resin has passed the liquid state, bending or alteration of the FRP bar is not possible due to the inability of the fibers to reorient within the resin matrix.

2.4.1 Gel time and peak exothermic temperature

To set the rate of the manufacturing process, the gel time and the peak exothermic temperature of the thermoset resin have to be evaluated. The gel time is the time from the initial mixing of the reactants of a thermoset plastic composition to the time when solidification commences, under conditions approximating the conditions of use. The peak exothermic temperature is the maximum temperature reached by a reacting thermoset plastic composition during the curing process. Errors in estimating a correct rate may cause internal defects in the cured material, which would ultimately induce lower durability and mechanical properties of the cured material. Figure 2.4 shows an example of thermocracking due to undercuring of the resin. Gel time and peak exothermic temperature are measured according to the ASTM D 2471, "Standard Test Method for Gel Time and Peak Exothermic Temperature of Reacting of Thermosetting Resins." For this test method, the typical test setup and a sample test output are shown in Figures 2.5(a) and 2.5(b). The bath simulates the oven environment and the resin samples are used to investigate the behavior of the resin in the oven.

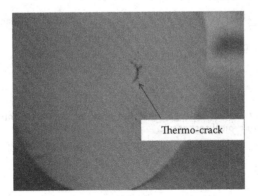

Thermo-crack

Figure 2.4 Thermocracking due to undercuring of the resin.

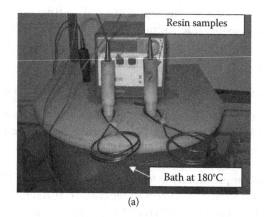

(a)

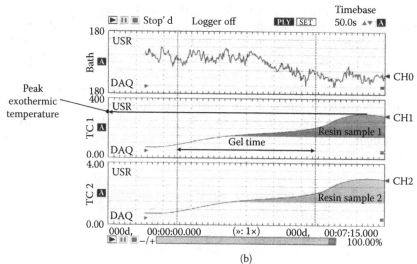

(b)

Figure 2.5 Test method for resin gel time and peak exothermic temperature: (a) test setup; (b) typical test result.

REFERENCES

1. ACI Committee 440. Guide for the design and construction of structural concrete reinforced with FRP bars. ACI 440.IR-06. American Concrete Institute, Farmington Hills, MI (2006).
2. *FIB Bulletin 40*, FRP rainforcement in RC structures. International Federation for Structural Concrete (FIB—Fédération Internationale du Béton), Lausanne, Switzerland (2007).
3. CNR Advisory Committee on Technical Recommendations for Construction. Guide for the design and construction of concrete structures with

fiber- reinforced polymer bars, CNR-DT 203/2006. National Research Council, Rome, Italy (2006).

4. L. Nicolais, A. Borzacchiello, and S. M. Lee. *Wiley encyclopedia of composites,* 2nd ed., Hoboken, NJ: John Wiley & Sons, ISBN (printed set): 978-0-470-12828-2 (2012).

5. T. Deak and T. Czigany. Investigation of basalt fiber reinforced polyamide composites. *Materials Science, Testing and Informatics* IV (589): 7–12 (2008).

6. M. C. Wang, Z. G. Zhang, et al. Chemical durability and mechanical properties of alkali-proof basalt fiber and its reinforced epoxy composites. *Journal of Reinforced Plastics and Composites* 27 (4), 393–407 (2008).

7. J. Sim, C. Park, and D. Y. Moon. Characteristics of basalt fiber as a strengthening material for concrete structure. *Composites:* Part B 36:504–512 (2005).

8. M. Di Ludovico, A. Prota, and G. Manfredi. Structural upgrade using basalt fibers for concrete confinement. *Journal of Composites for Construction* 14 (5): 541–552 (2010).

Chapter 3

FRP bar properties

3.1 PHYSICAL AND MECHANICAL PROPERTIES OF FRP BARS

Typical physical and mechanical properties of three common types of fiber-reinforced polymer (FRP) bars are discussed as follows. From a reference perspective, the most authoritative and comprehensive source of information on physical and mechanical properties of composites is the *Wiley Encyclopedia of Composites* [1].

Density. The density of FRP bars ranges between about one-sixth to one-fourth that of steel, which reduces transportation costs and makes FRP bars easy to handle at the job site [2,3]. Typical values of density of FRP bars are listed in Table 3.1.

Coefficient of thermal expansion (CTE). Longitudinal and transverse CTEs of FRP bars depend on the fiber type, resin, and volume fraction of the constituents [3]. The polymeric matrices and the glass fibers can be considered isotropic, while carbon and aramid fibers are orthotropic. Carbon FRP (CFRP) and aramid FRP (AFRP) bars typically contract in the longitudinal direction when temperature increases and dilate when temperature decreases (negative CTEs in the longitudinal direction). Typical longitudinal and transverse CTEs of FRP bars are listed in Table 3.1.

Tensile behavior. The tensile behavior of FRP bars is characterized by a stress–strain relationship that is linear elastic up to failure [4,5]. If compared to steel bars, FRP bars offer higher tensile strength but lower ultimate tensile strain (no yielding) and lower tensile modulus of elasticity. The tensile strength of an FRP bar varies with its diameter, while the longitudinal modulus does not change appreciably [2,3]. This phenomenon is primarily due to the effects of shear lag [6]. Typical tensile properties of FRP bars are listed in Table 3.1.

Compressive behavior. Testing of FRP bars in compression is complicated by the occurrence of fiber microbuckling due to the anisotropic and nonhomogeneous nature of the FRP material and can lead to inaccurate measurements [6]. Therefore, standard test methods are not well established yet.

Table 3.1 Typical FRP bar properties

Bar type	Density 10^3 lb/yd^3	CTE		Tensile strength ksi	Tensile modulus msi	Rupture strain %
		Long. 10^{-6}/°F	Transv. 10^{-6}/°F			
GFRP	3.63 ÷ 6.11	0.098 ÷ 0.17	0.35 ÷ 0.40	70 ÷ 230	5.1 ÷ 7.4	1.2 ÷ 3.1
CFRP	4.35 ÷ 4.67	−0.19 ÷ 0.0	1.2 ÷ 1.7	87 ÷ 535	15.9 ÷ 84.0	0.5 ÷ 1.7
AFRP	3.63 ÷ 4.11	−0.097 ÷ −0.32	0.99 ÷ 1.3	250 ÷ 368	6.0 ÷ 18.2	1.9 ÷ 4.4

Notes: 1 lb/yd^3 = 0.593 kg/m^3; 1/°F = 1/(9/5*°C + 32); 1 ksi = 6.89 N/mm^2; 1 msi = 0.645 m^2.

The compressive strength reduces by up to 45%, 80%, and 22% with respect to the values in tension for the case of glass FRP (GFRP), AFRP, and CFRP bars, respectively [4,7]. The compressive modulus of elasticity is approximately 80% for GFRP, 85% for CFRP, and 100% for AFRP of the tensile modulus of elasticity for the same product [7,8]. Different modes of failure (transverse tensile failure, fiber microbuckling, or shear failure) may characterize the response of an FRP bar in compression, depending on the type of fiber, fiber volume fraction, and type of resin [4,7,9].

Transverse shear behavior. The behavior of FRP bars under transverse shear loading is mostly influenced by the properties of the matrix. FRP bars are generally weak in transverse shear. The shear strength can be increased by braiding or winding additional fibers in the direction transverse to the longitudinal one. Typical transverse shear strength of FRP bars ranges between 4.3 and 7.3 kilopounds (kip) per square inch (ksi) (30 to 50 MPa) [10].

Creep rupture. When subjected to a sustained tensile load, FRP bars undergo progressive deformations, which may ultimately lead to failure (creep rupture) after a period of time called "endurance time." It should be noted that the term "endurance limit" represents the level of tensile stress below which the element would never fail irrespective of the load duration. The creep-rupture endurance time depends on the ratio of the sustained tensile stress to the short-term strength and the environmental conditions: The higher the sustained stress-to-strength ratio is and the harsher the environmental conditions (high temperature, high exposure to UV light, and high alkalinity) are, the shorter the endurance time is [11–14]. An experimental study was conducted on 0.25 in. (6 mm) diameter smooth carbon, aramid, and glass FRP bars. The bars were tested at different load levels at room temperature in laboratory conditions. Test results showed that the percentage of initial tensile strength retained, linearly extrapolated at the 50-year endurance time, was about 30%, 50%, and 95% for GFRP, AFRP, and CFRP, respectively [11]. Other tests confirmed that CFRP bars exhibit a better behavior under sustained-loads bars compared with GFRP bars [12–14].

Fatigue behavior. Carbon, aramid, and glass fibers are generally not prone to fatigue failure. In the last decades, data have been generated to characterize the fatigue behavior of stand-alone FRP materials. These data, however, were obtained for specimens made for aerospace applications. Results indicated that the plot of the ratio of the stress and the initial static strength versus the logarithm of the number of cycles at failure shows a continued downward trend (per decade of logarithmic life) of about 10%, 5%–6%, and 5%–8% for GFRP, AFRP, and CFRP, respectively [15–20]. The fatigue strength of CFRP bars encased in concrete was observed to be dependent on the moisture, the environmental temperature, the cyclic loading frequencies, and the ratio of maximum-to-minimum cyclic stress [14,21,22]. A recent study [23] showed that full-scale GFRP reinforced concrete bridge deck specimens exhibited better fatigue performance and longer fatigue life compared to their steel counterparts.

Durability. The mechanical properties of FRP bars are influenced by the environment. The presence of water, alkaline or acidic solutions, saline solutions, ultraviolet exposure, and high temperature may affect the tensile and bond properties of FRP bars.

Data from short-term experiments on bare bars (most of the time unstressed) subjected to alkaline environments are available. However, the extrapolation of these data to field conditions and expected lifetimes is difficult [24–26]. High pH degrades the tensile strength and modulus of GFRP bars [27]. The degradation is accelerated by high temperature and long exposure time. The reduction in tensile strength and modulus in GFRP bars (stressed or unstressed) ranges from 0% to 75% and between 0% and 20%, respectively [28–33,35,36]. In the case of AFRP bars (stressed or unstressed), tensile strength and stiffness reduce between 10% and 50%, and 0% and 20% of the initial value, respectively [37–39]. In the case of unstressed CFRP, strength and stiffness have been reported to each decrease between 0% and 20% [39]. Bars embedded in concrete at various temperatures and with good fiber–resin combinations show only limited degradation, which, however, increases with temperature and stress level [28–31,33–35,40,41]. Direct exposure of FRP bars to sunlight (UV rays) and moisture has a detrimental effect on the tensile strength. Strength reduces from 0% to 40% of its initial value in the case of GFRP bars, 0% to 30% for AFRP bars, and 0% to 20% for CFRP bars [34,42]. Although FRP bars embedded in concrete are not exposed to UV while in service, UV light may cause degradation during storage.

A field study conducted by the Intelligent Sensing for Innovative Structures (ISIS) Canada Research Network collected data with respect to the durability of GFRP bars in concrete exposed to natural environments [43,44]. Concrete cores containing GFRP bars were extracted from five selected structures: a 5-year-old harbor wharf and four 6- to 8-year-old reinforced concrete (RC) bridges. The GFRP bars were analyzed for their physical and chemical composition at the microscopic level. The experimental results were compared with the ones obtained from control GFRP bars preserved under controlled laboratory conditions [45]. The results of the analyses showed that there is no degradation of the GFRP bars in the real-life concrete structures. The analyses also indicated:

- No alkali ingress was observed in the GFRP bars from the concrete pore solution.
- The matrix in all GFRP bars was intact and unaltered from its original state.
- Neither hydrolysis nor significant changes in the glass transition temperature of the matrix took place after exposure for 5 to 8 years to the combined effects of the alkaline environment in the concrete and the external natural environment.

Bond-to-concrete behavior. Bond between FRP and concrete depends on the design, manufacturing process, mechanical properties of the bar itself, and the environmental conditions [46–50]. Bond stresses at the FRP bar/concrete interface are transferred by chemical bond (adhesion resistance of the interface), friction, and mechanical interlock due to irregularity of the interface. In the FRP bar, bond stresses are transferred through the resin to the reinforcement fibers. The bond behavior of an FRP bar is, therefore, limited by the shear strength of the resin [51–58].

3.2 TEST METHODS

Due to differences in the physical and mechanical behavior of FRP materials compared to steel, unique test methods for FRP bars are required. Standards development organizations have developed consensus-based test methods for FRP reinforcement for use in structural concrete.

The American Society for Testing and Materials (ASTM) International and the International Organization for Standardization (ISO) offer standardized test methods related to the use of FRP composites in structural engineering. Testing procedures directly have also been developed by the American Concrete Institute (ACI). The document "Guide Test Methods for Fiber-Reinforced Polymer (FRP) for Reinforcing or Strengthening Concrete and Masonry Structures" [59], prepared by the ACI Committee 440, provides model test methods for the short-term and long-term mechanical, thermomechanical, and durability testing of FRP bars not yet standardized by ASTM. These recommended test methods are based on the knowledge gained from research results and literature worldwide.

3.2.1 ASTM test methods

Table 3.2 provides a list of the test methods applicable to FRP bars for use in nonprestressed concrete as available in ASTM. Of these methods,

Table 3.2 ASTM test methods for FRP reinforcement

Property	ASTM test methods
Cross-sectional area	D7205 and D7205M
Longitudinal tensile strength and modulus	D7205 and D7205M
Shear strength	D7617 and D7617M
Creep properties	D7337 and D7337M
Flexural properties	D790 and D4476
Coefficient of thermal expansion (CTE)	E831 and D696
Glass transition temperature	E1356, E1640, D648, and E2092
Volume fraction	D3171, D2584, and C882

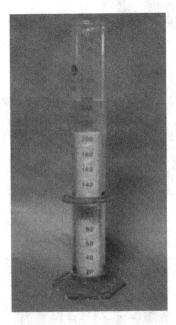

Figure 3.1 Measurement of the cross-sectional area of an FRP bar by the Archimedes principle.

the three that were specifically developed for FRP bars to be used in concrete are briefly discussed in the following.

ASTM D7205/D7205M, Tensile Properties of Fiber Reinforced Polymer Matrix Composite Bars. The area of the cross section of the FRP bar specimen is measured by applying the Archimedes principle (Figure 3.1). A bar sample of known length is immersed in a graduated cylinder filled with water. The increase of volume after immersing the sample is divided by the sample length to calculate the cross-sectional area. At least five measurements per lot of bars are required.

The FRP bar specimen of a given diameter and with length of at least 40 diameters is tested to failure. The tensile strength of the bar is measured by dividing the tensile capacity (load at failure) by the cross-sectional area. The tensile modulus is calculated from the difference between the stress–strain curve values at 20% and 50% of the tensile capacity, provided that the stress–strain curve is linear during this load range. The ultimate strain is the strain measured when the ultimate tensile capacity is reached or it can be calculated from the ultimate tensile capacity and modulus of elasticity. Photographic views of typical test setup and typical failure mode of a GFRP bar are shown in Figure 3.2(a–c). At the least, five specimens are required.

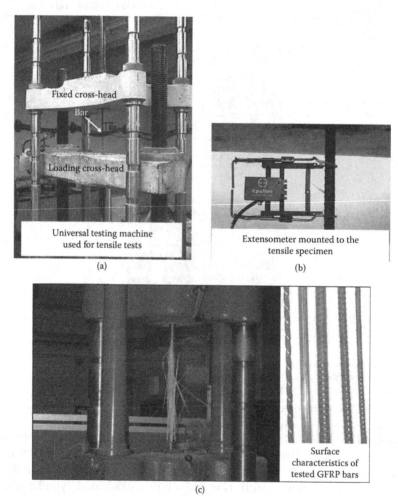

Figure 3.2 (a) Typical tensile test setup; (b) close-up view of the extensometer used to measure the axial deformation; (c) typical failure mode of a GFRP bar specimen.

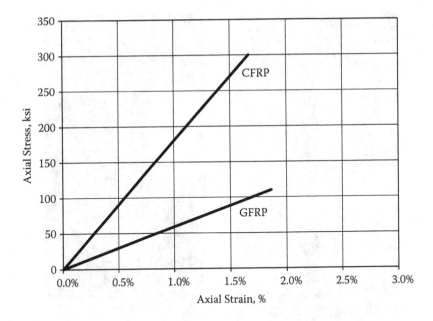

Figure 3.3 Typical tensile stress–strain curve of no. 3 GFRP and CFRP bars.

A typical stress–strain curve of no. 3 GFRP and CFRP bars is shown in Figure 3.3.

Measuring mechanical properties of large FRP bars poses unique challenges when using ASTM D7205. For traditional wedge-type grips, the difficulty stems from the significant transverse stresses that must be applied to the bar as a prerequisite to apply large enough longitudinal tensile loads. The local stress triaxiality due to stress concentrations near the gripping sections usually leads to premature failure at such locations, instead of the desired test gauge section. For unidirectional FRP composites, whose transverse strength is much less than the longitudinal strength, premature damage in the gripping region occurs frequently and depends on the method of gripping. The mechanisms of load transfer vary between different grip designs but they can be categorized into two distinct groups. The first group transfers load by applying a clamping force and relies on the interface interaction. If this clamping force is applied uniformly, then a high shear stress will develop at the loaded end. The second group relies on the underlying principle of the traditional wedge grip with various augmentations to mitigate the shear stress concentrations on the loaded ends.

An ASTM-recommended system is included in Annex A1 of ASTM D7205 and involves the use of steel pipes filled with expansive grout or polymer resin mixed with sand, as shown in Figure 3.4(c). The configuration

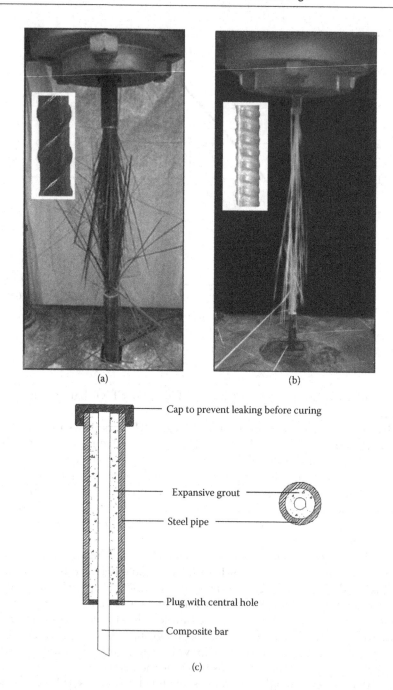

Figure 3.4 Successful axial tension tests: (a) 1.5 in. (38.1 mm) diameter bar; (b) 3/8 in. (9.5 mm) diameter bar; and (c) anchor schematics.

of the anchor is such that the grout forms a cylindrical shell around the specimen. After curing, the expansive grout exerts a pressure on the specimen and decreases the likelihood of precipitating failure in the grip region. The clamping pressure exerted on the specimen is dependent upon the confinement provided by the steel pipe and the properties of the grout layer, including the thickness, modulus of elasticity, Poisson's ratio, and coefficient of linear expansion. One advantage of the expansive grout-based grip is that one can adjust the grout layer thickness, level of confinement, and specimen embedment length (gripping length) to achieve the desired gripping behavior. To achieve such a versatile design, accurate assessment of the gripping pressure as a function of steel pipe dimensions and grout thickness (volume) is needed.

A study by Schesser et al. [60] developed a method to determine the grout material properties, including modulus of elasticity and the coefficient of linear expansion using the ASTM-recommended configuration for actual bar testing. Based on these parameters and an analytical solution, a design procedure was derived to dimension the anchor for any type of bar commercially available as demonstrated by successfully testing more than 100 specimens (Figure 3.4). The measured strength values were remarkably consistent with a coefficient of variance (CV) less than 5%. The design of expansive grout-based grips includes determination of the minimum gripping length, optimum confinement pipe dimensions, and minimum grout material volume.

ASTM D7617/D7617M, Standard Test Method for Transverse Shear Strength of Fiber Reinforced Polymer Matrix Composite Bars. The shear strength is measured by forcing the FRP bar specimen to fail due to transverse shear. Typical transverse shear test setup is shown in Figure 3.5. At least five specimens are required.

ASTM D7337/D7337M, Standard Test Method for Tensile Creep Rupture of Fiber Reinforced Polymer Matrix Composite Bars. The load-induced, time-dependent tensile strains of the FRP bar corresponding to at least five levels of load (ranging between 20% and 80% of the tensile capacity) are measured at certain ages (such as after 1, 10, 100, and 1000 hours) and in correspondence of at least five selected levels of load. The empirical strain values are plotted with respect to time (that is expressed on a logarithmic scale) and an approximation line from the graph data is extrapolated by means of the least-square method. The load ratio at 1 million hours, as determined from the calculated approximation line, is the creep-rupture load ratio. The load corresponding to this creep-rupture load ratio is the million-hour creep-rupture capacity. The million-hour creep-rupture strength is calculated by dividing this load capacity by the cross-sectional area. At least five specimens at each level of load are required.

Figure 3.5 Typical transverse shear test setup.

3.2.2 ACI 440 test methods

This section provides an overview of the test methods applicable to FRP bars for use in nonprestressed concrete that are outlined in ACI 440.3R "Guide Test Methods for Fiber-Reinforced Polymer (FRP) for Reinforcing or Strengthening Concrete and Masonry Structures" [59] and summarized in Table 3.3.

Test method for bond strength of FRP bars by pullout testing (test method B.3). Various types of test methods are available for the determination of different bond values of FRP reinforcement in concrete structures, as shown schematically in Figure 3.6 [61–64]. In this method, the bond strength of the FRP bar specimen is measured by pullout testing. The direct pullout test consists of measuring the pullout strength of an FRP bar embedded in a concrete block through a predetermined bonded length. A bond breaker is applied along the first five diameters on the pulling side of the bar, to minimize any undesired confinement effect that may affect the bond characteristics. Schematic and photographic views of a typical test setup and typical failure mode of a GFRP bar are shown in Figure 3.7(a–d). Typical free and

Table 3.3 ACI test methods for FRP reinforcement

Property	ACI 440 test method	Relationship to ASTM standards
Bond properties	B.3	ASTM C234 has been withdrawn. The only remaining ASTM test method for bond of steel bars to concrete is the beam-end test method (A944), which has not been modified for use with FRP bars.
Bent bar capacity	B.5	No existing ASTM test method is available.
Durability properties	B.6	No existing ASTM test method is available.
Fatigue properties	B.7	Method provides specific information on anchoring bars in the test fixtures and on attaching elongation measuring devices to the bars not available in ASTM D3479. Method also requires specific calculations that are not provided in the ASTM method.
Relaxation properties	B.9	Method provides specific information on anchoring bars in the test fixtures and on attaching elongation measuring devices to the bars not available in ASTM D2990 and E328. Method also requires specific calculations that are not provided in the ASTM method.
Corner radius	B.12	No existing ASTM test method is available.

loaded end bond-slip curves of a no. 4 GFRP bar are shown in Figure 3.8. The number of test specimens for each test condition should not be less than five. Other types of pullout tests include the rod–rod pullout test (Figures 3.9 and 3.10) and the beam-type pullout test (Figures 3.11 and 3.12). For both of these tests, typical test setups and results are shown in the referenced figures.

Test method for strength of FRP bent bars and stirrups at bend locations (test method B.5). The ultimate load capacity of the FRP stirrup specimen is measured by testing in tension the straight portion of an FRP C-shaped stirrup whose bent ends are embedded in two concrete blocks [65]. For assuring stability during testing, two FRP C-shaped stirrups are tested simultaneously and the failure is forced to occur at one of the two ends. Figures 3.13(a) and 3.13(b) show the test setup and the typical failure of an FRP bent bar specimen, respectively. The ultimate tensile capacity of the bent FRP stirrup is measured and compared to the ultimate tensile strength to obtain the strength-reduction factor due to bend effects. At the least, five specimens are required.

Accelerated test method for alkali resistance of FRP bars (test method B.6). The alkali resistance of the FRP bar specimen has been the subject of intense research [66–69]. In this test, it is measured considering three differ-ent conditionings. A first set of bar specimens is tested in tension after being

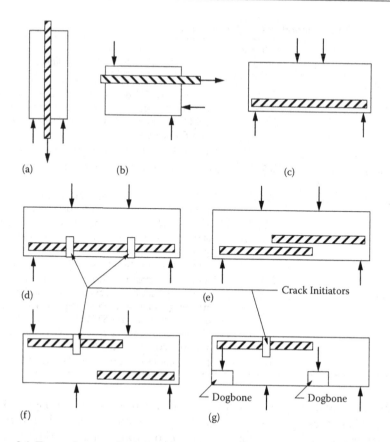

Figure 3.6 Types of test methods for different bond values of FRP reinforcement in concrete: (a) pullout specimen; (b) beam-end specimen; (c) simple beam specimen; (d) hinged beam-end specimen; (e) splice specimen; (f) cantilever beam
specimen (without dogbones); and (g) cantilever beam specimen (with dogbones) (ACI 440-3R).

aged in an alkaline solution[1] (with a pH included between 12.6 and 13) at
140°F (60°C) for exposure times of 1, 2, 3, 4, and 6 months. A second set
of bar specimens is subjected to a constant tensile load for exposure times
of 1, 2, 3, 4, and 6 months while immersed in the alkaline solution inside an
environmental cabinet and having a constant temperature of 140°F (60°C).
A third set of bar specimens embedded in concrete is subjected to the same
conditioning as the second set. The magnitude of the sustained load is
meant to be representative of the service conditions or such to induce a tensile strain of at least 2000 microstrain. At the end of the three series of tests,

---

[1] The suggested composition of the alkaline solution consists of 118.5 g of $Ca(OH)_2$, 0.9 g of
NaOH, and 4.2 g of KOH in 1 L of deionized water.

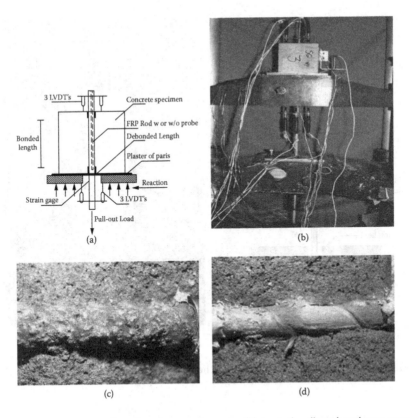

Figure 3.7 (a) Direct pullout bond test schematic; (b) typical pullout bond test setup; (c, d) typical failure modes of a GFRP bar specimen.

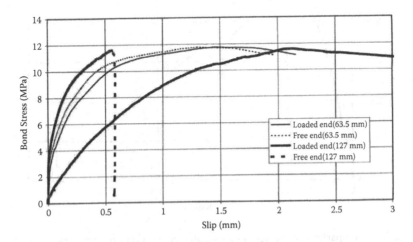

Figure 3.8 Typical free and loaded end bond-slip curves of a no. 4 GFRP bar.

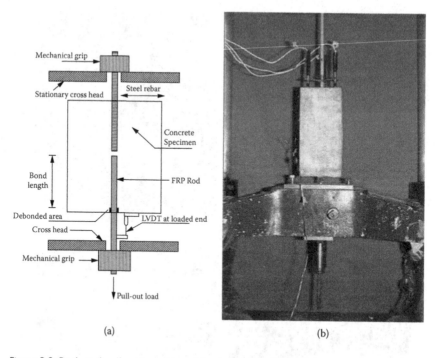

Figure 3.9 Rod–rod pullout bond test: (a) schematic and (b) photographic views.

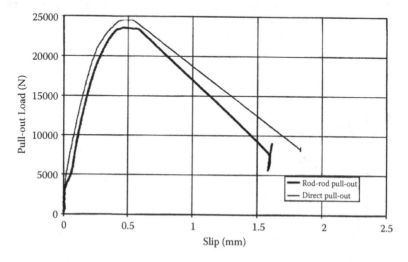

Figure 3.10 Comparison of the test results obtained by direct pullout versus rod–rod pullout on no. 4 GFRP bar.

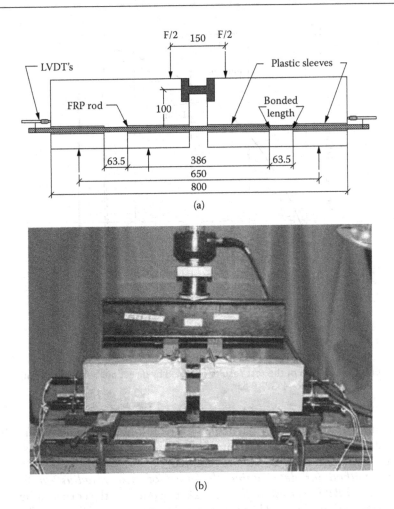

Figure 3.11 Beam-type pullout bond test: (a) schematic and (b) photographic views.

for all specimens, the mass change and the tensile capacity retention are computed. The tensile capacity retention corresponds to the ratio between the ultimate tensile capacity of the conditioned FRP bar specimen and the ultimate tensile capacity of a benchmark nonconditioned FRP bar. At least five specimens are required for each type of conditioning.

Test method for tensile fatigue of FRP bars (test method B.7). The tensile fatigue of the FRP bar is measured by applying a tensile load of magnitude included between 20% and 60% of the tensile capacity, and frequency included between 1 and 10 Hz. The FRP bar specimen is expected to fail in the gauge length at a cycle count of between 10^3 and 2×10^6 cycles. The numbers of repeated loading cycles required to fail the FRP bar are used

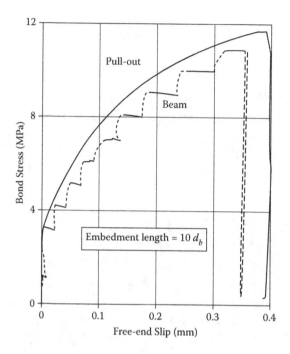

Figure 3.12 Comparison of the test results obtained by direct pullout versus beam-type pullout on no. 4 GFRP bar.

to create S–N curves for a particular set of testing conditions where the principal variable is the maximum value of the repeated load. At least five specimens at each level of load are required.

Test method for determining the effect of corner radius on tensile strength of FRP bars (test method B.12). The effect of the corner radius on tensile strength of the FRP bar is measured using a unique fixture. The test consists in pulling the straight sides of a reverse-U shaped FRP bar while the fixture, reacting against the testing machine frame, allows transferring the load to the bent portion. A photographic view of the test setup, a schematic view, and a typical failure mode of a bent GFRP bar are shown in Figure (3.14). At the least, five specimens are required.

3.3 PRODUCT CERTIFICATION AND QUALITY ASSURANCE

Product certification and quality assurance during FRP manufacturing are critical to ensure that what is specified by the engineer eventually ends up at the construction site for the contractor to install. Permitted constituent

(a)   (b)

Figure 3.13 Views of the (a) test setup and (b) a failed specimen.

materials, limits on constituent volumes, and minimum reinforcements for CFRP and GFRP bars to be used as nonprestressed concrete are indicated in ACI 440.6, "Specification for carbon and glass fiber-reinforced polymer bar materials for concrete reinforcement" [70]. As partially apparent in its title, this standard covers GFRP and CFRP bars made with thermosetting resins only (excluding polyesters).

For the determination of each of the mechanical and durability properties for product certification and quality assurance, ACI 440.6 prescribes the testing of at least 25 samples obtained in groups of five from five different production lots. The following sections summarize the requirements of this standard.

3.3.1 Constituent materials

The test method of reference for measuring the constituent in an FRP bar sample is the ASTM D3171, "Standard test method for constituent content of composite materials." With this test method, the fiber volume is

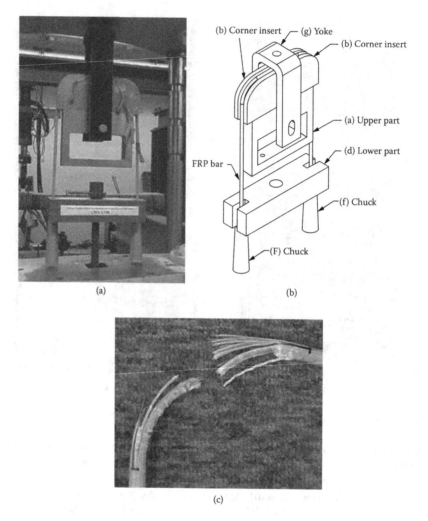

Figure 3.14 (a) Typical test setup; (b) assembled test fixture sketch; (c) typical failure mode of a bent GFRP.

determined by chemical matrix digestion, in which the matrix is chemically dissolved and the fibers weighed and calculated from substituent weights and densities.

A different test method is generally used to measure the fiber content. The reference is the ASTM D 2583, "Standard test method for ignition loss of cured reinforced resins." This test allows determining the ignition loss (or resin content) and the reinforcement content of the cured reinforced resin by heating a bar sample at 1500°F (815°C). What remains of a GFRP bar sample at the end of the test is shown in Figure 3.15.

Figure 3.15 Typical GFRP bar sample at the end of the ignition loss test.

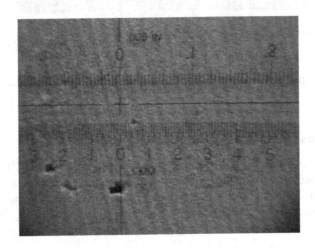

Figure 3.16 Voids in the cured material due to lack of glass.

Inaccurate proportions between resin and reinforcement may cause internal defects in the cured material, which would ultimately induce lower durability and mechanical properties. An example of void due to lack of reinforcing glass is shown in Figure 3.16.

According to ACI440-6, polyester cannot be used as the base polymer in FRP bar manufacturing because of the high content of styrene.[1] Styrene can be added to any polymer resin during processing, but its content cannot exceed 10% by mass of the polymer resin.

---

[1] Styrene has been recently described by the US National Toxicology Program as a possible carcinogen for human health.

Figure 3.17 Barcol hardness test setup and apparatus.

Other requirements on fillers and additives are also provided in ACI 440-6.

3.3.2 Glass transition temperature (T_G)

ACI 440.6 requires the T_G of the resin not to be smaller than 212°F (100°C). Two test methods are specified for determining the T_G of a resin: ASTM E 1356, "Standard test method for assignment of the glass transition temperature by differential scanning calorimetry," and ASTM E 1640, "Standard test method for assignment of the glass transition temperature by dynamic mechanical analysis."

The Barcol hardness test complements the measurement of T_G as an indirect measure of the degree of cure of a composite (Figure 3.17). The test method of reference is the ASTM D 2583, "Standard test method for indentation hardness of rigid plastics by means of a Barcol impressor." This test characterizes the indentation hardness of materials through the depth of penetration of an indentor and gives indications of the quality of the resin curing.

3.3.3 Bar size

ACI 440.6 defines the standard sizes and the nominal diameters for FRP bars to be used for designation and design (Table 3.4). The test method

Table 3.4 Size designation, minimum inside bend radius of bent bars, and minimum guaranteed tensile strength of FRP round bars

Bar size designation	Nominal diameter (in.)	Nominal area (in.^{2})	Bend radius (in.)	Minimum guaranteed tensile strength (ksi)	
				GFRP	CFRP
2	0.250	0.05	3/4	110	230
3	0.375	0.11	11/8	105	190
4	0.500	0.20	11/2	100	170
5	0.625	0.31	17/8	95	160
6	0.750	0.44	21/4	90	160
7	0.875	0.60	25/8	85	N/A
8	1.000	0.79	3	80	N/A
9	1.125	1.00	41/2	75	N/A
10	1.270	1.27	5	70	N/A

Notes: 1 in. = 25.4 mm; 1 in.^{2} = 645 mm^2; 1 ksi = 6.89 N/mm^2; N/A = not available.

Table 3.5 Minimum guaranteed values of tensile elastic modulus, tensile ultimate strain, transverse shear strength, and bond strength

Property	Bar type	
	GFRP	CFRP
Modulus of elasticity (ksi)	5,700	18,000
Ultimate tensile strain (%)	1.2	0.5
Transverse shear strength (ksi)	18	18
Bond strength (ksi)	1.4	1.4

Note: 1 ksi = 6.89 N/mm^2.

of reference for measuring the diameter of an FRP bar is the one given in ASTM D7205, "Standard test method for tensile properties of fiber reinforced polymer matrix composite bars."

3.3.4 Mechanical properties

ACI 440.6 defines the minimum values of tensile strength, tensile modulus of elasticity, tensile ultimate strain, transverse shear strength, and bond strength that have to be guaranteed by the manufacturer for product certification. The minimum values of tensile strength are reported in Table 3.4. Minimum values of tensile modulus of elasticity, tensile ultimate strain, transverse shear strength, and bond strength are listed in Table 3.5.

3.3.5 Durability properties

To evaluate the durability properties of GFRP and CFRP bars, ACI 440.6 requires the evaluation of moisture absorption, resistance to alkaline environment, and longitudinal wicking for product certification.

Moisture content. ACI 440.6 requires that the average of the individual measurements be not less than 1.0%. The moisture content of a plastic is intimately related to the durability and mechanical properties of the FRP bar. The test method of reference is the ASTM D 570, "Standard test method for water absorption of plastics." The objective of this test method is to determine the relative rate of absorption of water by plastic when immersed in water. A photographic view of samples being tested is shown in Figure 3.18. An alternative test method that can be used for moisture content measurement is ASTM D5229, "Standard test method for moisture absorption properties and equilibrium conditioning of polymer matrix composite materials."

Resistance to alkaline environment. ACI 440.6 has not defined yet the minimum strength retention values to characterize the resistance to alkaline environments. The test method of reference is the one discussed in test method B.6 of ACI 440.3R.

Longitudinal wicking. To guarantee the integrity of the composite material, ACI 440.6 does not allow continuous voids in the resin. The presence of voids and cracks is detrimental. These defects may be due to shrinkage of the resin during processing or poor consolidation of fibers and resin matrix during production. Hollow fibers in the cured material are allowed. The test method of reference to measure the longitudinal wicking is the ASTM D 5117, "Standard test method for dye penetration of solid fiberglass reinforced pultruded stock." A typical case of voids in the bar is due to the presence of hollow fibers shown in Figure 3.19.

Figure 3.18 Water absorption test.

Figure 3.19 Voids due to the presence of hollow glass fibers.

3.3.6 Bent bars

Only FRP bars that are made with thermosetting resin are allowed by ACI 440.6. For such bars, the bends can be made only while the resin is still in its liquid state. ACI 440.6 defines the minimum values of inside bend radius for GFRP and CFRP bent bars (Table 3.4). Minimum values of the strength of bends have not yet been established.

3.4 PERFORMANCE OF FRP RC UNDER FIRE CONDITIONS

The performance of an FRP RC member under fire strictly relies on the temperature the embedded bars experience. When the surface of an FRP bar embedded in the concrete exposed to fire approaches the glass transition temperature (T_g) of the resin matrix, the resin begins to decompose and eventually becomes unable to transfer the stresses from the concrete to the fibers because of the loss of bond between the FRP bar and the surrounding concrete. When the temperature of the bar exceeds T_g, the individual fibers can continue to carry the load if adequate bond is provided in the zones not exposed to fire (*cold anchorage region*). The RC member collapse ultimately occurs when the temperature of the fibers reaches their *critical temperature*.

The behavior of FRP bars under fire conditions is affected by several parameters, namely fiber type, matrix type, bar diameter, fiber volume fraction, manufacturing process, and surface treatment. Available research

[71–74] shows that both CFRP and GFRP bars when tested in tension under temperatures in the proximity of T_g [about 150 to 250°F (65 to 120°C)] experience a small reduction in tensile strength and elastic modulus, while a reduction larger than 50% is typically reached at temperatures higher than approximately 620°F (325°C) and 480°F (250°C) for GFRP and CFRP, respectively. Both CFRP and GFRP bars predictably experience an almost full loss of bond strength with the surrounding concrete at temperatures in the range of T_g, since bond fully depends upon the mechanical properties of the resin at the surface of the bars.

Guidance on the performance of FRP RC members under fire conditions is still limited. Because of their different behavior at elevated temperatures, the guidelines proposed by ACI 216 [75] for determining the fire resistance of steel RC members are not applicable in the case of FRP reinforcement.

A recent publication by Nigro et al. [76] proposes a new methodology to perform fire safety checks for bending moment capacity of unprotected FRP RC flexural members exposed to fire on the side of the fibers under tension. This conceptual approach is discussed in Chapter 4 (Section 4.12) and its application is explained in Chapter 6 (Step 10).

REFERENCES

1. L. Nicolais, A. Borzacchiello, and S. M. Lee. *Wiley encyclopedia of composites,* 2nd ed. Hoboken, NJ: John Wiley & Sons ISBN (printed set): 978-0-470-12828-2 (2012).
2. C. E. Bakis. FRP reinforcement: Materials and manufacturing. In *fiber-reinforced plastic (FRP) for concrete structures: Properties and applications,* ed. A. Nanni, 13–58. New York: Elsevier Science Publishers (1993).
3. L. C. Bank. Properties of FRP reinforcements for concrete. In *fiber-reinforced plastic (FRP) for concrete structures: Properties and applications,* ed. A. Nanni, 59–86. New York: Elsevier Science Publishers (1993).
4. W. P. Wu. Thermomechanical properties of fiber reinforced plastic (FRP) bars, PhD dissertation, West Virginia University, Morgantown, WV (1990).
5. S. Faza and H. GangaRao. Glass FRP reinforcing bars for concrete. In *fiber-reinforced plastic (FRP) for concrete structures: Properties and applications,* ed. A. Nanni, 42, 167–188. Amsterdam: Elsevier (1993).
6. V. O. Volkersen. Die Nietkraftverteilung in zugbeanspruchten Nievtverbindungen mit konstanten Laschenquerschnitten. *Luftfahrforschung* 15:41–47 (1938).
7. P. K. Mallick. *Fiber reinforced composites, materials, manufacturing, and design.* New York: Marcell Dekker, Inc. 1988.
8. M. R. Ehsani. Glass-fiber reinforcing bars. In *Alternative materials for the reinforcement and prestressing of concrete,* ed. J. L. Clarke, 35–54. London: Blackie Academic & Professional (1993).
9. C. C. Choo, I. E. Harik, and H. Gesund. Minimum reinforcement ratio for fiber-reinforced polymer reinforced concrete rectangular columns. *ACI Structural Journal* 103 (3): 460–466 (2006).

10. fib. Bulletin 40. FRP reinforcement in RC structures. International Federation for Structural Concrete (FIB—Fédération Internationale du Béton), Lausanne, Switzerland (2007).

11. T. Yamaguchi, Y. Kato, T. Nishimura, and T. Uomoto. Creep rupture of FRP rods made of aramid, carbon and glass fibers. *Proceedings of the Third International Symposium on Non-Metallic (FRP) Reinforcement for Concrete Structures (FRPRCS-3)*, Japan Concrete Institute, Tokyo, Japan, 2: 179–186 (1997).

12. N. Ando, H. Matsukawa, A. Hattori, and A. Mashima. 1997. Experimental studies on the long-term tensile properties of FRP tendons. *Proceedings of the Third International Symposium on Non-Metallic (FRP) Reinforcement for Concrete Structures (FRPRCS-3)*, Japan Concrete Institute, Tokyo, Japan, 2:203–210 (1997).

13. H. Seki, K. Sekijima, and T. Konno. Test method on creep of continuous fiber reinforcing materials. *Proceedings of the Third International Symposium on Non-Metallic (FRP) Reinforcement for Concrete Structures (FRPRCS-3)*, Japan Concrete Institute, Tokyo, Japan, 2:195–202 (1997).

14. H. Saadatmanesh and F. E. Tannous. Relaxation, creep, and fatigue behavior of carbon fiber-reinforced plastic tendons. *ACI Materials Journal* 96 (2): 143–153 (1999).

15. H. Saadatmanesh and F. E. Tannous. Long-term behavior of aramid fiber-reinforced plastic tendons. *ACI Materials Journal* 96 (3): 297–305 (1999).

16. M. Roylance and O. Roylance. Effect of moisture on the fatigue resistance of aramid-epoxy composite. *Polymer Engineering & Science* 22:988–993 (1982).

17. J. F. Mandell. Fatigue behavior of fiber resin composites. *Development of Fiber Reinforced Plastic* 2:67–107 (1982).

18. P. T. Curtis. The fatigue behavior of fibrous composite materials. *Journal of Strain Analysis* 24 (4): 235–244 (1989).

19. J. F. Mandell and U. Meier. Effects of stress ratio frequency and loading time on the tensile fatigue of glass-reinforced epoxy. Long term behavior of composites. ASTM STP 813, ASTM International, West Conshohocken, PA, 55–77 (1983).

20. T. Odagiri, K. Matsumoto, and H. Nakai. Fatigue and relaxation characteristics of continuous aramid fiber reinforced plastic rods. *Proceedings of the Third International Symposium on Non-Metallic (FRP) Reinforcement for Concrete Structures (FRPRCS-3)*, Japan Concrete Institute, Tokyo, Japan, 2:227–234 (1997).

21. A. H. Rahman and C. Y. Kingsley. Fatigue behavior of a fiber-reinforced-plastic grid as reinforcement for concrete. *Proceedings of the First International Conference on Composites in Infrastructure (ICCI-96)*, ed. H. Saadatmanesh and M. R. Ehsani, 427–439. Tucson, AZ, (1996).

22. R. Adimi, H. Rahman, B. Benmokrane, and K. Kobayashi. Effect of temperature and loading frequency on the fatigue life of a CFRP bar in concrete. *Proceedings of the Second International Conference on Composites in Infrastructure (ICCI-98)*, Tucson, AZ, 2:203–210 (1998).

23. A. El-Ragaby, E. El-Salakawy, and B. Benmokrane. Fatigue life evaluation of concrete bridge deck slabs reinforced with glass FRP composite bars. *ASCE Journal of Composites for Construction* 11 (3): 258–268 (2007).

24. T. R Gentry. Life assessment of glass-fiber reinforced composites in Portland cement concrete. Paper no. 077. *Proceedings 16th Annual Meeting of the American Society for Composites,* Blacksburg, VA, Sept. 10–12 (2001).

25. J. Clarke and P. Sheard. Designing durable FRP reinforced concrete structures. *Proceedings of the First International Conference on Durability of Composites for Construction,* ed. B. Benmokrane and H. Rahman, Sherbrooke, Québec, Canada, 13–24 (1998).

26. A. Kamal and M. Boulfiza. Durability of GFRP rebars in simulated concrete solutions under accelerated aging conditions. *Journal of Composites for Construction* 15 (4): 473–481 (2011).

27. M. L. Porter and B. A. Barnes. Accelerated aging degradation of glass fiber composites. *Second International Conference on Composites in Infrastructure,* V. II, ed. H. Saadatmanesh and M. R. Eshani, University of Arizona, Tucson, 446–459 (1998).

28. A. Coomarasamy and S. Goodman. Investigation of the durability characteristics of fiber reinforced polymer (FRP) materials in concrete environment. American Society for Composites—Twelfth Technical Conference, Dearborn, MI 1997.

29. H. V. S. GangaRao and P. V. Vijay. Aging of structural composites under varying environmental conditions. *Proceedings of the Third International Symposium on Non-Metallic (FRP) Reinforcement for Concrete Structures (FRPRCS-3),* Japan Concrete Institute, Tokyo, Japan, 2:91–98 (1997).

30. M. L. Porter, J. Mehus, K. A. Young, E. F. O'Neil, and B. A. Barnes. Aging for fiber reinforcement in concrete. *Proceedings of the Third International Symposium on Non-Metallic (FRP) Reinforcement for Concrete Structures (FRPRCS-3),* Japan Concrete Institute, Tokyo, Japan, 2:59–66 (1997).

31. C. E. Bakis, A. J. Freimanis, D. Gremel, and A. Nanni. Effect of resin material on bond and tensile properties of unconditioned and conditioned FRP reinforcement rods. *Proceedings of the First International Conference on Durability of Composites for Construction,* ed. B. Benmokrane and H. Rahman, Sherbrooke, Québec, 525–535 (1998).

32. F. E. Tannous and H. Saadatmanesh. Durability of AR-glass fiber reinforced plastic bars. *Journal of Composites for Construction* 3 (1): 12–19 (1999).

33. T. Uomoto. Durability of FRP as reinforcement for concrete structures. *Proceedings of the Third International Conference on Advanced Composite Materials in Bridges and Structures,* ACMBS-3, ed. J. L. Humar and A. G. Razaqpur, Canadian Society for Civil Engineering, Montreal, Québec, Canada, 3–17 (2000).

34. K. Takewaka and M. Khin. Deterioration of stress-rupture of FRP rods in alkaline solution simulating as concrete environment. In *Advanced composite materials in bridges and structures,* ed. M. M. El-Badry, Canadian Society for Civil Engineering, Montréal, Québec, Canada, 649–664 (1996).

35. L. C. Bank and M. Puterman. Microscopic study of surface degradation of glass fiber-reinforced polymer rods embedded in concrete castings subjected to environmental conditioning, high temperature and environmental effects on polymeric composites. ASTM STP 1302, ed. T. S. Gates and A.-H. Zureick, ASTM International, West Conshohocken, PA, 2:191–205 (1997).

36. A. Mukherjee and S. I. Arwikar. Performance of glass fiber-reinforced polymer reinforcing bars in tropical environments—Part II: Microstructural tests. *ACI Structural Journal* 102 (6): 816–822 (2005).

37. F. S. Rostasy. On durability of FRP in aggressive environments. *Proceedings of the Third International Symposium on Non-Metallic (FRP) Reinforcement for Concrete Structures (FRPRCS-3)*, Japan Concrete Institute, Tokyo, Japan 2:107–114 (1997).

38. R. Sen, M. Shahawy, J. Rosas, and S. Sukumar. Durability of aramid pretensioned elements in a marine environment. *ACI Structural Journal* 95 (5): 578–587 (1998).

39. I. Sasaki, I. Nishizaki, H. Sakamoto, K. Katawaki, and Y. Kawamoto. Durability evaluation of FRP cables by exposure tests. *Proceedings of the Third International Symposium on Non-Metallic (FRP) Reinforcement for Concrete Structures (FRPRCS-3)*, Japan Concrete Institute, Tokyo, Japan, 2:131–137 (1997).

40. F. Micelli and A. Nanni. Durability of FRP rods for concrete structures. *Construction and Building Materials* 18:491–503 (2004).

41. H. Y. Kim, Y. H. Park, Y. J. You, and C. K. Moon. Short-term durability test for GFRP rods under various environmental conditions. *Composite Structures* 83 (1): 37–47 (2008).

42. J. Bootle, F. Burzesi, and L. Fiorini. Design guidelines. In *ASM handbook, composites*. ASM International, Material Park, OH, 21:388–395 (2001).

43. A. A. Mufti, N. Banthia, B. Benmokrane, M. Boulfiza, and J. Newhook. Durability of GFRP composite rods in field structures. *Concrete International* 29 (2): 37–42 (2007).

44. A. A. Mufti. Report on studies of GFRP durability in concrete from field bridge demonstration projects. *Proceeding 3rd International. Conference on Composites in Construction (CCC 2005)*, ed. P. Hamelin, D. Bigaud, E. Ferrier, and E. Jacqelin, Université Claude Bernard, Lyon, France, 889–895 (2005).

45. G. Nkurunziza, A. Debaiky, P. Cousin, and B. Benmokrane. Durability of GFRP bars: A critical review of the literature. *Progress in Structural Engineering and Materials* 7 (4): 194–209 (2005).

46. S.U. Al-Dulaijan, M. Al-Zahrani, A. Nanni, C. E. Bakis, and T. E. Boothby. Effect of environmental pre-conditioning on bond of FRP reinforcement to concrete. *Journal of Reinforced Plastics and Composites* 20: 881–900 (2001).

47. A. Nanni, J. Nenninger, K. Ash, and J. Liu. Experimental bond behavior of hybrid rods for concrete reinforcement. *Structural Engineering and Mechanics* 5 (4): 339–354 (1997).

48. C. E. Bakis, S. U. Al-Dulaijan, A. Nanni, T. E. Boothby, and M. M. Al-Zahrani. Effect of cyclic loading on bond behavior of GFRP rods embedded in concrete beams. *Journal of Composites Technology and Research* 20 (1): 29–37 (1998).

49. L. C. Bank, M. Puterman, and A. Katz. The effect of material degradation on bond properties of FRP reinforcing bars in concrete. *ACI Materials Journal* 95 (3): 232–243 (1998).

50. A. J. Freimanis, C. E. Bakis, A. Nanni, and D. Gremel. A comparison of pull-out and tensile behaviors of FRP reinforcement for concrete. *Proceedings, ICCI-98*, Tucson, AZ, 2:52–65 (Jan. 5–7, 1998).

51. M. Pecce, G. Manfrei, R. Realfonzo, and E. Cosenza. Experimental and analytical evaluation of bond properties of GFRP bars. *Journal of Materials in Civil Engineering* 13 (4): 282–290 (2001).

52. S. Faza and H. GangaRao. Bending and bond behavior of concrete beams reinforced with plastic rebars. *Transportation Research Record* no. 1290, 2:185–193 (1990).

53. M. R. Ehsani, H. Saadatmanesh, and S. Tao. Design recommendations for bond of GFRP rebars to concrete. *Journal of Structural Engineering* 122 (3): 247–254 (1996).

54. B. Benmokrane, B. Zhang, K. Laoubi, B. Tighiouart, and I. Lord. Mechanical and bond properties of new generation of carbon fiber reinforced polymer reinforcing bars for concrete structures. *Canadian Journal of Civil Engineering* 29 (2): 338–343 (2002).

55. C. Shield, C. French, and J. Hanus. Bond of GFRP rebar for consideration in bridge decks. *Fiber Reinforced Polymer Reinforcement for Reinforced Concrete Structures, Fourth International Symposium*, SP-188, ed. C. W. Dolan, S. H. Rizkalla, and A. Nanni, American Concrete Institute, Farmington Hills, MI, 393–406 (1999).

56. C. Mosley. Bond performance of fiber reinforced plastic (FRP) reinforcement in concrete. MS thesis, Purdue University, West Lafayette, IN (2002).

57. B. W. Wambeke and C. K. Shield. Development length of glass fiber-reinforced polymer bars in concrete. *ACI Structural Journal* 103 (1): 11–17 (2006).

58. B. Tighiouart, B. Benmokrane, and P. Mukhopadhyaya. Bond strength of glass FRP rebar splices in beam under static loading. *Construction and Building Material* 13 (7): 383–392 (1999).

59. ACI Committee 440. Guide test methods for fiber-reinforced polymers (FRPs) for reinforcing or strengthening concrete and masonry structures, ACI 440.3R-12. American Concrete Institute, Farmington Hills, MI (2012).

60. D. Schesser, Q. D. Yang, A. Nanni, and J. W. Giancaspro. Expansive grout-based gripping systems for tensile testing of large diameter composite bars. *ASCE Journal of Materials in Civil Engineering* DOI: 10.1061/(ASCE)MT.1943-5533.0000807, Feb. (2013).

61. M. Al-Zahrani, A. Nanni, S. U. Al-Dulaijan, and C. E Bakis. Bond of FRP to concrete for rods with axisymmetric deformations. *Proceedings, ACMBS-II,* Montreal, Canada, 853–860 (1996).

62. B. Benmokrane, P. Wang, T. M. Ton-That, H. Rahman and J.-F. Robert. Durability of GFRP reinforcing bars in concrete environment. *ASCE Journal of Composites for Construction* 6 (3): 143–155 (2002).

63. O. Chaallal and B. Benmokrane. Pullout and bond of glass-fiber rods embedded in concrete RILEM. *Journal of Materials and Structures* 26 (157): 167–175 (1993).

64. A. J. Freimanis, C. E. Bakis, A. Nanni, and D. Gremel. A comparison of pull-out and tensile behaviors of FRP reinforcement for concrete. *Proceedings, ICCI-98*, Tucson, AZ, 2:52-65 (Jan. 5–7, 1998).

65. R. Morphy, E. Shehata, and S. Rizkalla. Bent effect on strength of CFRP stirrups. *Proceedings of the Third International Symposium on Non-Metallic (FRP) Reinforcement for Concrete Structures*, Sapporo, Japan, Oct. 19–26 (1997).

66. B. Benmokrane, P. Wang, and T. M. Ton-That. Durability of GFRP reinforcing bars in alkaline environments. *Proceedings of the CICE 2001, International Conference on FRP Composites in Civil Engineering*, Hong Kong, 1527–1534 (Dec. 12–15, 2001).

67. B. Benmokrane, B. Zhang, K. Laoubi, B. Tighiouart, and I. Lord. Mechanical and bond properties of new generation of carbon fiber reinforced polymer reinforcing bars for concrete structures. *Canadian Journal of Civil Engineering* 29 (2): 338–343 (2002).

68. T. R. Gentry, L. C. Bank, A. Barkatt, and L. Prian. Accelerated test methods to determine the long-term behavior of composite highway structures subject to environmental loading. *Journal of Composites Technology and Research* 20 (1): 38–50 (1998).

69. G. Nkurunziza, R. Masmoudi, and B. Benmokrane. Effect of sustained tensile stress and temperature on residual strength of GFRP composite bars. *Proceedings Second International Conference on Durability of Fiber Reinforced Polymer (FRP) Composites for Construction*, CDCC 2002, Montreal, Quebec, Canada, 347–358 (May 29–31, 2002).

70. ACI Committee 440. Specification for carbon and glass fiber-reinforced polymer bar materials for concrete reinforcement, ACI 440.6-08. American Concrete Institute, Farmington Hills, MI (2008).

71. L. A. Bisby, M. Green, V. K. R. Kodur. Response to fire of concrete structures that incorporate FRP. *Progress in Structural Engineering and Materials* 7(2): 136–149 (2005).

72. L. A. Bisby, M. Green, V. K. R. Kodur. Evaluating the fire endurance of concrete slabs reinforced with FRP bars: Considerations for a holistic approach. *Composites Part B: Engineering* 38(5–6): 547–558 (2007).

73. E. Nigro, G. Cefarelli, A. Bilotta, G. Manfredi, E. Cosenza. Mechanical Behavior of Concrete Slabs Reinforced with FRP Bars in Case of Fire: Experimental Investigation and Numerical Simulation. *3rd FIB International Congress* 15pp (2010).

74. C. Maluk, L. Bisby, G. Terrasi, M. Green. Bond Strength Degradation for CFRP and Steel reinforcing Bars in Concrete at Elevated Temperature. *ACI SP-297, American Concrete Institute* pp. 2.1–2.36 (2011).

75. *ACI Committee 216*. Code Requirements for Determining Fire Resistance of Concrete and Masonry Construction Assemblies. Farmington Hills, MI (2007).

76. E. Nigro, G. Cefarelli, A. Bilotta, G. Manfredi, E. Cosenza. Guidelines for flexural resistance of FRP reinforced concrete slabs and beams in fire. *Composites Part B: Engineering*, 58: 103–112 (2014).

Part II

Analysis and design

Chapter 4

Flexural members

NOTATION

A_{cp} = area enclosed by outside perimeter of concrete cross section, in.^{2} (mm^2)

A_f = area of FRP reinforcement in^2. (mm^2)

A_{fb} = area of flexural reinforcement producing balanced failure, in.^{2} (mm^2)

$A_{f,bar}$ = area of one FRP bar, in.^{2} (mm^2)

$A_{f,i}$ = area of reinforcement in i-th layer, in.^{2} (mm^2)

$A_{f,min}$ = minimum area of FRP reinforcement needed to prevent failure of flexural members upon cracking, in.^{2} (mm^2)

$A_{f,sh}$ = area of shrinkage and temperature FRP reinforcement per linear foot, in.^{2} (mm^2)

A_{fv} = amount of FRP shear reinforcement within spacing s, in.^{2} (mm^2)

$A_{fv,min}$ = minimum amount of FRP shear reinforcement within spacing s, in.^{2} (mm^2)

A_{oh} = area enclosed by centerline of the outermost closed transverse torsional reinforcement, in.^{2} (mm^2)

A_s = area of tension steel reinforcement, in.^{2} (mm^2)

A_t = area of one leg of a closed stirrup resisting torsion within spacing s, in.^{2} (mm^2)

A_{vf} = area of shear friction reinforcement perpendicular to the plane of shear

a = depth of equivalent rectangular stress block, in. (mm)

b = width of rectangular cross section, in. (mm)

b_o = perimeter of critical section for slabs and footings, in. (mm)

b_{eff} = effective width of the slab, in. (mm)

b_w = width of the web, in. (mm)

C = Compressive force, lb (N)

C_E = environmental reduction factor for various fiber types and exposure conditions

c	=	distance from extreme compression fiber to the neutral axis, in. (mm)
c	=	spacing or cover dimension
c_b	=	distance from extreme compression fiber to neutral axis at balanced strain condition, in. (mm)
c_{cr}	=	cracked neutral axis depth, in. (mm)
D	=	diameter of circular cross section, in. (mm)
d	=	distance from extreme compression fiber to centroid of tension reinforcement, in. (mm)
d_b	=	diameter of reinforcing bar, in. (mm)
d_c	=	thickness of concrete cover measured from extreme tension fiber to center of bar or wire location closest thereto, in. (mm)
di	=	distance from centroid of ith layer of longitudinal reinforcement to geometric centroid of cross section, in. (mm)
e	=	ratio of ε_{fu} over ε_{cu}
d_f	=	effective depth of the FRP reinforcement, in. (mm)
E_c	=	modulus of elasticity of concrete, psi (MPa)
E_f	=	design or guaranteed modulus of elasticity of FRP defined as mean modulus of sample of test specimens ($E_f = E_{f,ave}$), psi (MPa)
$E_{f,ave}$	=	average modulus of elasticity of FRP, psi (MPa)
E_s	=	modulus of elasticity of steel, psi (MPa)
f'_c	=	specified compressive strength of concrete, psi (MPa)
f_r	=	modulus of rupture of concrete, psi (MPa)
f_f	=	stress in FRP reinforcement in tension, psi (MPa)
f_{fb}	=	strength of bent portion of FRP bar, psi (MPa)
f_{fd}	=	design tensile strength, psi (MPa)
f_{fe}	=	bar stress that can be developed for embedment length l_e, psi (MPa)
f_{fr}	=	required bar stress, psi (MPa)
$f_{f,s}$	=	stress level induced in FRP by sustained loads, psi (MPa)
f_{fu}	=	design tensile strength of FRP, considering reductions for service environment ($f_{fu} = C_E f^*_{fu}$), psi (MPa)
f^*_{fu}	=	guaranteed tensile strength of FRP bar, defined as mean tensile strength of sample of test specimens minus three times standard deviation ($f^*_{fu} = f_{fu,ave} - 3\sigma$), psi (MPa)
f_{fv}	=	tensile strength of FRP for shear design, taken as smallest of design tensile strength f_{fu}, strength of bent portion of FRP stirrups f_{fb}, or stress corresponding to $0.004E_f$, psi (MPa)
f_s	=	allowable stress in steel reinforcement, psi (MPa)
$f_{u,ave}$	=	mean tensile strength of sample of test specimens, psi (MPa)
f_{vf}	=	shear friction stress in reinforcement
f_y	=	specified yield stress of nonprestressed steel reinforcement, psi (MPa)
h	=	overall height of rectangular member, in. (mm)
I	=	moment of inertia, in.^{4} (mm^4)

I_{cr}	=	moment of inertia of transformed cracked section, in.^{4} (mm^4)
I_e	=	effective moment of inertia, in.^{4} (mm^4)
I_g	=	gross moment of inertia, in.^{4} (mm^4)
K_1	=	parameter accounting for boundary conditions
k	=	ratio of depth of neutral axis to reinforcement depth
k_b	=	bond-dependent coefficient
k_m	=	neutral axis depth to reinforcement depth ratio at midspan
l	=	span length of member, ft (m)
l_a	=	additional embedment length at support or at point of inflection, in. (mm)
l_{bhf}	=	basic development length of FRP standard hook in tension, in. (mm)
l_d	=	development length, in. (mm)
l_e	=	embedded length of reinforcing bar, in. (mm)
f_{ct}	=	tensile strength of concrete, psi (MPa)
f_{ft}	=	tensile strength of transverse reinforcement, psi (MPa)
$k_{creep-R}$	=	creep rupture factor
$l_{d,fi,t,T>Tcr}$ =		embedment length of a bar with a temperature exceeding 122°F (50°C)
l_{thf}	=	length of tail beyond hook in FRP bar, in. (mm)
M	=	maximum positive moment lb-in.(N-mm)
m	=	non-dimensional moment parameter
M_a	=	maximum moment in member at stage deflection is computed, lb-in. (N-mm)
M_C, M_T =		contributions to nominal moment capacity for circular section, lb-in. (N-mm)
M_{cr}	=	cracking moment, lb-in. (N-mm)
M_n	=	nominal moment capacity, lb-in. (N-mm)
M_{nb}	=	nominal moment strength corresponding to balanced failure, lb-in. (N-mm)
M_s	=	moment due to sustained load, lb-in. (N-mm)
M_u	=	factored moment at section, lb-in. (N-mm)
n_f	=	ratio of modulus of elasticity of FRP bars to modulus of elasticity of concrete
p_{cp}	=	outside perimeter of concrete cross section, lb-in. (N-mm)
r_b	=	internal radius of bend in FRP reinforcement, in. (mm)
s	=	stirrup spacing or pitch of continuous spirals, and longitudinal FRP bar spacing, in. (mm)
T	=	temperature, °C
T_1, T_2	=	tensile force corresponding to A_1 and A_2, lb (N)
$T_{f,i}$	=	tensile force in i^{th} layer, lb (N)
T, T_{max}	=	tensile force and maximum tensile force, lb (N)
T_n	=	nominal torsional moment strength, lb-in. (N-mm)
T_g	=	glass transition temperature, °F (°C)

T_u	=	factored torsional moment at section, lb-in. (N-mm)
t	=	time, minutes
t_{slab}	=	slab thickness, in. (mm)
u	=	average bond stress acting on the surface of FRP bar, psi (MPa)
V_c	=	nominal shear strength provided by concrete, lb (N)
V_f	=	shear resistance provided by FRP stirrups, lb (N)
V_n	=	nominal shear strength at section, lb (N)
V_s	=	shear resistance provided by steel stirrups, lb (N)
V_u	=	factored shear force at section, lb (N)
w	=	maximum crack width, in. (mm)
w_c	=	maximum crack width at the tension face of a flexural member, in., (mm)
x, x_b	=	distance of N.A. from compression edge and at balanced condition, in. (mm)
y_t	=	distance from centroidal axis of gross section, neglecting reinforcement, to tension face, in. (mm)
α	=	angle of inclination of stirrups or spirals
α	=	top bar modification factor
α	=	ratio of x over d
α_1	=	ratio of average stress of equivalent rectangular stress block to f'_c
α_L	=	longitudinal coefficient of thermal expansion, 1/°F (1/°C)
α_T	=	transverse coefficient of thermal expansion, 1/°F (1/°C)
β	=	ratio of distance from neutral axis to extreme tension fiber to distance from neutral axis to center of tensile reinforcement
β_1	=	factor relating depth of equivalent stress block to neutral axis depth
β_2	=	factor representing the influence of the load duration and repetition
β_d	=	reduction coefficient used in calculating deflection
γ	=	ratio of d over h or of d over D for columns
$\Delta_{(cp+sh)}$	=	additional deflection due to creep and shrinkage under sustained loads, in. (mm)
$(\Delta i)_{sus}$	=	immediate deflection due to sustained loads, in. (mm)
$(\Delta/l)_{max}$	=	limiting deflection-span ratio
ε_c	=	strain in concrete
ε_{cu}	=	ultimate strain in concrete
ε_f	=	strain in FRP reinforcement
ε_{fi}	=	reinforcement strain in i^{th} layer
ε_o	=	maximum strain of unconfined concrete corresponding to f_c
ε_{fu}	=	design rupture strain of FRP reinforcement
ε^*_{fu}	=	guaranteed rupture strain of FRP reinforcement defined as the mean tensile strain at failure of sample of test specimens minus three times standard deviation ($\varepsilon^*_{fu} = \varepsilon_{u,ave} - 3\sigma$)

ε_m	=	reinforcement tensile strain at midspan
$\varepsilon_{u,ave}$	=	mean tensile strain at rupture of sample of test specimens
ε_v	=	shear friction strain in reinforcement
ε_y	=	design yield strain
η	=	ratio of distance from extreme compression fiber to centroid of tension reinforcement (d) to overall height of flexural member (h)
λ	=	multiplier for additional long-term deflection
μ	=	coefficient of subgrade friction for calculation of shrinkage and temperature reinforcement
ξ	=	time-dependent factor for sustained load
ρ'	=	ratio of steel compression reinforcement, $\rho' = A_s'/bd$
ρ_b	=	FRP reinforcement ratio producing balanced strain conditions
ρ_f	=	average deterioration factors for modulus of elasticity at a specific temperature T in °C
ρ_f	=	FRP reinforcement ratio
ρ_f'	=	ratio of FRP compression reinforcement
ρ_{fv}	=	ratio of FRP shear reinforcement
$\rho_{f,ts}$	=	reinforcement ratio for temperature and shrinkage FRP reinforcement
ρ_{min}	=	minimum reinforcement ratio for steel
σ	=	standard deviation
σ_c	=	compressive stress, psi (MPa)
χ	=	curvature
ϕ	=	strength reduction factor
ω_f	=	tension reinforcement index
ω_{fb}	=	tension reinforcement index corresponding to balanced failure

4.1 INTRODUCTION

In this chapter, the design of slabs and beams is discussed. Following the conventional design procedure, this chapter first investigates the structural analysis of flexural members and elaborates on the parameters that define the input and determine the output of such analysis methods. Next, flexural design with fiber-reinforced polymer (FRP) bars is discussed and it is demonstrated how their mechanical behavior can divert the design process from methods that are well established for steel bars. Flexural design is completed by detailing and explaining the serviceability provisions of FRP reinforced concrete (FRP RC) flexural members that, when compared to steel RC, play a more prominent role in the overall design process. Shear design of flexural members with or without FRP transverse reinforcement concludes this chapter. Chapters 6–8 and 10 propose design examples that clarify the topics covered in this chapter.

4.2 STRUCTURAL ANALYSIS

FRP RC slabs and beams are designed to resist the maximum effects induced by the factored and the service loads. These maximum effects can be computed based on the elastic or plastic structural analyses. When applicable, approximate analysis methods are also available. Frame analysis to compute maximum design bending moments and shear forces in concrete slabs and beams is outside the scope of this book.

4.2.1 Loading conditions for ultimate and serviceability limit states

For the design of flexural members, the following load combinations are generally considered [1,2]:

$$U = 1.2D + 1.6L \tag{4.1}$$

$$U = 1.2D + 1.6 \, (L_r \text{ or } S \text{ or } R) + (1.0L \text{ or } 0.8W) \tag{4.2}$$

$$U = 1.2D + 1.6W + 1.0L + 0.5 \, (L_r \text{ or } S \text{ or } R) \tag{4.3}$$

where D is the dead load, L is the live load, L_r is the roof live load, S is the roof snow load, R is the roof rain load, and W is the roof wind load.

Traditionally, the load combinations for serviceability limit states use a load factor of 1.0 on all service loads [1].

4.2.2 Concrete properties

Several models interpreting the behavior of concrete in compression are available in the technical literature. The compressive stress–strain diagram for normal strength concrete proposed by Todeschini, Bianchini, and Kesler [3] is represented in Figure 4.1. f'_c is the design concrete compressive

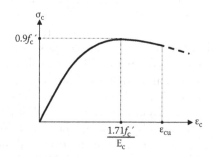

Figure 4.1 Todeschini and colleagues' stress–strain curve for concrete in compression. (Todeschini, Bianchini, and Kesler. *ACI Journal Proceedings* 61 (6): 701–716, 1964.)

strength and ε_{cu} is the maximum usable concrete compressive strain and is assumed equal to 0.003.

For ultimate strength calculations controlled by concrete crushing, ACI 318-11 [2] allows the approximation of the stress–strain curve to an equivalent rectangular stress, or "stress block," distribution as discussed in Section 4.5. In the design examples discussed in the chapters of Part 3, the stress–strain curve proposed by Todeschini et al. [3] is adopted when concrete crushing does not control failure.

Modulus of elasticity. The modulus of elasticity of concrete varies with concrete compressive strength (f'_c), concrete age, properties of cement and aggregates, and rate of loading. Based on statistical analysis of experimental data available for concrete with unit weights, w, varying between 90 and 155 pcf (1442 and 2483 kg/m^3), ACI 318-11 [2] provides the following empirical equation for computing the modulus of elasticity:

$$E_c = 33 \cdot w^{1.5} \sqrt{f'_c} \tag{4.4}$$

[or $E_c = 0.43 \cdot w^{1.5} \sqrt{f'_c}$ in SI units]

This equation is representative of the secant modulus for a compressive stress at service load. The secant modulus of elasticity in tension is generally assumed to be the same as in compression for computing deflections under service conditions.

For normal-weight concrete weighing 145 pcf (2323 kg/m^3), the following simplified equation is suggested by ACI 318-11:

$$E_c = 57000 \sqrt{f'_c} \tag{4.5}$$

[or $E_c = 4700 \sqrt{f'_c}$ in SI units]

It must be noted that, generally, creep and shrinkage over time cause a reduction of the secant modulus in compression inducing larger deflections. These effects are not taken into consideration in either Equation (4.4) or (4.5).

Tensile strength. ACI 318-11 suggests the following equation to compute the concrete tensile strength:

$$f_{ct} = 7.5\lambda \sqrt{f'_c} \tag{4.6}$$

[or $f_{ct} = 0.6\lambda \sqrt{f'_c}$ in SI units]

where λ is a modification factor equal to 1.0 for normal-weight concrete or 0.75 for lightweight concrete.

4.2.3 Cross-sectional properties

The analysis of a reinforced concrete frame should reflect the potential for cracking and inelastic action for each member. However, this approach would require selecting different stiffnesses for all members. For this reason, when analyzing concrete frames, the gross sectional properties are generally used.

Transformed section. The elastic-beam theory of de Saint-Venant can be applied to FRP reinforced concrete mechanics only if the FRP concrete section is transformed to an all-concrete section (this is generally preferred to an all-FRP section) [1]. In fact, an area of FRP bars equal to A_f can be considered equivalent to an area of concrete equal to $n_f A_f$, where the modular ratio, n_f, is equal to

$$n_f = \frac{E_f}{E_c} \tag{4.7}$$

When computing the neutral axis depth below the top fiber of the cross-section, c_g, of uncracked reinforced concrete, the contribution of the FRP reinforcement can, generally, be neglected. In this case, the neutral axis intersects the geometrical centroid of the gross section.

In a cracked section, the neutral axis intersects the mechanical centroid of the section. For a rectangular cross section, the cracked neutral axis depth, c_{cr}, can be computed as follows:

$$c_{cr} = kd_f = \left(\sqrt{2\rho_f n_f + \left(\rho_f n_f \right)^2} - \rho_f n_f \right) \cdot d_f \tag{4.8}$$

where ρ_f is the FRP reinforcement ratio and d_f is the effective depth of the FRP reinforcement.

T-sections. A T-beam section is generally considered when a beam is built integrally with the slab. ACI 318-11 defines the effective width of the slab, b_{eff}, acting as a T-beam flange as the minimum of: (a) one-quarter of the beam span length, (b) eight times the slab thickness, or (c) one-half the clear distance to the next beam.

For a spandrel beam (or reverse L-beam), ACI 318-11 mandates the effective flange width, b_{eff}, not to exceed: (a) 1/12 of the beam span length, (b) six times the slab thickness, or (c) half the clear distance to the next beam.

COMMENTARY

Equation (4.8) can be used for T-sections when the cracked neutral axis is included within the slab thickness, t_{slab}, which occurs when the following condition applies:

$$b_{eff} \frac{t_{slab}^2}{2} \geq n_f A_f (d_f - t_{slab}) \tag{4.9}$$

Table 4.1 Recommended minimum thickness of one-way slabs and beams

	Minimum thickness, h			
	Simply supported	One-end continuous	Both-ends continuous	Cantilever
One-way slabs	l/13	l/17	l/22	l/5.5
Beams	l/10	l/12	l/16	l/4

4.3 INITIAL MEMBER PROPORTIONING

For steel reinforced concrete flexural members, ACI 318-11 mandates minimum values of member thickness that satisfy given deflection-to-span ratios. The applicability of this approach for FRP reinforced concrete members is still under scrutiny. However, ACI 440.1R-06 [4] recommends minimum thickness values for indirect control of deflections in one-way slabs and beams, which can be used for initial member proportioning only (Table 4.1).

Deflections in FRP RC members tend to be larger than in their steel RC counterparts, due to the lower tensile modulus of elasticity of commercially available FRP reinforcing bars. For this reason, FRP reinforced members are generally subjected to higher depth requirements for comparable span lengths.

The values in Table 4.1 were derived by Ospina, Alexander, and Cheng [5] based on the following equation:

$$\frac{l}{h} = \frac{48\eta}{5K_1}\left(\frac{1-k}{\varepsilon_f}\right)\left(\frac{\Delta}{l}\right)_{max} \tag{4.10}$$

where ε_f is the reinforcement tensile strain at midspan, h is the height of the cross section, k is the neutral axis depth to reinforcement depth ratio at midspan, η is the reinforcement depth to member thickness ratio, and K_1 is a constant that depends on the loading and support conditions. For example, K_1 can be taken as 1.0, 0.8, 0.6, and 2.4 for uniformly loaded, simply supported, one-end continuous, both-ends continuous, and cantilevered spans, respectively. The ratio η may be assumed to range between 0.85 and 0.95.

COMMENTARY

When a linear distribution of the strain over a member's cross section is assumed, the maximum immediate elastic deflection at midspan, Δ_{max}, of a one-way flexural member under uniform distributed load can be computed as

$$\Delta_{max} = K_1 \frac{5}{48} \cdot \frac{Ml^2}{E_c I} \tag{4.11}$$

---

where M is the maximum positive moment, l is the span length, E_c is the modulus of elasticity of concrete, and I is the moment of inertia. The term $M/E_c I$ is the curvature of the cross section at midspan, $\chi_{m,cr}$, which can be computed as follows when the elastic cracked conditions are assumed:

$$\chi_{m,cr} = \frac{\varepsilon_m}{d(1-k_m)}$$

(4.12)

where ε_m is the reinforcement tensile strain at midspan, d is the flexural reinforcement effective depth, and k_m is the neutral axis depth to reinforcement depth ratio at midspan. Rearranging terms and setting $\eta = d/h$, the maximum deflection at midspan to span length ratio, Δ_m/l, is given by the following equation:

$$\frac{\Delta_m}{l} = \frac{5}{48} \cdot \frac{K_1}{\eta} \cdot \frac{\varepsilon_m}{1-K_m} \cdot \frac{l}{h}$$

(4.13)

where η may be assumed to range between 0.85 and 0.95. Equation (4.13) can be used to estimate the minimum thickness of the flexural member by posing $\Delta_m = \Delta_{max}$.

In the attempt to account for the effect of tension stiffening of concrete, Ospina and Gross [6] rewrote Equation (4.13) in the following fashion:

$$\frac{\Delta_m}{l} = \frac{5}{48} \cdot \frac{K_1}{\eta} \cdot l \cdot \left[(1-\xi)\chi_{m,g} + \xi \cdot \chi_{m,cr} \right]$$

(4.14)

where $\chi_{m,g}$ is the cross-sectional curvature at midspan for uncracked conditions that can be computed as follows:

$$\chi_{m,g} = \frac{2f_{ct}}{E_c h}$$

(4.15)

ξ is a constant to allow expressing the maximum deflection as a function of the average curvature instead of the curvature at a cracked cross section and its expression is reported in the following:

$$\xi = 1 - \beta_1 \beta_2 \left(\frac{M_{cr}}{M} \right)^2$$

(4.16)

where β_1 is a factor accounting for bond quality of the bars ($\beta_1 = 1$ for high bond bars), β_2 is a factor representing the influence of the load duration and repetition ($\beta_2 = 1$ for first loading), and M_{cr} is the cracking moment. Imposing $E_c = 57,000\sqrt{(f'_c)}$ and $f_r = 7.5\sqrt{(f'_c)}$ (in SI units: $4700\sqrt{(f'_c}$ and $0.62\sqrt{(f'_c)})$), and rearranging terms, the maximum deflection can be written as follows:

$$\frac{\Delta_m}{l} = \frac{5}{48} \cdot \frac{K_l}{\eta} \cdot \left[(1-\xi)\frac{15\eta}{57000} + \xi \cdot \frac{\varepsilon_m}{(1-k_m)} \right] \cdot \frac{l}{h} \qquad (4.17)$$

$$\left[\text{or } \frac{\Delta_m}{l} = \frac{5}{48} \cdot \frac{K_l}{\eta} \cdot \left[(1-\xi)\frac{1.24\eta}{4700} + \xi \cdot \frac{\varepsilon_m}{(1-k_m)} \right] \cdot \frac{l}{h} \text{ in SI Units} \right]$$

Posing $\Delta_m = \Delta_{m,limit}$ in Equation (4.17), the minimum thickness of the flexural member that satisfies the deflection requirement can be estimated.

4.4 FRP DESIGN PROPERTIES

The FRP tensile properties to be used in design equations are reduced by ACI 440.1R-06 to include the detrimental effect of long-term exposure to various environments. The design tensile strength, f_{fu}, and the design ultimate tensile strain, ε_{fu}, are calculated as

$$f_{fu} = C_E f_{fu}^* \qquad (4.18)$$

and

$$\varepsilon_{fu} = C_E \varepsilon_{fu}^* \qquad (4.19)$$

where C_E is the environmental reduction factor, and f_{fu}^* and ε_{fu}^* are the manufacturer's guaranteed tensile strength and ultimate strain, respectively, and are defined as the mean value of the sample population minus three times the standard deviation. The values of C_E recommended by ACI-440.1R-06 are summarized in Table 4.2.

Table 4.2 Environmental reduction factor for various fibers and exposure conditions

Exposure condition	Fiber type	C_E
Concrete not exposed to earth and weather	Carbon	1.0
	Glass	0.8
	Aramid	0.9
Concrete exposed to earth and weather	Carbon	0.9
	Glass	0.7
	Aramid	0.8

Although the effect of temperature is included in the values of C_E, ACI 440.1R-06 does not recommend using FRP bars in environments with a service temperature higher than the glass transition temperature (T_g) of the resin used for their manufacturing.

4.5 BENDING MOMENT CAPACITY

As it has been demonstrated experimentally, irrespectively of the reinforcing material used (steel or FRP), the basic assumptions for the flexural theory of RC members analyzed using the strength design method can be summarized as follows:

1. Plane sections remain plane; this means that shear deformations can be disregarded.
2. Perfect bond exists between reinforcing bars and the surrounding concrete; in other words, the strain in the reinforcement is equal to the strain in the concrete at the same level.
3. Stresses in both concrete and reinforcement are computed based on the strain level reached in each material using the appropriate constitutive laws for concrete and reinforcing bars. In particular, for the case of concrete, up to the serviceability limit state, a linear–elastic relationship will be used; past the linear elastic point up to crushing, either the Todeschini model or the equivalent stress block can be used. Regardless of the limit state considered, steel and FRP reinforcing bars are considered elastic–plastic and linear–elastic, respectively.

Although not formally needed, the following two assumptions are usually introduced to simplify the calculation process with little loss in the accuracy of the final results:

4. The tensile strength of the concrete is neglected.
5. The concrete is assumed to fail when it reaches a maximum preset compressive strain.

The basic safety relationship at the ultimate limit state can be written as

$$\phi M_n \geq M_u \tag{4.20}$$

In Equation (4.20), ϕM_n is the factored bending moment capacity of the member and is a function of the member geometry, the location of the reinforcement, and the mechanical properties of the materials; the term "factored" means that the nominal calculated bending moment capacity has been reduced by the safety factors associated with the materials or the failure mode depending upon the calculation procedures followed.

The second term of Equation (4.20), M_u, is the factored bending moment resulting from the analysis of the member and is a function of the member

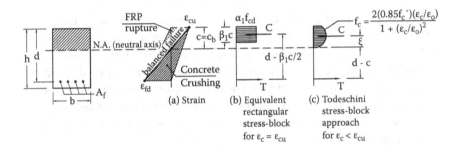

Figure 4.2 Failure mode regions for FRP RC flexural members.

geometry, stiffness and boundary conditions, and the applied loads; the term "factored" means that the calculated bending moment associated to a specified loading condition has been amplified by the safety factors related to the acting loads. M_u comes from the structural analysis performed on the system being studied.

4.5.1 Failure mode and flexural capacity

Balanced failure. Both concrete crushing and FRP rupture are possible failure modes. When the failure mode is controlled by the simultaneous occurrence of concrete crushing and FRP rupture, it is termed "balanced failure." The neutral axis position for balanced failure, $c = c_b$, can easily be determined from strain compatibility as follows (Figure 4.2):

$$c_b = \frac{\varepsilon_{cu}}{\varepsilon_{cu} + \varepsilon_{fu}} d \tag{4.21}$$

where ε_{fu} is the design tensile strength of the FRP reinforcement.

The position of the neutral axis corresponding to balanced failure is used as the basis to establish the member failure mode. When the position of the neutral axis at ultimate, c, is larger than c_b, the failure is controlled by the crushing of the concrete; conversely, when c is less than c_b, the failure is initiated by rupture of the FRP reinforcement.

COMMENTARY

Traditional steel reinforced flexural members are designed to display concrete failure when the strain in the steel has passed its yielding limit. In this way, the member is said to be "under-reinforced." Such behavior corresponds to a ductile failure mode with signs of the incipient collapse in the form of extensive cracking and large deflections visible on the flexural member.

Figures 4.3(a) and (b) show the normalized bending moment (defined as the ratio of the bending moment acting on the beam at a given load stage, M, to the member flexural capacity at the ultimate limit state, M_n) and the normalized neutral axis depth (defined as the ratio of the neutral axis depth [axis of zero strain] measured from the extreme compression fiber, c, to the beam's depth, h) plotted as a function of the beam curvature defined as $\chi = \varepsilon/y$, where ε represents the strain at a distance y from the neutral axis at the load stage considered.

Several considerations can be drawn by the observation of such diagrams. Line OA in Figure 4.3(a) represents all points where the beam is uncracked and both bending moment and neutral axis depth can be found without appreciable error by neglecting the presence of the steel reinforcing bars. In this case, the beam can be considered homogeneous with the neutral axis located at $c = h/2$ (therefore, the normalized neutral axis depth $= h/2/h = 0.5$). At point A, the beam cracks; from point A to B, the neutral axis depth and the stiffness of the member rapidly decrease to a point of stability. At this stage, the steel is linear–elastic and the concrete is practically linear–elastic. In this curvature range, corresponding to service conditions, the internal moment magnitude increases because of an increase in the tensile and compressive forces. At point B, the steel reaches the yield point. From points B to C, the steel is yielded and the concrete gradually moves to the inelastic and postpeak range. The neutral axis position further shifts toward the top fiber as the internal moment magnitude increases (due to an increase of the internal couple arm), while the total compressive and tensile forces remain constant. At point C, the compressive strain reaches the value of ε_{cu}, which corresponds to the assumed concrete crushing, and failure occurs.

If the steel does not yield (as in the case of Figure 4.3b), the neutral axis after cracking has occurred moves downward toward the tensile side of the beam as the increase of moment is given by the increase of the compressive and tensile forces. From this last figure, it is clear how the neutral axis position for "over-reinforced" beams sways around the mid-depth of the beam.

The behavior of FRP RC members is affected by the presence of reinforcement that does not yield and is to be considered liner–elastic up to failure. As opposed to traditional RC structures, where failure is always controlled by crushing of the concrete before or after yielding of the steel, members reinforced with FRP bars may display either concrete crushing or FRP rupture as the governing failure mode (Figure 4.4). As a result, the material (concrete or FRP) that first reaches its ultimate strain is the one that dictates the member failure mode.

Similar considerations to what has been observed for steel RC members in Figure 4.3 may be provided here for members reinforced with FRP bars. With reference to Figure 4.4, line OA represents points where the beam is uncracked; in this range, the constituent materials are linear–elastic, the neutral axis depth is located at $h/2$, and the bending moment corresponding to the cracking of the concrete (point A) can be found based on the Navier's equation as follows:

$$M_{cr} = \frac{f_r I_g}{y_t} \tag{4.22}$$

where y_t is the distance from the centroidal axis of the gross section neglecting the tensile reinforcement and I_g is the gross moment of inertia.

From point A to B, the beam is cracked; this range corresponds to service condition where the concrete is still a linear–elastic material. Such assumption can be considered acceptable up to a maximum compressive strain in the concrete of $0.45 \cdot f'_c / E_c$. In this curvature range, the stiffness of the beam greatly decreases as it is indicated by the slope of line AB (compare with line AB of Figure 4.3a) due to the reduced modulus of elasticity of FRP bars compared to steel bars. From point B to C, as the member flexural capacity is approached, the neutral axis position remains basically constant (Figure 4.4a) or slightly changes direction and shifts downward (Figure 4.4b). This is necessary to maintain the equilibrium of the horizontal forces: As the value of the internal tensile force increases, the total compressive force also needs to increase. At point C, the tensile strain in the FRP reaches its ultimate value and the failure is controlled by the rupture of the FRP bars (Figure 4.4a) or the maximum concrete compressive strain, ε_{cu}, is attained and the failure is controlled by the crushing of the concrete (Figure 4.4b).

The lack of yielding of the FRP reinforcement that produces an overall less ductile behavior of FRP RC members as compared to steel RC members must be compensated for with an increase in the safety factor used for design. This increased safety can take the form of more stringent strength-reduction factors or larger material safety factors depending upon the design methodology followed.

The moment-curvature diagram for the four beams shown in Figures 4.3 and 4.4 is depicted in Figure 4.5. These diagrams are based on the material properties summarized in Table 4.3.

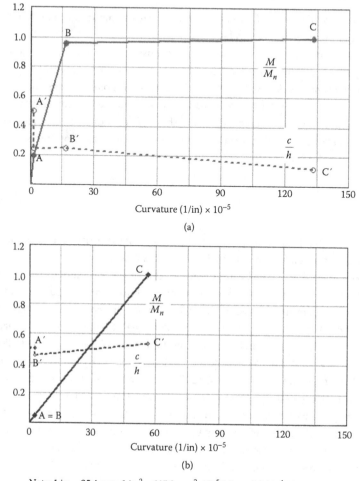

Note: 1 in. = 25.4 mm; 1 in.² = 645.2 mm²; 10^{-5} 1/in. = 3.9 10^{-4} 1/m

Figure 4.3 Moment-curvature diagram for RC beams reinforced with steel bars: (a) *under-reinforced member (b = 8 in., d = 16 in., A_s = 0.80 in².); (b) over-reinforced member (b = 16 in., d = 8 in., A_s= 5.67 inch.²).*

When failure is initiated by the crushing of the concrete ($c > c_b$), the stress distribution in the concrete can be approximated with the equivalent rectangular stress block having the parameters α_1 and β_1 defined below (see also Figure 4.2b):

- α_1 = 0.85 is the ratio of average stress of equivalent rectangular stress block to f'c.
- β_1 is the factor taken as 0.85 for concrete strength f'c up to and including 4000 psi (28 MPa). For strength above 4000 psi (28 MPa), this factor

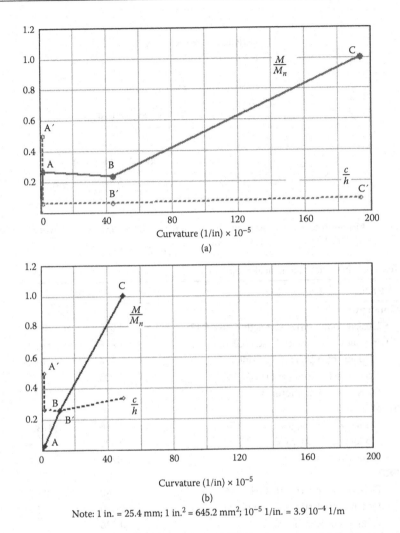

Figure 4.4 Moment-curvature diagram for RC beams reinforced with FRP bars: (a) *FRP rupture (b = 8 in., d = 16 in., A_f = 0.35 in².)*; (b) *concrete crushing (b = 8 in., d = 16 in., A_f = 5.55 in².)*

Note: 1 in. = 25.4 mm; 1 in.² = 645.2 mm²; 10^{-5} 1/in. = 3.9 10^{-4} 1/m

is reduced continuously at a rate of 0.05 per each 1000 psi (7 MPa) of strength in excess of 4000 psi (28 MPa), but is not taken less than 0.65.

The strain in the FRP reinforcement is less than that corresponding to the bar rupture and it is determined from strain compatibility as follows

$$\varepsilon_f = \frac{d-c}{c}\varepsilon_{cu} \leq \varepsilon_{fu} \qquad (4.23)$$

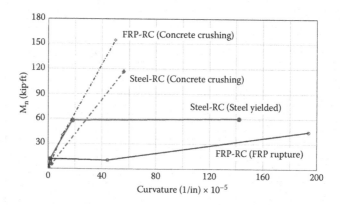

Figure 4.5 Moment-curvature diagram for steel and GFRP beams. *Note:* I ft-kip = 1.36 KN-m; 10^{-5} I/in. = 3.9 10^{-4} I/m.

Table 4.3 Material mechanical properties

Material	Value	
Concrete		
• Design compressive strength, f'_c, psi (MPa)	4,000	(27.6)
• Modulus of elasticity, E_c, ksi (GPa)	3,605	(24.9)
• Modulus of rupture, f_r, psi (MPa)	474	(3.3)
• Compressive strain at crushing, ε_{cu}	0.003	
• Stress block parameter, β_1	0.85	
Steel		
• Design yield strength, f_y, ksi (MPa)	60	(413.7)
• Modulus of elasticity, E_s, ksi (GPa)	29,000	(199.9)
• Design yield strain, ε_y	0.00207	
FRP		
• Design tensile strength, f_{fd}, ksi (MPa)	140	(965)
• Modulus of elasticity, E_f, ksi (GPa)	5,000	(34.5)
• Design tensile strain, ε_{fd}	0.028	

The result of the compressive and tensile forces in both concrete and FRP reinforcement can be expressed as

$$C = \alpha_1 f'_c \beta_1 cb \tag{4.24}$$

$$T = A_f E_f \varepsilon_f \tag{4.25}$$

where b is the width of the flexural member. By substituting Equation (4.23) into Equation (4.24) and imposing the equilibrium condition, $C = T$, the following second-order equation is derived:

$$\alpha c^2 + \beta c + \gamma = 0 \tag{4.26}$$

where the coefficients of the unknown term, c, are

$$\alpha = \alpha_1 f_c' \beta_1 b$$

$$\beta = A_f E_f \varepsilon_{cu}$$

$$\gamma = A_f E_f \varepsilon_{cu} d \tag{4.27}$$

The only solution with physical meaning of Equation (4.26) gives the position of the neutral axis as follows:

$$c = \frac{-\beta + \sqrt{\beta^2 - 4\alpha\gamma}}{2\alpha} \tag{4.28}$$

The nominal bending moment capacity of the member is calculated as

$$M_n = \left(\alpha_1 f_c' \beta_1 cb\right)\left(d - \frac{\beta_1 c}{2}\right) \tag{4.29}$$

Figure 4.6 shows an example of glass fiber-reinforced polymer (GFRP) RC type slab tested under four-point bending. Figure 4.6(a) depicts the fairly large deflection while Figure 4.6(b) portrays the compression failure mode.

FRP rupture: rigorous approach. When failure is initiated by rupture of the FRP reinforcement ($c < c_b$), the compressive strain in the concrete has not reached the ultimate value ε_{cu}. Todeschini et al. [3] developed the stress-block approach reported in Figure 4.2(c) for concrete columns reinforced with high-strength steel that can be used for any ε_c value assumed by the compressive strain in the concrete and calculated using strain compatibility as follows:

$$\varepsilon_c = \frac{c}{d-c}\varepsilon_{fu} \leq \varepsilon_{cu} \tag{4.30}$$

The result of the compression and tensile forces in both concrete and FRP reinforcement can be expressed as

$$C = 0.85 f_c' \frac{bc}{\varepsilon_c/\varepsilon_o} \text{ in } \left[1 + \left(\varepsilon_c/\varepsilon_o\right)^2\right] \tag{4.31}$$

$$T = A_f f_{fu} \tag{4.32}$$

where ε_o represents the concrete strain at maximum strength as determined from cylinder tests calibrated based on the results reported by Todeschini et al. [3] as

$$\varepsilon_o = 1.71\frac{f_c'}{E_c} \tag{4.33}$$

(a) Slab under load showing the large deflection

(b) Compression failure

Figure 4.6 Testing of GFRP-RC slab at the Universitá di Napoli–Federico II, Naples, Italy. (Courtesy of Prof. Andrea Prota.)

Since a closed-form solution of the equation obtained by replacing Equations (4.30) and (4.33) into Equation (4.31) and imposing the equilibrium condition, $C = T$, is impractical, a numerical procedure may be adopted by assuming tentative values for the neutral axis depth, c, until the equilibrium condition obtained by equating Equations (4.31) and (4.32) is met.

Once the position of the neutral axis, c, is known, the nominal bending moment capacity can be calculated as

$$M_n = A_f f_{fu} (d - c + \xi) \qquad (4.34)$$

where ξ represents the distance from the neutral axis to the centroid of the stress block (Figure 4.2c) determined as

$$\xi = \frac{2c \left[\varepsilon_c / \varepsilon_o - \tan^{-1} (\varepsilon_c / \varepsilon_o) \right]}{(\varepsilon_c / \varepsilon_o) \ln \left[1 + (\varepsilon_c / \varepsilon_o)^2 \right]} \qquad (4.35)$$

with $\varepsilon_c / \varepsilon_o$ is in radians when computing $\tan^{-1}(\varepsilon_c / \varepsilon_o)$.

Available guidelines on concrete members reinforced with FRP bars allow for the use of simplified approaches to compute the bending moment capacity with little loss in the accuracy of the final result.

FRP rupture: the ACI 440.1R-06 approach. ACI 440.1R-06 allows using the equivalent rectangular stress block also for the case of FRP rupture irrespectively of the maximum strain reached by the concrete in compression. This means that the compression resultant in the concrete, C, may be expressed as

$$C = \alpha_1 f_c' \beta_1 bc \qquad (4.36)$$

where α_1 is 0.85, β_1 is the parameter of the rectangular stress block, and c is the position of the neutral axis at failure. The force in the FRP reinforcement can be evaluated as

$$T = A_f \varepsilon_{fu} E_f \qquad (4.37)$$

where A_f is the total FRP area; ε_{fu} and E_f represent the design tensile strain and modulus of elasticity of the FRP reinforcement, respectively.

The proposed equation for the bending moment capacity is expressed as follows:

$$M_n = A_f f_{fu} \left(d - \frac{\beta_1 c_b}{2} \right) \qquad (4.38)$$

Note that ACI 440.1R-06 assumes a conservative estimation of the bending moment capacity by replacing the neutral axis depth at failure with the value at balanced condition, $c = c_b$, expressed by Equation (4.21).

Figure 4.7 shows an example of an FRP RC beam tested under four-point bending. Figure 4.7a depicts the crack pattern while Figure 4.7b portrays the tensile failure mode.

(a) Immediately prior to failure with marked-up cracks

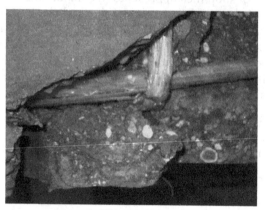

(b) Detail of failed GFRP bar in tension

Figure 4.7 Testing of GFRP-RC beam at the Universitá di Napoli–Federico II, Naples, Italy. (Courtesy of Prof. Andrea Prota.)

COMMENTARY

When FRP rupture controls failure, two approaches have been investigated in order to improve the ductility of FRP RC flexural members. The first is to use hybrid FRP bars fabricated by combining two or more different fiber types to simulate the elastic–plastic behavior of the steel bars [7,8]. This approach has shown some success in research studies but has resulted in limited practical applications because of the complicated and costly manufacturing process of the hybrid bars [9]. The second approach is to improve the pseudoductility of the concrete by adding steel or polypropylene fibers to the concrete mixture [9,10].

4.5.2 Nominal bending moment capacity of bond-critical sections

One of the main assumptions at the basis of reinforced concrete mechanics is a "perfect" bond between the reinforcing bars and the surrounding concrete. Therefore, adequate embedment length of the FRP reinforcement is required to avoid bond failure. ACI 440.1R-06 defines as bond critical those sections where the maximum achievable stress in the FRP reinforcement is limited by the following equation:

$$f_{fe} = \frac{\sqrt{f'_c}}{\alpha}\left(13.6\frac{l_e}{d_b} + \frac{C}{d_b}\frac{l_e}{d_b} + 340\right) \le f_{fu} \tag{4.39}$$

$$\left[\text{or } f_{fe} = \frac{0.083\sqrt{f'_c}}{\alpha}\left(13.6\frac{l_e}{d_b} + \frac{C}{d_b}\frac{l_e}{d_b} + 340\right) \le f_{fu} \text{ in SI units}\right]$$

where l_e is the length of FRP bar embedded in concrete, d_d is the bar diameter, C is the lesser of the cover to the center of the bar or one-half of the center-on-center spacing of the bars being developed, and α is a factor to account for the bar location. α is taken equal to 1.5 for bars with more than 12 in. (300 mm) of concrete cast below, otherwise it is taken equal to 1.0. In other words, the developable FRP bar stress, f_{fe}, cannot be smaller than the design FRP tensile stress, $f_f \le f_{fu}$, as defined in Equation (4.40), to prevent bond failure. Otherwise, corrective measures must be taken, such as increasing the embedment length, increasing the number of bars, replacing the planned bars with smaller diameter ones for the same reinforcement area, or recomputing the nominal moment capacity based on bond strength limitation.

When bond limits the stress that can be developed in reinforcement, the two possible failure modes are concrete crushing or bond failure. The failure mode is concrete crushing when the condition stated in Equation (4.40) is verified:

$$f_{fe} \ge f_f = \sqrt{\frac{(E_f\varepsilon_{cu})^2}{4} + \frac{0.85\beta_1 f'_c}{\rho_f}E_f\varepsilon_{cu}} - 0.5E_f\varepsilon_{cu} \tag{4.40}$$

In this case, the nominal bending moment capacity can be computed as per ACI 440.1R.06 Equation (8.5) reported below:

$$M_n = \rho_f f_f\left(1 - 0.59\frac{\rho_f f_f}{f'_c}\right)bd^2 \tag{4.41}$$

Conversely, the bending moment capacity for failure controlled by bond can be computed based on the following equation:

$$M_n = A_f f_{fe} \left(d_f - \frac{\beta_1 c_b}{2} \right)$$

(4.42)

COMMENTARY

Equation (4.39) is based on the work by Wambeke and Shield [11]. They reviewed a database of 269 beam bond tests, including beam-end tests, notch-beam tests, and splice tests. The majority of the bars included in the study were GFRP. A linear regression of the normalized average bond stress, u, versus the normalized cover and embedment length was performed and resulted in the following relationship:

$$\frac{u}{\sqrt{f_c'}} = 4.0 + 0.3 \frac{C}{d_b} + 100 \frac{d_b}{l_e}$$

(4.43)

By considering the free-body diagram of an FRP bar of diameter d_b and area $A_{f,bar}$ embedded in concrete for a length equal to l_e, the equilibrium of forces was written as follows:

$$l_e \pi d_b u = A_{f,bar} f_f$$

(4.44)

Wambeke and Shield [11], then, solved Equations (4.43) and (4.44) for the achievable bar stress given the existing embedment length and cover obtaining Equation (4.3).

ACI 440.1R-06 recommends avoiding embedment lengths shorter than $20d_b$, and limiting the ratio C/d_b to 3.5 to prevent pullout failure. The factor α is generally taken equal to 1.5 for bars with more than 12 in. (305 mm) of concrete cast below them ("top bars"), or 1.0 when the bars are in the bottom 12 in. (305 mm) of the formwork when the concrete is cast. When there is insufficient embedment length to develop full anchorage of a bar, a bent bar may be used.

4.5.3 Minimum FRP reinforcement

ACI 440.1R-06 prescribes that at every section of a flexural member where tensile reinforcement is required by analysis, A_f provided should not be less than the area given by

$$A_f = 4.9 \frac{\sqrt{f_c'}}{f_{fu}} b_w d$$

(4.45)

$$[\text{or } A_f = 0.41 \frac{\sqrt{f_c'}}{f_{fu}} b_w d \text{ in SI units}]$$

and not less than $(330/f_{fu})b_w d$ [or $(2.3/f_{fu})\, b_w d$ in SI units], where b_w and d are the cross-section web width and the distance from the extreme compression fiber to the centroid of tension reinforcement, respectively.

It has been shown that the requirements of Equation (4.45) may become unrealistic for large concrete cross sections. It is therefore suggested that Equation (4.45) need not be applied in a member where at every cross section, the area of tensile reinforcement provided is at least one-third greater than that required by analysis.

COMMENTARY

Traditional steel RC flexural members designed in compliance with code provisions must exhibit additional strength capacity beyond concrete cracking. This requirement is to prevent a brittle failure. Provisions for a minimum amount of reinforcement were originally established for members that, for architectural or other reasons, were larger in cross section than required by strength. If the amount of tensile reinforcement is too small, the ultimate bending moment capacity computed using the cracked section analysis may be less than that of the corresponding concrete section computed from its modulus of rupture, neglecting the presence of the existing reinforcement (Equation 4.22).

Similar considerations may be applied to FRP reinforced concrete members designed to fail by FRP rupture to prevent failure upon concrete cracking; for failures controlled by concrete crushing, the amount of FRP reinforcement automatically satisfies the minimum requirement.

4.5.4 Maximum FRP reinforcement

Provision 10.3.5 in ACI 318-11 limits the minimum tensile strain at failure in the longitudinal steel reinforcement of flexural members to a value of 0.004, which corresponds to roughly twice the yield stain of a Grade 60 steel (420 MPa). This strain limit is to ensure that the failure of the steel-RC structural member will always be ductile. Even though this limit loses relevance in the case of FRP reinforcement, it may be argued that irrespective of the fact that FRP bars do not yield, this strain threshold would at least ensure some visible level of distress in terms of deflection and crack width for a flexural member approaching failure. According to this provision, the maximum reinforcement ratio, ρ_{max}, for an FRP-RC member in flexure would be:

$$\rho_{max} = 0.85\beta_1 \frac{f'_c}{0.004 E_f} \frac{0.003}{0.003 + 0.004} = 91.1\beta_1 \frac{f'_c}{E_f}$$

It should be noted that this value of ρ_{max} cannot be physically attained in GFRP RC flexural members as it would be impossible to fit the many bars in a cross section.

COMMENTARY

Considering a worst case scenario of low compressive strength concrete, compute ρ_{max} and the corresponding number of bars for a slab and a beam using the following data:

Slab	Beam
Size:	Size:
$\quad b = 12.0$ in.	$\quad b = 12.0$ in.
$\quad d = 6.0$ in.	$\quad d = 24.0$ in.
Concrete:	
$\quad f'_c = 3.0$ ksi	
$\quad \beta_1 = 0.85$	
Reinforcement:	
$\quad E_f = 6,000$ ksi	
$\quad \rho_{max} = 0.387$	
$\quad \#6 = 0.44$ in.^{2}	$\quad \#9 = 1.00$ in.^{2}
Solution:	Solution:
$\quad A_f = 2.78$ in.^{2}	$\quad A_f = 11.15$ in.^{2}
$\quad$ Or 6#6 bars which would barely fit in the given width (clear spacing of 1.25 in.)	$\quad$ Or 11#9 bars which would not fit in the given width

---

4.5.5 Examples—Flexural strength

The examples presented here (US customary only) are intended to show an application of the algorithms discussed in the preceding sections. More exhaustive design examples are given in Chapters 6, 7, and 10 that deal with one-way slab, beam, and isolated footing, respectively.

Design equations aim at (a) calculating the flexural strength of an existing beam or (b) calculating design parameters such as size and reinforcement area for an applied bending moment. A critical parameter in all these calculations is the tension reinforcement index, ω_f, defined as

$$\omega_f = \rho_f \frac{f_{fu}}{f_c'} = \frac{A_f}{bd}\frac{f_{fu}}{f_c'}$$

In *case a*, ω_f is known and the strength can be defined as a function of it. In *case b*, unknown parameters are derived as functions of ω_f that may be known or unknown at first. Another dimensionless parameter is defined as

$$e = \frac{\varepsilon_{fu}}{\varepsilon_{cu}}; \varepsilon_{cu} = 0.003$$

Case a—calculation of nominal flexural strength of an existing member
(a1) Failure is initiated by concrete crushing if

$$\omega_f \geq \omega_{fb} = \frac{0.85\beta_1}{1+e}$$

In this case the stress level in the reinforcement can be calculated as

$$f = \frac{f_f}{f_{fu}} = \frac{\sqrt{1 + \dfrac{3.4\beta_1 e}{\omega_f}} - 1}{2e} \leq 1$$

And the nominal flexural strength is

$$M_n = \omega_f'\left(1 - \frac{\omega_f'}{1.7}\right)f_c'bd^2$$

where

$$\omega_f' = f\omega_f$$

Example 4.1

Calculate the nominal flexural strength of the following beam:

Concrete:	$f'_c = 4.0$ ksi
	$\beta_1 = 0.85$
Reinforcement:	$f_{fu} = 60$ ksi
	$E_f = 6{,}000$ ksi
	$\varepsilon_{fu} = 0.010$
	$A_f = 4\#10 = 5.08$ in.^{2}
Size:	$b = 16.0$ in.
	$h = 25.0$ in.
	$c = 3.0$ in.
	$d = h-c = 22.0$ in.

Continued

Solution: $e = 3.333$
$\omega_f = 0.2165$
$\omega_{fb} = 0.1667 < \omega_f$: Failure by concrete crushing
$f = 0.862$
$\omega'_f = 0.1839$
$M_n = 5,145$ kip-in. $= 428.8$ ft-kip

Example 4.2

Calculate the nominal flexural strength of the beam in Example 4.1, if $f'_c = 5.0$ ksi $(\beta_1 = 0.80)$.

Solution: $e = 3.333$
$\omega_f = 0.1732$
$\omega_{fb} = 0.1569 < \omega_f$: Failure by concrete crushing
$f = 0.946$
$\omega'_f = 0.1638$
$M_n = 5,731$ kip-in. $= 477.6$ ft-kip

This example demonstrates that when failure is governed by concrete crushing, the flexural strength of FRP RC members is sensitive to the strength of concrete. This is in contrast to the general notion about the traditional steel RC flexural members, whose strength is largely perceived to be independent of the concrete strength.

Example 4.3

Calculate M_{nb}, the nominal flexural strength of the beam in Example 4.1, if the area of its reinforcement is changed so that concrete crushing and FRP rupture occur simultaneously (balanced failure).

Solution: $M_{nb} = M_n(\omega_f = \omega_{fb})$
$f = 1.0$
$\omega'_f = \omega_{fb} = 0.1667$
$M_{nb} = 4658.1$ kip-in. $= 388.2$ ft-kip
$A_{fb} = 3.91$ in.^{2}

(a2) Failure is initiated by FRP rupture if

$$\omega_f \leq \omega_{fb} = \frac{0.85\beta_1}{1+e}$$

In this case two methods are available to calculate the nominal strength:

1. Exact method. The equivalent stress block for concrete must be calculated. The first step is to compute maximum strain of concrete corresponding to f'_c as

$$\varepsilon'_c = 1.7 \frac{f'_c}{E_c}$$

Then, an initial c, the depth of the compressive part of the section, is assumed and the strain on the compressive edge of the beam is calculated as

$$\varepsilon_c = \frac{c}{d-c} \varepsilon_{fu}$$

The parameters regarding the intensity (α_1) and depth of the stress block (β_1) are related as

$$\alpha_1 \beta_1 = \left(\frac{\varepsilon_c}{\varepsilon'_c} \right) - \frac{1}{3} \left(\frac{\varepsilon_c}{\varepsilon'_c} \right)^2$$

A new value for c can be calculated as

$$c = \frac{f_{fu}}{\alpha_1 \beta_1 f'_c} \frac{A_f}{b} = \frac{\omega_f}{\alpha_1 \beta_1} d$$

The previous three steps are repeated until convergence is achieved. Then:

$$M_n = \omega_f \left(1 - \frac{\beta_1 c}{2d} \right) f'_c b d^2 = \omega_f \left(1 - \frac{\omega_f}{2\alpha_1} \right) f'_c b d^2$$

where

$$\beta_1 = \frac{4\varepsilon'_c - \varepsilon_c}{6\varepsilon'_c - 2\varepsilon_c}$$

Selecting an appropriate value for the first guess of c can expedite the procedure. Noting that for this mode of failure,

$$c \le c_b = \frac{d}{1+e}$$

c_b is the compressive depth of the section corresponding to the balanced failure. Therefore, observing the preceding condition in selecting c in the first iteration is highly recommendable.

Figure 4.8 shows an example of FRP RC large beam tested under four-point bending. Figure 4.8(a) depicts the crack pattern while Figure 4.8(b) portrays the tensile and compressive failures that happened simultaneously (balanced failure).

(a) Immediately prior to failure with marked-up cracks

(b) Detail of failed concrete in compression and GFRP bar in tension

Figure 4.8 Testing of GFRP-RC beam at Missouri University of Science and Technology, Rolla, Missouri. (Courtesy of Prof. Fabio Matta.)

Example 4.4

Calculate the nominal flexural strength of the beam in Example 4.1, if $A_f = 2\#10 = 2.54$ in.^{2}.

Solution:	$\omega_f = 0.1082$
	$\omega_{fb} = 0.1667 > \omega_f$: Failure by FRP rupture
	E_c (ksi) $= 57\sqrt{f'_c}$ (psi) $= 3605$ ksi
	$\varepsilon'_c = 0.00189$
	$c_b = 5.08$ in.
Iteration 1:	$c = 4.0$ in. : First guess
	$\varepsilon_c = 0.00222$
	$\alpha_1\beta_1 = 0.715$
	$c = 3.3$ in.

Continued

Iteration 2:	$c = 3.3$ in.
	$\varepsilon_c = 0.00176$
	$\alpha_1\beta_1 = 0.643$
	$c = 3.7$ in.
Iteration 3:	$c = 3.7$ in.
	$\varepsilon_c = 0.00202$
	$\alpha_1\beta_1 = 0.688$
	$c = 3.5$ in. : Convergence is achieved
	$\beta_1 = 0.759$
	$\alpha_1 = 0.906$
	$M_n = 3{,}151$ kip-in. $= 262.6$ ft-kip

2. Approximate method

From the previous example it is obvious that the exact value of c is not an important factor in the overall accuracy of the solution. Therefore, when FRP rupture governs the failure, it is allowed to assume $c = c_b$ conservatively, which results in

$$M_n = \omega_f \left(1 - \frac{\omega_{fb}}{1.7}\right) f'_c b d^2$$

Example 4.5

Calculate the nominal flexural strength of the beam in Example 4.4, using the approximate method.

Solution:	$\omega_f = 0.1082$
	$\omega_{fb} = 0.1667 > \omega_f$: Failure by FRP rupture
	$M_n = 3{,}023$ kip-in. $= 251.9$ ft-kip

Example 4.6

Repeat Example 4.5 with $f'_c = 5.0$ ksi.

Solution:	$\omega_f = 0.0866$
	$\omega_{fb} = 0.1569 > \omega_f$: Failure by FRP rupture
	$M_n = 3{,}044$ kip-in. $= 253.6$ ft-kip

This example shows that, in the FRP rupture mode, the strength of concrete has little effect on the flexural strength of the member.

Example 4.7

In previous examples, the assumed rupture stress of the FRP reinforcement is equal to the yielding stress of ordinary reinforcing steel

($f_{fu} = f_y = 60$ ksi). Compare the nominal strength of the beams in previous examples with their steel reinforced counterparts.

Solution:	Example	FRP RC	Steel RC*
	1: Concrete crushing	$M_n = 428.8$ ft-kip	$M_n = 487.6$ ft-kip
	2: Concrete crushing	$M_n = 477.6$ ft-kip	$M_n = 501.9$ ft-kip
	3: Balanced failure	$M_{nb} = 388.2$ ft-kip	$M_{nb} = 826.1$ ft-kip**
	4: FRP rupture	$M_n = 262.6$ ft-kip	$M_n = 261.6$ ft-kip
	5: FRP rupture	$M_n = 251.9$ ft-kip	$M_n = 261.6$ ft-kip
	6: FRP rupture	$M_n = 253.6$ ft-kip	$M_n = 265.2$ ft-kip

*   Note that for steel RC beams in all cases $M_n \leq M_{nb}$ and therefore all beams fail in the same mode of steel yielding followed by concrete crushing.
** M_{nb} for steel RC has a reinforcement ratio 2.6 times greater than that of GFRP RC.

From this comparison, it also appears that FRP reinforcement is at its most effective when the failure is governed by FRP rupture. However, for reasons of safety and serviceability, which are elaborated in their respective chapters, this is not always the most desirable mode of failure.

Case b—calculation of design parameters of a rectangular flexural member

(b1) Applied moment and dimensions are known, reinforcement area is unknown:

This is probably the most common case, as in order to perform the structural analysis and obtain the internal moments, it is normally required to assume the dimensions of the members. Since the reinforcement area and subsequently the failure mode are not known beforehand, the formulae for **Case a** need to be altered as

$$m = \frac{M_n}{0.85f'_c bd^2}$$

$$\omega'_f = 0.85\left(1 - \sqrt{1 - 2m}\right)$$

$$f = \frac{\dfrac{0.85\beta_1}{\omega'_f} - 1}{e}$$

$$\omega_f = \begin{cases} \dfrac{\omega'_f}{f} & : if\ f \leq 1 \\ \dfrac{0.85m}{1 - \dfrac{\omega_{fb}}{1.7}} & : if\ f > 1 \end{cases}$$

Example 4.8

Design a beam for the nominal flexural strength of $M_n = 5400$ kip-in. = 450 ft-kip. Other properties are as follows:

Concrete:	$f'_c = 4.0$ ksi
	$\beta_1 = 0.85$
Reinforcement:	$f_{fu} = 60$ ksi
	$E_f = 6000$ ksi
	$\varepsilon_{fu} = 0.010$
Size:	$b = 16.0$ in.
	$h = 25.0$ in.
	$c = 3.0$ in.
	$d = h-c = 22.0$ in.
Solution:	$m = 0.2051$
	$\omega'_f = 0.1972$
	$e = 3.333$
	$f = 0.799 < 1$: Failure by concrete crushing
	$\omega_f = 0.2468$
	$A_f = 5.79$ in.^{2} : Use 5#10

Example 4.9

Design a beam for the nominal flexural strength of $M_n = 3600$ kip-in. = 300 ft-kip. Other properties are similar to those in Example 4.8.

Solution:	$m = 0.1367$
	$\omega'_f = 0.1255$
	$e = 3.333$
	$f = 1.427 > 1$: Failure by FRP rupture
	$\omega_{fb} = 0.1667$
	$\omega_f = 0.1289$
	$A_f = 3.02$ in.^{2} : Use 3#9

(b2) Applied moment and the stress level in reinforcement are known; dimensions and the reinforcement area are unknown:

In this case, if a predetermined value is assigned to the stress level $(f < 1)$, then:

$$\omega'_f = \frac{0.85\beta_1}{1+ef}$$

The reinforcement area and the dimensions can be calculated from

$$\omega_f = \frac{\omega'_f}{f}$$

$$bd^2 = \frac{M_n}{\omega_f'\left(1-\frac{\omega_f'}{1.7}\right)f_c'}$$

Example 4.10

The calculated stress level in Example 4.8 is $f = 0.799 \approx 0.80$. Repeat the example for the stress levels of $f = 0.90$ and $f = 0.70$. Assume that the effective depth of the beam, d, is unknown but the width of the beam is unchanged ($b = 16.0$ in.).

Solution:	$e = 3.333$	$f = 0.90$	$f = 0.80$	$f = 0.70$
		$\omega_f' = 0.1806$	$\omega_f' = 0.1976$	$\omega_f' = 0.2168$
		$\omega_f = 0.2007$	$\omega_f = 0.2470$	$\omega_f = 0.3097$
		$bd^2 = 8362$ in.^{3}	$bd^2 = 7752$ in.^{3}	$bd^2 = 7138$ in.^{3}
		$d = 22.9$ in.	$d = 22.0$ in.	$d = 21.1$ in.
		$A_f = 4.89$ in.^{2}	$A_f = 5.80$ in.^{2}	$A_f = 6.98$ in.^{2}
		Use 4#10	Use 5#10	Use 6#10

As expected, the stress level decreases if the required reinforcement area increases or the effective depth decreases.

(b3) Applied moment and reinforcement ratio are known; dimensions are unknown:

If, for practical reasons or to impose a certain mode of failure, a predetermined reinforcement ratio is considered, then the design procedure is almost identical to **Case a.**

If $\omega_f \geq \omega_{fb}$, the failure is governed by concrete crushing and

$$f = \frac{f_f}{f_{fu}} = \frac{\sqrt{1+\dfrac{3.4\beta_1 e}{\omega_f}}-1}{2e} \leq 1$$

$$\omega_f' = f\omega_f$$

$$bd^2 = \frac{M_n}{\omega'f\left(1-\frac{\omega_f'}{1.7}\right)f_c'}$$

Otherwise, the failure is governed by FRP rupture and

$$bd^2 = \frac{M_n}{\omega_f\left(1-\frac{\omega_{fb}}{1.7}\right)f_c'}$$

Example 4.11

The calculated ratio in Example 4.8 is $\rho_f = 1.65\%$. Repeat the example for the reinforcement ratios of $\rho_f = 1.0\%$ and $\rho_f = 2.0\%$. Assume that the effective depth of the beam, d, is unknown but the width of the beam is unchanged ($b = 16.0$ in.).

Solution:	$\omega_b = 0.1667$	$\rho_f = 1.0\%$	$\rho_f = 1.65\%$	$\rho_f = 2.0\%$
		$\omega_f = 0.150$	$\omega_f = 0.2475$	$\omega_f = 0.300$
		FRP rupture	Concrete crushing	Concrete crushing
		$f = 1.00$	$f = 0.798$	$f = 0.713$
		—	$\omega'_f = 0.1975$	$\omega'_f = 0.2139$
		$bd^2 = 9978$ in.^{3}	$bd^2 = 7736$ in.^{3}	$bd^2 = 7218$ in.^{3}
		$d = 25.0$ in.	$d = 22.0$ in.	$d = 21.2$ in.
		$A_f = 4.0$ in.^{2}	$A_f = 5.81$ in.^{2}	$A_f = 6.80$ in.^{2}
		Use 4#9	Use 5#10	Use 6#10

4.6 STRENGTH-REDUCTION FACTORS FOR FLEXURE

4.6.1 ACI 440.1R-06 approach

The relationship suggested by ACI 440.1R-06 for the strength-reduction factor shown in Figure 4.9 is given in the following equation:

$$\phi = \begin{cases} 0.55 & \text{for} & \rho_f \leq \rho_b \\ 0.3 + 0.25\dfrac{\rho_f}{\rho_b} & \text{for} & \rho_b < \rho_f < 1.4\rho_b \\ 0.65 & \text{for} & \rho_f \geq 1.4\rho_b \end{cases} \tag{4.46}$$

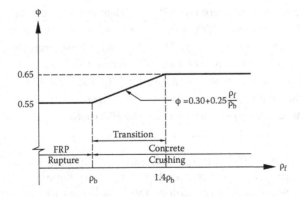

Figure 4.9 Strength-reduction factor as per ACI 440.1R-06.

where $\rho_f = A_f/bd$ and ρ_b represent the FRP reinforcement ratio and the FRP reinforcement ratio at balanced failure condition, respectively. The latter may be expressed as A_{fb}/bd, where A_{fb} represents the area of flexural reinforcement producing balanced failure. In such conditions, the result of compressive and tensile forces in both concrete and FRP reinforcement can be calculated as

$$C = 0.85f_c'\beta_1 c_b b$$

$$T = A_{fb}f_{fu} = \rho_b bd f_{fu}$$

(4.47)

where c_b is the neutral axis depth at balance. Replacing the expression of c_b from Equation (4.21) into Equation (4.47) and equating $C = T$, the following relationship for ρ_b can be determined (note that ε_{fu} has been replaced with f_{fu}/E_f):

$$\rho_b = 0.85\beta_1 \frac{f_c'}{f_{fu}} \frac{\varepsilon_{cu}}{\varepsilon_{cu} + \varepsilon_{fu}}$$

(4.48)

For bond-critical sections, a ϕ-factor of 0.55 is recommended.

COMMENTARY

The reasoning behind the development of Equation (4.46) is presented here. If $\rho_f \leq \rho_b$ failure of the concrete cross section is initiated by the rupture of the FRP reinforcement; if $\rho_f > \rho_b$ (or better, $\rho_f > 1.4\,\rho_b$), failure is controlled by crushing of the concrete. The corresponding ϕ-factors are 0.55 and 0.65, respectively, and there is a linear transition between the two failure modes. The value of 1.4 was adopted to account for the variability of the concrete compressive strength in ensuring the predicted failure (higher than specified concrete strength would produce FRP rupture). Such value was determined as $1/0.75 = 1.333$, rounded to 1.4, where the 0.75 coefficient represented the threshold set by ACI 318 prior to the 2002 edition of the code to ensure concrete crushing after extensive steel yielding. The 0.65 value of the strength-reduction factor has been derived from the steel RC tradition to ensure the same structural reliability of under-reinforced systems [12]. The value of 0.55 is based on ACI 440 Committee's consensus and represents a further reduction to penalize the less "ductile" behavior shown by FRP rupture.

Similarly to ACI 318-11, in order to introduce a unified strength-reduction factor applicable to any member subject to flexure (e.g., columns for which ρ_b is not used), the failure mode is linked to the strain in the extreme tensile layer of reinforcement (ε_f in this study as opposed to ε_t in

ACI 318-11). For fiber-reinforced concrete (FRC) RC flexural members, the relationship between reinforcing ratio and its tensile strain is

$$\rho_f = 0.85\beta_1 \frac{f'_c}{f_f} \frac{\varepsilon_{cu}}{\varepsilon_{cu} + \varepsilon_f} \tag{4.49}$$

Therefore,

$$\frac{\rho_f}{\rho_{fb}} = \frac{f_{fu}}{f_f} \frac{\varepsilon_{cu} + \varepsilon_{fu}}{\varepsilon_{cu} + \varepsilon_f} = \frac{\varepsilon_{fu}}{\varepsilon_f} \frac{\varepsilon_{cu} + \varepsilon_{fu}}{\varepsilon_{cu} + \varepsilon_f} \tag{4.50}$$

And the strength-reduction factor can be reformulated as

$$0.55 \leq \phi = 0.30 + 0.25 \frac{\varepsilon_{fu}}{\varepsilon_f} \frac{\varepsilon_{cu} + \varepsilon_{fu}}{\varepsilon_{cu} + \varepsilon_f} \leq 0.65 \tag{4.51}$$

The strength-reduction factor expressed by Equation (4.51) is calibrated by targeting a level of reliability of at least 3.5. This target reliability, however, is not always obtainable even by ordinary steel RC flexural members [13].

4.6.2 New approach

An attempt by Jawaheri Zadeh and Nanni [14] to equalize the reliability indices of the two member types (steel RC vs. FRP RC) leads to ϕ with a lower limit of 0.70 (instead of 0.55) and an upper limit of 0.75 (instead of 0.65). Based on a conservative interpretation of these values, a new ϕ may be formulated as

$$0.65 \leq \phi = 1.15 - \frac{\varepsilon_f}{2\varepsilon_{fu}} \leq 0.75 \tag{4.52}$$

It should be noted that ACI 318-11 also imposes a strength-reduction factor of $\phi = 0.65$ on compression-controlled sections, which is lower than the value (0.75) proposed by Equation (4.52). Nevertheless, a reliability-based calibration of the strength-reduction factors of steel RC members [13] confirms the adequacy of $\phi = 0.75$. Figure 4.10 compares the two ϕ-factors (Equation 4.51 vs. Equation 4.52) for different levels of maximum tensile strain if $\varepsilon_{fu} = 0.01$.

In the design examples presented in Chapters 6–8 and 10, the values of ϕM_n computed according to both methods (i.e., ACI 440.1R-06 and Jawaheri Zadeh and Nanni [14]) are shown.

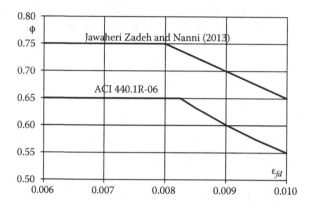

Figure 4.10 Comparison of the strength-reduction factors ($\varepsilon_{fu} = 0.01$).

4.6.3 Examples—Flexural strength-reduction factor

Example 4.12

Calculate the strength reduction factors and the ultimate strength of the cross sections in Examples 4.1–4.4.

Solution:

Equation (4.52) can be rewritten in terms of f, the stress level in FRP reinforcement, as $0.65 \leq \phi = 1.15\text{-}f/2 \leq 0.75$.

Example	Stress level	ϕ-factor	Design strength
1:	$f = 0.862$	$\phi = 0.719$	$\phi M_n = 308.3$ ft-kip
2:	$f = 0.946$	$\phi = 0.678$	$\phi M_n = 323.3$ ft-kip
3:	$f = 1$	$\phi = 0.65$	$\phi M_{nb} = 252.3$ ft-kip
4:	$f = 1$	$\phi = 0.65$	$\phi M_n = 170.7$ ft-kip

COMMENTARY

Validation of strength-reduction factors can be attained by the reliability analysis that links the probability of failure to the load and safety factors, providing a basis for their calibration to achieve desired levels of safety. Conventionally, the reliability index is defined as an indicator of the probability of failure of a member with a resistance of R against the loads it may experience during its lifetime, Q, with both Q and R being random variables.

This, however, poses a few obvious difficulties due to the presence of load parameters in the calculations:

- Compared to resistance, statistical parameters of loading are far more difficult to obtain due to the vast number of factors affecting load.
- Load and resistance, being of different natures, follow different statistical distributions that, especially in the case of multiple load cases, make the problem of calculating the reliability index less tractable.
- Reliability analysis has to be performed for several types of loads and load combinations.
- For each load combination, covering the whole range of plausible loadings makes the calculations cumbersome, especially when more than two loads are involved.

The idea is to calibrate strength-reduction factors of the elements reinforced with new materials—not by setting them against loads, but rather by comparing them to elements of the same capacity that are made of better established and better known materials. In other words, if the current strength factors for steel RC and its associated load factors are taken as a design standard (as there is little doubt about its performance when designed according to code), how should the safety factors of FRP reinforced members be proportioned so that the same level of safety is attained? This section shows how this concept applies to FRP RC and proposes revised strength-reduction factors for use in design.

Comparative reliability. Assume two structural elements 1 (e.g., steel RC beam) and 2 (e.g., FRP RC beam) that are equal in their ultimate design capacity or:

$$\phi_1 N_1 = \phi_2 N_2 \tag{4.53}$$

where ϕ_i indicates the strength-reduction factor and N_i is the nominal strength (design strength) of each element ($i = 1, 2$). Resistance is a lognormal random variable defined by its mean, μ, and standard deviation, σ, or coefficient of variation $\delta = \sigma/\mu$. These parameters can be obtained by testing. The nominal resistance (or capacity), N, is typically a conservative estimate below the mean value of resistance. The bias factor, λ_i, is defined as the ratio of the mean value, μ_i, to the nominal value, N_i, of a random variable—in this case, the resistance of elements 1 and, therefore,

$$\lambda_1 = \frac{\mu_1}{N_1}; \lambda_2 = \frac{\mu_2}{N_2} \tag{4.54}$$

Table 4.4 (first row) shows these parameters for ordinary steel RC beams subject to the limit state of flexure.

For a structural element, safety is measured in terms of the reliability index, β, which is a function of the statistical parameters of the resistance of an element and the loads applied to it:

$$\beta = f\left[(\phi, \lambda_R, \delta_R), (\gamma, \lambda_Q, \delta_Q)_1, (\gamma, \lambda_Q, \delta_Q)_2, \ldots, (\gamma, \lambda_Q, \delta_Q)_n\right] \qquad (4.55)$$

where the index R denotes the resistance, Q the load, and 1 to n the different types of loading that are applied to the element (dead, live, seismic, etc.) in the load combination for which the reliability index is calculated. γ_1 to γ_1 are, then, the load factors that constitute the combination. Traditionally, the reduction factors are calibrated by targeting a preset level of reliability for all the design combinations:

$$\beta = \beta_T \qquad (4.56)$$

Table 4.4 also shows this target reliability for ordinary RC beams subject to flexure.

The essence of the concept of comparative reliability is to calibrate the unknown strength-reduction factor of a newly introduced element (element 2 in this case) by equalizing its reliability index to that of the element 1, whose reduction factor and reliability level are agreed upon. In other words, it solves this set of equations by eliminating the load parameters:

$$\begin{cases} \beta_1 = f\left[(\phi_1, \lambda_1, \delta_1), (\gamma, \lambda_Q, \delta_Q)_1, (\gamma, \lambda_Q, \delta_Q)_2, \ldots, (\gamma, \lambda_Q, \delta_Q)_n\right] = \beta_T \\ \beta_2 = f\left[(\phi_2, \lambda_2, \delta_2), (\gamma, \lambda_Q, \delta_Q)_1, (\gamma, \lambda_Q, \delta_Q)_2, \ldots, (\gamma, \lambda_Q, \delta_Q)_n\right] = \beta_T \\ \phi_1 N_1 = \phi_2 N_2 \end{cases} \qquad (4.57)$$

Jawaheri Zadeh and Nanni [14] found a solution for this set as

$$\frac{\ln\left[\left(\dfrac{\phi_1}{\phi_2}\dfrac{\lambda_2}{\lambda_1}\right)\right]}{\sqrt{\delta_1^2 + \delta_2^2}} = \frac{\delta_2 - \delta_1}{\delta_1 + \delta_2}\beta_T \qquad (4.58)$$

where the unknown ϕ_2 has to be calculated from other known parameters.

Table 4.4 (second and third rows) summarizes the statistical data (bias factor and coefficient of variation) obtained from tests discussed in detail in Gulbrandsen [12], its assumptions (minimum target reliabilities), and strength-reduction factors adopted by ACI 440.IR-06. To avoid discontinuity, ACI 440.IR-06 inserts a transitional range between the two modes (ρ_{fb} $\leq \rho_f \leq 1.4 \, \rho_{fb}$) in which the reduction factor is linearly interpolated between 0.55 and 0.65 based on the reinforcement ratio, ρ_f.

Taking advantage of the concept of comparative reliability, the strength-reduction factors can be calibrated proportional to those of a steel RC beam whose statistical parameters are given in Table 4.4. In other words, λ_1 and δ_1 correspond to the first row of Table 4.4, while λ_2 and δ_2, depending on the failure mode, are taken from the second and third rows of the same table, respectively. As a result, the calculated values of the strength-reduction factor, ϕ, according to which, $\phi = 0.70$ for tension-controlled sections and $\phi = 0.75$ for compression-controlled sections, are recommendable, respectively.

These new factors may be formulated, similarly to ACI 440.IR-06, in terms of reinforcement ratio; however, a more general approach as followed by ACI 318-11 is more desirable as it unifies the strength-reduction factors of columns and flexural members by describing them as a function of the maximum tensile strain in the reinforcement. As ACI 440.IR-06 is silent about columns, this method allows generalizing the flexural reduction factors for cases involving axial force. To that end, the failure mode shall be linked to the strain in the extreme tensile layer of reinforcement (ε_f as opposed to ε_t in ACI 318-11) instead of the reinforcement ratio, ρ_f, which is not the only decisive factor in determining the failure mode when axial forces interfere. Equation (4.52) is the proposed strength-reduction factor as a function of the tensile strain in the reinforcement.

Table 4.4 Strength-reduction and statistical parameters of cast-in-place steel RC and FRP RC beams

Steel RC	ϕ^a	ϕ^a	$\beta_T{}^b$	Bias $(\lambda)^b$	CoV$(\delta)^b$
Flexure	0.65	0.90	3.5	1.190	0.089
FRP RC	ϕ^c	ϕ^c	$\beta_T{}^d$	Bias $(\lambda)^d$	CoV$(\delta)^d$
FRP rupture	0.55	0.65	3.5	1.11	0.157
Concrete crushing	0.65	0.65	3.5	1.19	0.158

[a] ACI 318-11. ACI Committee 318. Building code requirements for reinforced concrete, ACI 318-11. American Concrete Institute, Farmington Hills, MI (2011).
[b] A. S. Nowak and M. M. Szerszen. ACI Structural Journal 100 (3): 383–391 (2003).
[c] ACI 440.IR-06. ACI Committee 440. Guide for the design and construction of structural concrete reinforced with FRP bars. 440.IR-06. American Concrete Institute, Farmington Hills, MI (2006).
[d] P. Gulbrandsen. Reliability analysis of the flexural capacity of fiber reinforced polymer bars in concrete beams. Master thesis, University of Minnesota, MN, p. 80 (2005).

4.7 ANCHORAGE AND DEVELOPMENT LENGTH

Development length of straight bars. The development length, l_d, of FRP straight bars is defined in ACI 440.1R-06 as the bond length necessary to develop the minimum of f_{fu}, the design tensile stress; f_f, as defined in Equation (4.40); or f_{fe} as defined in Equation (4.39). l_d is given by the following equation:

$$l_d = \frac{\alpha \dfrac{f_{fr}}{\sqrt{f_c'}} - 340}{13.6 + \dfrac{C}{d_b}} d_b$$
(4.59)

$$\left[\text{or } l_d = \frac{\alpha \dfrac{f_{fr}}{0.083\sqrt{f_c'}} - 340}{13.6 + \dfrac{C}{d_b}} d_b \text{ in SI units} \right]$$

where d_d is the bar diameter, C is the lesser of the cover to the center of the bar or one-half of the center-on-center spacing of the bars being developed, and α is a factor to account for the bar location. α is taken equal to 1.5 for bars with more than 12 in. (300 mm) of concrete cast below, otherwise α is taken equal to 1.0. For the case of columns, α is always equal to 1.5.

ACI 440.1R-06 recommends the following criterion for development of positive reinforcement at points of inflection and simple supports:

$$l_d \le \frac{\phi M_n}{V_u} + l_a$$
(4.60)

where M_n is the effective nominal bending moment capacity at the critical cross section, V_u is the factored shear force at that section, and l_a is the embedment length (a) beyond the center of the support or (b) at a point of inflection. For the latter, l_a is the larger of the effective depth of the member or $12d_b$.

Development length of bent bars. Because of the constituents (i.e., thermosetting resin) and manufacturing method (i.e., pultrusion), it is difficult, if not impossible, to create bends at the end of a long FRP bar. Bent bars are typically spliced to longitudinal bars at their ends, where a hook anchorage is required. ACI 440.1R-06 recommends the following development length, l_{bhf} for hooked bars:

$$l_{bhf} = \begin{cases} 2000\dfrac{d_b}{\sqrt{f_c'}} \text{ for } f_{fu} \leq 75{,}000 \text{ psi} \\[2ex] \dfrac{f_{fu}}{37.5}\dfrac{d_b}{\sqrt{f_c'}} \text{ for } 75{,}000 \text{ psi} < f_{fu} < 150.000 \text{ psi} \\[2ex] 4000\dfrac{d_b}{\sqrt{f_c'}} \text{ for } f_{fu} \geq 150{,}000 \text{ psi} \end{cases} \qquad (4.61)$$

$$\text{or } l_{bhf} = \begin{cases} 165\dfrac{d_b}{\sqrt{f_c}} \text{ for } f_{fu} \leq 520 \text{ MPa} \\[2ex] \dfrac{f_{fu}}{3.1}\dfrac{d_b}{\sqrt{f_c}} \text{ for } 520 \text{ MPa} < f_{fu} < 1040 \text{ MPa in SI units} \\[2ex] 330\dfrac{d_b}{\sqrt{f_c}} \text{ for } f_{fu} \geq 1040 \text{ MPa} \end{cases}$$

l_{bhf} is measured from the critical section to the farthest point on the bar and should not be less than $12d_b$ or 9 in. (230 mm). Also, the tail length of the hook should not be shorter than $12d_b$ and the radius of the bend should not be less than $3d_b$.

Lap splices. ACI 440.1R-06 recommends that all splices be considered class B splices and that a tension lap splice of $1.3l_d$ be used.

COMMENTARY

The production of bar bends is a challenge to FRP bar manufacturers, and scarce information is available in the technical literature related to the method of production and available dimensions. Standard bends available from one manufacturer are given as an example in Table 4.5.

Table 4.5 Standard bends of ASLAN 100 GFRP bar

4.8 SPECIAL CONSIDERATIONS

4.8.1 Multiple layers of reinforcement

Because FRP reinforcement is linear–elastic to failure, when multiple layers of FRP reinforcement are used, rupture of the bars at the outermost layer controls overall reinforcement failure; in such a case, the strain compatibility approach should be used to determine the flexural resistance of the member. If different FRP bars (in terms of size or material) are used in a multiple-layer configuration in the same concrete component, failure may not occur in the outermost layer of reinforcement; in fact, failure will occur in the FRP bar that first reaches its ultimate tensile strain irrespectively of its position within the concrete cross section. This latter case, however, has little significance as concrete sections with different FRP bars have rare, if any, practical application.

The methodology followed with traditional steel reinforcement of locating the centroid of the layers of steel bars and determining a result of magnitude equal to the total area times the yield strength cannot be used with FRP bars because each layer of FRP reinforcement is subjected to different levels of strain and, therefore, different values of the tensile stress due to the linear–elastic behavior of the FRP reinforcement. Nonetheless, the strain compatibility approach should also be used with traditional steel reinforcement if the strain in one of the layers does not reach the yielding strain.

A rectangular cross section with multiple layers of FRP reinforcement as indicated in Figure 4.11 is selected. Assume that the FRP reinforcement is made of the same material and bar size so that failure of the reinforcement

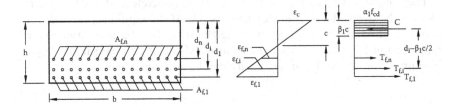

Figure 4.11 Flexural member with multiple layers of FRP reinforcement.

is controlled by the layer located farthest from the extreme compression fiber. Also assume the rectangular stress block approach applies (a similar conclusion could be drawn if the rigorous method is selected).

Under these conditions, the neutral axis depth at balanced failure is determined from Equation (4.21) with d replaced with d_1. A numerical procedure is initiated by assuming a tentative value for the neutral axis c. Assume first that concrete crushing is the controlling failure mode ($c > c_b$); this means that the extreme fiber in the concrete has reached the value of $\varepsilon_c = \varepsilon_{cu}$, while the strain in the ith layer of reinforcement can be found as

$$\varepsilon_{f,i} = \frac{d_i - c}{c}\varepsilon_{cu} < \varepsilon_{fu} \tag{4.62}$$

The compressive force in the concrete, C, and the tensile force in the reinforcement, T, can be calculated as

$$C = \alpha_1 f'_c \beta_1 cb$$

$$T = \sum_{i=1}^{n} T_{f,i} = \sum_{i=1}^{n} A_{f,i}\varepsilon_{f,i}E_f \tag{4.63}$$

Assume, now, that FRP rupture is the controlling failure mode ($c < c_b$); this means that extreme fiber in the concrete has reached a strain $\varepsilon_c < \varepsilon_{cu}$, while the layer of reinforcement located farthest from the extreme compression fiber has reached the design tensile strain of $\varepsilon_{f,1} = \varepsilon_{fd}$. The strain in the concrete and in each ith layer of reinforcement can be determined from similar triangles as

$$\varepsilon_c = \frac{c}{d_1 - c}\varepsilon_{fu} < \varepsilon_{cu}$$

$$\varepsilon_{f,i} = \frac{d_i - c}{d_1 - c}\varepsilon_{fu} \leq \varepsilon_{fd} \tag{4.64}$$

The compressive and tensile forces in both concrete and reinforcement can still be determined from Equation (4.63). If the equilibrium condition is not satisfied, the procedure involving the selection of a new value of c is repeated until $T - C = 0$. Once the neutral axis depth, c, that satisfies equilibrium is found, the flexural capacity can be computed as

$$M_n = \sum_{i=1}^{n} A_{f,i} \varepsilon_{f,i} E_f \left(d_i - \frac{\beta_1 c}{2} \right) \tag{4.65}$$

4.8.2 Redistribution of moments

Traditional under-reinforced steel RC continuous flexural members undergo a significant amount of inelastic deformations prior to failure, mostly due to the yielding of the steel reinforcement. For this particular reason, such members exhibit a ductile behavior at ultimate condition and redistribution of moment is allowed with benefits in the economy of reinforcement. Conversely, moment redistribution is not attainable in over-reinforced beams for which the yielding strain is never reached in the steel reinforcement and inelastic deformations are not fully developed. Moment redistribution is not allowed for FRP reinforced continuous flexural members.

COMMENTARY

The typical result from moment redistribution in a continuous flexural member is a reduction of maximum negative moments at the supports and an increase of positive moments at midspan from those computed by elastic analysis. Sometimes, the opposite mechanism may be preferable (i.e., reduction of positive moments and increase of negative moments) to narrow the envelope of maximum positive and negative bending moments at any section within the continuous span as determined from different loading configurations [15].

Redistribution of moments in statically indeterminate concrete structures is not allowed for when FRP reinforcement is used because plastic hinges cannot form due to the linear–elastic behavior up to failure of the FRP reinforcement. A pseudoductile behavior has been observed in over-reinforced FRP beams (with amount of FRP reinforcement well above the balanced reinforcement ratio) due to the inelastic behavior of concrete prior to failure. Continuous FRP RC beams demonstrated moment redistribution when cracking and debonding between FRP and concrete

---

occurred. However, no moment redistribution occurred when either the midspan or middle support section reached its respective moment capacity due to the brittle nature of FRP reinforcement rupture or concrete crushing [16].

4.8.3 Compression FRP in flexural members

Differently from steel reinforced flexural members, the contribution of FRP bars in compression does not increase the strength nor reduce the effects of concrete creep of FRP reinforced flexural members due to the limited compressive strength and modulus of the FRP bars. For this reason, ACI 440.1R-06 does not recommend relying upon compression FRP reinforcement in flexural members, but allows its use for fabrication purposes.

The lack of effectiveness of FRP reinforcement in the compression zone of flexural members does not necessarily preclude its use in columns. In fact, Chapter 5 presents a methodology for the design of FRP RC columns subjected to combined axial load and flexure. For analysis purposes, this methodology is based on the assumption that the area of FRP bars in compression can be replaced with an equivalent area of concrete—as if the FRP bars were not present in the cross section.

COMMENTARY

Studies from Washa and Fluck [17] conducted on steel RC simply supported beams demonstrated that the effect of compression reinforcement is beneficial in reducing the long-term deflections under sustained loads. Such behavior is attributed to the creep of concrete that transfers a portion of the load from the concrete itself to the compression steel, thus reducing the overall stress in the concrete and the resulting deflection caused by the sustained load. Similar studies are not available for FRP RC members; however, because of the relatively low elastic modulus of some FRP products (particularly GFRP reinforcement) compared to steel and creep characteristics of FRP (which are resin dominated), it is believed that such beneficial effects on long-term deflection due to sustained loads are negligible.

In the technical literature, there is no direct measurement of time-dependent creep of FRP bar coupons subjected to compression. Experimental evidence obtained from GFRP pultruded shapes indicates that the ratio creep strain to initial elastic strain is low for stress levels up to 45% of the proportionality limit [18]. Accordingly, creep should not cause geometrical integrity problems at the stress levels typical of FRP bars in compression.

Cohn, Ghosh, and Parimi [19] reported an increase on the strength and ductility of under-reinforced steel RC beams with the increase of compression reinforcement. Such behavior, particularly important in seismic regions or if moment redistribution is desired, is due to the reduction of the depth of the compression stress block caused by the presence of compression reinforcement that allows for the neutral axis to shift toward the extreme compression fiber to maintain force equilibrium. As such, the strain in the tension reinforcement at failure increases, resulting in more ductile behavior. These authors also conducted tests on over-reinforced beams and demonstrated that the brittle failure associated to the crushing of the concrete before yielding of the steel can be mitigated by the presence of compression reinforcement.

In the case of FRP reinforced flexural members, over-reinforced sections exhibit a more ductile behavior than their under-reinforced counterparts. The presence of compression FRP reinforcement (assumed to have a higher stiffness than that of concrete) would increase the strain in the FRP tension reinforcement as seen in the case of over-reinforced steel RC, but the beneficial effect on the member ductility would not be as relevant since FRP is linear–elastic to failure. The addition of compression reinforcement could, instead, cause the section failure mode to shift from concrete crushing to FRP rupture.

4.9 SERVICEABILITY

The serviceability limit states for FRP reinforced concrete flexural members generally include crack width, maximum deflections, and maximum FRP stress levels to avoid FRP creep-rupture and fatigue. In many instances, serviceability criteria (crack width and deflections) may control the design of FRP reinforced one-way slabs or beams because of the relatively smaller stiffness of these members after cracking.

COMMENTARY

Replacing the steel reinforcement of a flexural member with an equal area of FRP reinforcement results in larger deflections and wider cracks [20,21]. However, the serviceability criteria for crack width and deflections are generally satisfied when a section is designed to achieve concrete crushing failure [22]. A section designed to reach a failure controlled by FRP bar rupture, vice versa, is almost unattainable as it would require a very small amount of reinforcement that, in turn, would not satisfy the service condition requirements.

4.9.1 Control of crack width

Crack width in FRP reinforced concrete is generally limited for aesthetic reasons and to prevent water leakage. Differently from the case of steel reinforced concrete, crack width limitations related to the potential corrosion of the reinforcement are not required for FRP reinforced concrete. ACI 440.1R-06 recommends the following limiting values: 0.020 in. (0.5 mm) for exterior exposure and 0.028 in. (0.7 mm) for interior exposure. More stringent crack width limits might be considered for the design of liquid-retaining structures.

The maximum crack width at the tension face of a flexural member, w_c, may be computed by Equation (4.66), which is based on the work of Frosch [23]:

$$w_c = 2\frac{f_f}{E_f}\beta k_b \sqrt{d_c^2 + \left(\frac{s}{2}\right)^2} \qquad (4.66)$$

where f_f is the reinforcement stress, d_c is the thickness of cover from tension face to center of closest bar, s is the bar spacing, β is the ratio between the distance from the neutral axis to the tension face of the member and the distance from the neutral axis to the centroid of the tensile reinforcement, and k_b is a bond coefficient that accounts for the bond characteristics of the reinforcement. Frosch's equation [23] is written in a more general fashion than Equation (4.66), so that it can be applied regardless of whether the reinforcement is steel or FRP. For the case of FRP reinforced concrete, Frosch's original equation is modified by including the bond coefficient k_b. k_b varies between 0.60 and 1.70 and depends on the FRP bar manufacturer, the fiber type, the resin formulations, and the surface treatments. Sand-coated bar surface treatments generally trend toward the lower bound of this range. ACI 440.1R-06 conservatively recommends the value of 1.4 when k_b is not based on experiments.

COMMENTARY

Reviewing the work of Broms [24], Frosch [23] had noted that the crack spacing, S_c, depends on the concrete cover and can be written as follows:

$$S_c = \Psi_s d^* \qquad (4.67)$$

where $d^* = \sqrt{d_c^2 + \left(\frac{s}{2}\right)^2}$ is the controlling cover distance and ψ_s is the crack spacing factor, which was assumed equal to 1.0 for minimum crack spacing, 1.5 for average crack spacing, and 2.0 for maximum crack spacing.

Frosch also noted that the crack width at the level of the reinforcement, w_c, depends on the strain level in the reinforcement, ε, and the crack spacing. Therefore, w_c can be written as

$$w_c = \varepsilon \cdot S_c = \frac{\sigma}{E} \cdot S_c \tag{4.68}$$

Substituting the expression for S_c in Equation (4.68) and assuming the case of maximum crack spacing ($\psi_s = 2.0$), Frosch obtained the expression in Equation (4.66). The coefficient β was introduced to account for the strain gradient and was defined as the ratio between the strain at the bottom of the member and the strain at the level of the reinforcement.

Frosch's equation can be rearranged to solve for the permissible bar spacing, s, as a function of the permissible maximum crack width, w_c:

$$s = \sqrt{\left(\frac{w_c E_f}{2 f_f \beta k_b}\right)^2 + d_c^2} \tag{4.69}$$

Given the difficulty and uncertainty in determining crack width, a design approach based on maximum bar spacing appears more logical. Frosch [23] also developed a simplified equation that was modified and adopted in ACI 318 as a new crack control equation to evaluate the maximum bar spacing for the 1999 revision of the building code:

$$s = 15\left(\frac{40,000}{f_s}\right) - 2.5c_c \leq 12\left(\frac{40,000}{f_s}\right) \tag{4.70}$$

$$[\text{or } s = 380\left(\frac{280}{f_s}\right) - 2.5c_c \leq 300\left(\frac{280}{f_s}\right) \text{ in SI units}]$$

where f_s is the steel reinforcement stress at service level in psi (MPa) and c_c is the clear concrete cover.

Equation (4.70) represents the starting point for the development of the new proposed model by Ospina and Bakis [25] for indirect flexural crack control of one-way concrete flexural members. Normalizing the first term in the left-hand side of Equation (4.70) by the crack width and the elastic modulus ratios, using 0.017 in. and 29,000 ksi (4 mm and 200 GPa) as reference values, and introducing the bond coefficient, k_b, Ospina and Bakis [25] obtained the following equation:

$$s = 1.2\left(\frac{E_f w_c}{f_f k_b}\right) - 2.5c_c \le 0.95\left(\frac{E_f w_c}{f_f k_b}\right) \tag{4.71}$$

In the design examples presented in Chapters 6–8, compliance with the serviceability criterion of limiting crack width is checked according to both methods (i.e., ACI 440.1R-06 and that of Ospina and Bakis [25]).

4.9.2 Control of deflections

Deflections in reinforced concrete slabs and beams are generally limited to prevent damage to nonstructural or other structural elements and to avoid disruptions of function. Limiting values for deflections of steel RC flexural members are defined in ACI 318-11.

As discussed in Section 4.3, minimum values of thickness of FRP reinforced flexural members are proposed by ACI 440.1R-06 for initial member proportioning only. The control of deflections consists in verifying that the sum of the immediate deflections due to the loads and the long-term deflection due to creep and shrinkage is smaller than a limiting value as given in the following equation:

$$\Delta_{max} = \Delta_{immediate} + \Delta_{creep/shrinkage} \le \Delta_{limit} \tag{4.72}$$

The immediate deflections are generally computed including live loads only, whereas the long-term deflections are computed including the dead loads and a percentage (typically 20%) of the live loads.

4.9.2.1 Elastic immediate deflections of one-way slabs and beams

ACI 440.1R-06 recommends using indirect procedures for initial member proportioning and then checking deflections explicitly. The formula to compute the maximum immediate elastic deflection of a one-way flexural member, Δ_{max}, under uniformly distributed load is reported in Equation (4.73). The maximum immediate elastic deflection of a one-way flexural member, Δ_{max}, can be computed as follows [26]:

$$\Delta_{max} = \frac{5l^2}{48E_c I_e}\left[M_0 + 0.1(M_1 + M_2)\right] \tag{4.73}$$

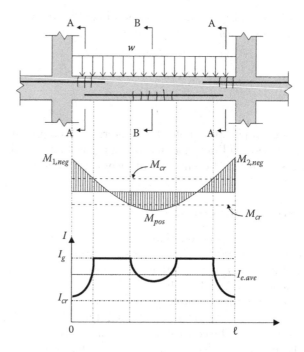

Figure 4.12 Moment of inertia for a continuous beam.

where M_0 is the midspan moment, M_1 and M_2 are the end moments, l is the span length, E_c is the modulus of elasticity of concrete, and I_e is the effective moment of inertia.

Equation (4.73) is based on the assumptions of linear distribution of the strains over the member cross section. For a flexural member subjected to various load cases, the deflection for each case is calculated separately and then algebraically added to the others to obtain the total.

Moment of inertia. When computing deflections of a flexural member, the magnitude of the flexural rigidity (defined as the product of the concrete modulus of elasticity and the moment of inertia, E_cI) must be determined. E_cI is not constant throughout the length of the element as I (the moment of inertia) is a section property that depends on the applied moment and the resulting cracking.

The span of a continuous reinforced concrete beam subjected to a uniformly distributed load is shown in Figure 4.12. If the applied load is such that bending moments do not exceed the cracking moment, the flexural rigidity is constant throughout the beam and can be computed using the uncracked or gross moment of inertia, I_g. As the load increases and the induced bending moments exceed the cracking moment, cracking occurs at the supports first and, eventually, at midspan. When a beam section cracks, a reduced or cracked moment of inertia results at various locations, as shown in the graph of Figure 4.12. It is therefore obvious that, for the purpose of simplifying design, an effective moment of inertia must be determined.

Effective moment of inertia for steel RC. To express the transition between the gross moment of inertia, I_g, and the cracked moment of inertia, I_{cr}, ACI 318-11 proposes to use the following equation (Branson's [26] equation) to calculate an effective moment of inertia, I_e, at any given cross section along the beam:

$$I_e = \left(\frac{M_{cr}}{M_a}\right)^3 I_g + \left[1 - \left(\frac{M_{cr}}{M_a}\right)^3\right] I_{cr} \leq I_g \tag{4.74}$$

where M_a is the unfactored moment at the section where the deflection is calculated.

As M_a increases, the effective moment of inertia, I_e, decreases. When M_a is close to the nominal flexural capacity of the cross section, I_e becomes close to the fully cracked moment of inertia. The relationship between the sectional rigidity and the applied bending moment is illustrated in its general fashion in Figure 4.13.

A single value of effective moment of inertia for the flexural member, $I_{e,av}$, can be derived when the variable I results from the variation in the extent of concrete cracking. For the case of a continuous beam such as the one shown in Figure 4.12, ACI Committee 435 [27] recommends the following equation:

$$I_{e,\,ave} = 0.70\,I_{e.\,midspan} + 0.15\,(I_{e,\,sup1} + I_{e,\,sup2}) \tag{4.75}$$

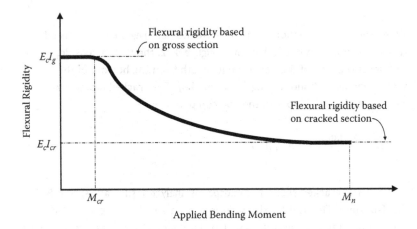

Figure 4.13 Flexural rigidity versus applied bending moment relationship.

For the case of a beam continuous only at one end, the following expression can be used:

$$I_{e,\,ave} = 0.85\,I_{e.\,midspan} + 0.15\,I_{e,\,sup1} \tag{4.76}$$

Effective moment of inertia for FRP RC. For FRP reinforced members, Branson's equation is adapted by including the reduction coefficient β_d as the multiplier of the first term in Equation (4.74). β_d is expressed by

$$\beta_d = \frac{1}{5}\left(\frac{\rho_f}{\rho_{fb}}\right) \leq 1.0 \tag{4.77}$$

β_d is a reduction coefficient related to the reduced tension stiffening of FRP when compared to steel reinforcement. This reduced tension stiffening can be attributed to the lower modulus of elasticity and different bond stresses for the FRP reinforcement as compared with those of steel.

The expression of the effective average moment of inertia for the flexural member, $I_{e,av}$, computed for traditional steel RC is also applicable to FRP RC.

COMMENTARY

With reference to the continuous T-beam in Figure 4.12, the gross moment of inertia can be computed as follows (Figure 4.14a):

$$I_g = \frac{b_{eff}t_{slab}^3}{12} + b_{eff}t_{slab}\left(d_g - \frac{t_{slab}}{2}\right)^2 + \frac{b_w\left(h - t_{slab}\right)^3}{12} + b_w\left(h - t_{slab}\right)\left[d_g - \left(h - t_{slab}\right)\right]^2 \tag{4.78}$$

When the cracked section in the negative moment region is considered, such as Section A-A of the T-beam in Figure 4.12, the neutral axis depth intersects the web of the beam and its depth from the bottom of the section can be computed using Equation (4.8). The cracked moment of inertia (see Figure 4.14b) can then be computed as

$$I_{cr} = \frac{b_w\left(kd_f\right)^3}{3} + n_f A_f d_f\left(1 - k\right)^2 \tag{4.79}$$

When the cracked section at midspan is analyzed, such as Section B-B of the T-beam in Figure 4.12, the section can be studied as rectangular if the cracked neutral axis depth falls within the depth of the flange. This occurs when the condition in Equation (4.9) is satisfied. In this case, the cracked moment of inertia can be computed as (Figure 4.14c):

$$I_{cr} = \frac{b_{eff}(kd_f)^3}{3} + n_f A_{f,tot} \frac{d_{f1}+d_{f2}}{2}(1-k)^2 \tag{4.80}$$

If Equation (4.9) is not satisfied (Figure 4.14d), the neutral axis depth is to be computed by solving the following equation for the unknown neutral axis depth, $x = c_{cr}$:

$$\frac{b_{eff}x^2}{3} - (b_{eff}-b_w)\frac{(x-t_{slab})^2}{2} = n_f A_{f,tot}\left(\frac{d_{f1}+d_{f2}}{2}-x\right) \tag{4.81}$$

Once the neutral axis depth is determined, the cracked moment of inertia of a T-section with N.A. through the web can be determined as

$$I_{cr} = \frac{b_{eff}c_{cr}^3}{3} - (b_{eff}-b_w)\frac{(c_{cr}-t_{slab})^3}{3} + n_f A_{f,tot}\left(\frac{d_{f1}+d_{f2}}{2}-c_{cr}\right)^2 \tag{4.82}$$

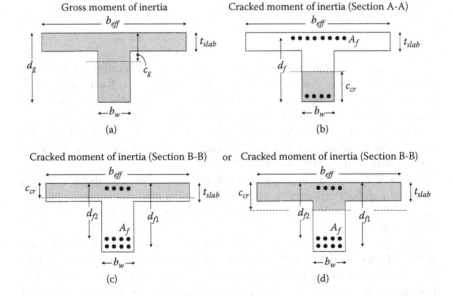

Figure 4.14 Gross and cracked moments of inertia of a T-beam.

4.9.2.2 Elastic immediate deflections according to Bischoff

As demonstrated by Bischoff [28], Branson's equation overestimates member stiffness when the I_g/I_{cr} ratio of the member is greater than about 3 or 4. This corresponds to most FRP reinforced concrete beams, which typically have an I_g/I_{cr} ratio between 5 and 25. It is for this reason that past research on deflection of FRP reinforced concrete beams has shown that Branson's equation underestimates deflection, particularly for lightly reinforced members with a high I_g/I_{cr} ratio.

Bischoff [28] and, more recently, Bischoff and Gross [29] proposed an alternative section-based expression for the effective moment of inertia I_e that works equally well for both steel and FRP reinforced concrete members without the need for empirical correction factors. Branson's original expression represents a weighted average of the uncracked and cracked member stiffness (EI), while Bischoff's proposed approach represents a weighted average of flexibility $(1/EI)$. The approach using a weighted average of flexibility represents better the deflection response of members with discrete cracks along their length [30].

The section-based expression proposed by Bischoff [28] is modified to include an additional factor γ to account for the variation in stiffness along the length of the member. The modified expression for the effective moment of inertia is given by

$$I_e = \frac{I_{cr}}{1 - \gamma \left(\dfrac{M_{cr}}{M_a}\right)^2 \left(1 - \dfrac{I_{cr}}{I_g}\right)} \leq I_g \tag{4.83}$$

This approach provides reasonable estimates of deflection for FRP RC beams and one-way slabs. The factor γ is dependent on load and boundary conditions and accounts for the length of the uncracked regions of the member and for the change in stiffness in the cracked regions. In lieu of a more comprehensive analysis, the value $\gamma = 1.72 - 0.72\,(M_{cr}/M_a)$ is suggested, which is the result from integrating the curvature over the length of a beam with a uniformly distributed load.

Unless stiffness values are obtained by more comprehensive analysis, immediate deflections can be computed with the effective moment of inertia given by Equation (4.83) using the maximum service load moment M_a in the member.

When $M_a > M_{cr}$, the assumed M_{cr} value has a significant effect on computed values of deflection. A lower M_{cr} can be used to account for the tensile stresses that develop in the concrete from restraint to shrinkage [30]. The use of a reduced M_{cr} also accounts for cases where the calculated maximum unfactored moment is only slightly less than the unrestrained M_{cr} (based on $f_r = 7.5\sqrt{f'_c}$), as factors such as shrinkage and temperature can still cause a cross section to crack over time.

An example on how to compute deflection using this method is presented in Chapter 7.

4.9.2.3 Elastic immediate deflections of two-way slabs

The availability of various finite element (FE) analysis computer programs makes the FE method the most popular for computing deflections of two-way reinforced concrete slabs. Unless FE software dedicated to the analysis and design of concrete structures is used, the flexural rigidity of a two-way member is computed based on the gross-sectional properties. Accounting for the concrete elements' cracking-induced reduction in stiffness is complex due to its dependence on the amount and orientation of the tension reinforcement and the distribution of the internal bending moments. A simplified way of accounting for cracking is to compute deflections conservatively considering the cracked conditions. This can be achieved by modifying the modulus of elasticity of concrete elements using the reduction factor I_{cr}/I_g. Moreover, to account for different reinforcement ratios along the in-plane x and y directions with, for example, $\rho_{sx} > \rho_{sy}$, the modulus of elasticity along the y direction of the concrete elements can be reduced by the ratio ρ_{sy}/ρ_{sx}.

Another viable method for a more realistic computation of deflections is the *crossing beam method* [27], which consists in treating the two-way slab as an orthogonal one-way system, in which the middle and column strips are treated as continuous beams of unit width. For each of these beams, midspan deflections can be computed by beam analogy using the elastic beam deflection equation. This approach allows the calculation of deflections by beam analogy, assuming the middle strips are supported by the perpendicular column strips.

The classical solution based on the *plate bending theory* [27] for elastic thin plates may also be used. This approach can compute the immediate deflections of uncracked two-way slab systems loaded uniformly. Closed-form solutions are, however, available only for a limited number of load and geometry cases. Modifications to the original theory can be considered to include the effect of cracking.

4.9.2.4 Concrete creep effects on deflections under sustained load

In general, sustained loads and time-dependent factors cause the flexural member's immediate deflections to increase. ACI 318-11 provides a deflection multiplier for the additional deflection due to sustained loading. Research has demonstrated that the approach used for estimating long-term deflections of steel reinforced members can also be applied to FRP reinforced members [4].

As noted in ACI 440.1R-06, the time-dependent deflection increase for FRP RC can be expected to be proportionally less than steel reinforced concrete. The creep-induced reduction of the concrete modulus of elasticity causes an overall reduction of the members' flexural rigidity, $E_c I$. The reduced modulus also causes an increase of the neutral axis depth of the cracked cross section, which, in turn, induces an increase of the moment of inertia. This second effect is typically more significant in FRP reinforced members because of the lower tensile modulus of the FRP reinforcement compared to steel.

The following equation is proposed by ACI 440.1R-06 to compute long-term deflections due to creep and shrinkage, $\Delta_{(cp+sh)}$:

$$\Delta_{(cp+sh)} = 0.6 \, \xi \, (\Delta_i)_{sus} \tag{4.84}$$

where $(\Delta_i)_{sus}$ is the immediate deflection due to sustained loads and ξ is a factor varying between 0 and 2 depending on the time period over which deflections are computed. Values of ξ are given by ACI 318-11. For 100 years, $\xi = 2$ is assumed.

4.9.3 FRP creep rupture and fatigue

Sustained loads can cause FRP bars to fail suddenly after a period of time defined as the endurance time. This phenomenon is known in literature as creep rupture (or static fatigue). ACI 440.1R-06 recommends limiting the stress level in the FRP reinforcement induced by sustained loads (dead loads and the sustained portion of the live load) to prevent creep rupture. The stress level in the FRP can be computed using the Navier equation:

$$f_{f,s} = \frac{M_s}{I_{cr}} n_f d_f (1-k) \leq k_{creep-R} f_{fu} \tag{4.85}$$

where M_s is the bending moment acting on the cross section where the FRP stress is computed, and $k_{creep-R}$ is the knock-down factor applied to the design tensile strength of the FRP bars to account for creep rupture. According to ACI 440.1R-06, $k_{creep-R}$ is equal to 0.20 for glass, 0.30 for aramid, and 0.55 for carbon FRP.

The provision given in Eq. (4.85) indirectly implies that the concrete in compression is still within its linear-elastic range. For this, it has to be checked and verified that the maximum concrete compressive stress (f_c) is smaller than $0.45 \, f'_c$ when the applied moment is M_s.

If the structure is subjected to fatigue regimes (i.e., repeated cycles of loading and unloading), ACI 440.1R-06 recommends limiting the FRP stress to the same threshold values adopted to prevent creep rupture, $k_{creep-R}$ f_{fu}. Equation (4.85) can be used with M_s equal to the moment due to sustained loads plus the maximum moment induced in a fatigue loading cycle.

4.10 SHEAR CAPACITY

When using FRP as shear reinforcement, one needs to recognize that FRP has a relatively low modulus of elasticity, FRP has a high tensile strength and no yield point, tensile strength of the bent portion of an FRP bar is significantly lower than the straight portion, and FRP has low dowel resistance [22].

COMMENTARY

The use of FRP as shear reinforcement has to be further explored to provide the foundation of a fully rational model to predict shear strength [31]. This is also the case for a deeper understanding of the contributions of concrete, aggregate interlock, and dowel effect of the longitudinal FRP reinforcement. At present, most of the shear design provisions incorporated in existing codes and guides are based on the design formulas of members reinforced with conventional steel, considering some modifications to account for the differences between FRP and steel reinforcement [32].

Compared with a steel reinforced section with equal areas of longitudinal reinforcement, a cross-section using FRP flexural reinforcement after cracking has a smaller depth to the neutral axis because of the lower axial stiffness (that is, product of reinforcement area and modulus of elasticity). The compression region of the cross-section is reduced, and the crack widths are wider. As a result, the shear resistance provided by both aggregate interlock and compressed concrete is smaller [33].

The contribution of longitudinal FRP reinforcement in terms of dowel action has not been determined. Because of the lower strength and stiffness of FRP bars in the transverse direction, however, it is assumed that their dowel action contribution is less than that of an equivalent steel area [33].

Recent research [31] showed that the presence of GFRP stirrups (similar to steel stirrups) enhances the concrete contribution after the formation of the first shear crack. The shear resistance is influenced by the spacing between the stirrups. A small spacing contributes to enhance the confinement of the concrete, to control the shear cracks, and to improve the aggregate interlocking.

The basic safety relationship at the ultimate limit state can be written as

$$\phi V_n \geq V_u \tag{4.86}$$

In Equation (4.86), ϕV_n is the factored shear capacity of the member and is a function of the member geometry, the spacing of the reinforcement, and the mechanical properties of the materials; the term "factored" means that the nominal calculated shear capacity has been reduced by the safety factors associated with the material or the failure mode depending upon the calculation procedures followed.

The second term of Equation (4.86), V_u, is the factored shear force resulting from the analysis of the member and is a function of the member geometry, stiffness and boundary conditions, and the applied loads; the term "factored" means that the calculated shear force associated to a specified loading condition has been amplified by the safety factors related to the acting loads. V_u comes from the structural analysis performed on the system being studied and follows ACI 318-11 provisions.

The nominal shear strength of a member, V_n, is the sum of the contributions of concrete, V_c, and FRP reinforcement, V_f:

$$V_n = V_c + V_f \tag{4.87}$$

An upper limit for V_n based on the magnitude of V_f is discussed in Section 4.10.3 in order to maintain the current shear strength-reduction factor of 0.75 adopted by ACI 440.1R-06.

4.10.1 Concrete contribution, V_c

According to ACI 440.1R-06, the concrete contribution to the shear capacity, V_c, of a member of rectangular section with FRP bars used as the main reinforcement can be evaluated according to the equation developed by Tureyen and Frosch [34]:

$$V_c = 5\sqrt{f_c'}b_w c \tag{4.88}$$

[or $V_c = \dfrac{5}{2}\sqrt{f_c'}b_w c$ in SI units]

where f_c' is the compressive strength of concrete (psi), b_w is the width of the beam web or the unit width of a slab, d is the effective depth of the FRP flexural reinforcement, and c is the compressive depth of the neutral axis in the fully cracked section computed per Equation (4.8).

A lower limit for V_c is discussed in Section 4.10.3 in order to maintain the current shear strength-reduction factor.

COMMENTARY

In their analysis, Tureyen and Frosch [34] considered a portion of a flexural member subjected to constant shear and computed the principal stress of an infinitesimal concrete element located in the compression zone where shear failure may initiate. This uncracked concrete element is subjected to shear and axial compression stresses.

When failure occurs, the principal tensile stress reaches the tensile strength of the concrete. Using the Mohr's circle, the following can be written:

$$\frac{\sigma}{2} - \sqrt{\tau^2 + \left(\frac{\sigma}{2}\right)^2} = -f_{ct} \tag{4.89}$$

Solving for τ, the expression of the tensile stress at failure can be obtained:

$$\tau = \sqrt{f_{ct}^2 + f_{ct}\sigma} \tag{4.90}$$

By considering the equilibrium of a member strip of infinitesimal width chosen at a flexural crack location, the shear capacity of the section due to the concrete itself was written in the following fashion:

$$V_c = \frac{2}{3} b_w c \sqrt{f_{ct}^2 + f_{ct}\frac{\sigma_m}{2}} \tag{4.91}$$

In fact, analyzing the free-body diagram of the member strip at a crack location, Tureyen and Frosch [34] considered the shear stresses generated from the flexural stresses due to $\Delta M = V \cdot \Delta x$ and noted that the maximum shear stress, achieved at the mid-depth of the compression zone, is equal to

$$\tau_{max} = \frac{3}{2} \frac{V}{b_w c} \tag{4.92}$$

Substituting $\tau_{max} = \tau$ and $\sigma = \sigma_m/2$ in Equation (4.90) and solving for V, Equation (4.91) was obtained.

The design Equation (4.89) was then arrived at after substituting $f_{ct} = 6\sqrt{f_c'}$ in Equation (4.91), factoring out $\sqrt{f_c'}$, and rearranging terms:

$$V_c = \left(\sqrt{16 + \frac{4\sigma_m}{3\sqrt{f_c'}}}\right)\sqrt{f_c'}b_w c = K\sqrt{f_c'}b_w c \tag{4.93}$$

> The theoretical and experimental values of the constant K were computed and compared based on a database of 370 beams. Based on these results, it was suggested to adopt a value of $K = 5$ for design purposes.

Design requirements. For the case of FRP reinforced concrete beams, shear reinforcement is not required when the following condition is verified:

$$V_u \leq \frac{\phi V_c}{2} \tag{4.94}$$

For the case of slabs and footings, this threshold is raised to ϕV_c. Typically, slabs and footings do not include shear reinforcement.

COMMENTARY

FRP RC members without shear reinforcement exhibit a size effect in shear whose extent and relevant parameters are similar to those of steel RC members [35]. Experimental evidence gained using smaller specimens may lead to overestimating strength when used for validation purposes and in practice on larger members without shear reinforcement. When used, longitudinal skin reinforcement or minimum shear reinforcement contributes to mitigating the size effect. In particular, the former option improves the flexural stiffness, allowing the formation of more closely spaced cracks. The size effect is exacerbated as the amount of flexural reinforcement and the maximum aggregate size are reduced, both resulting in smaller shear strength values. However, the ACI 440.1R-06 algorithm remains conservative due to an implicit strength-reduction factor that offsets size effect and is also effective for large beams (with effective depth limited to that in the research presented) with minimum FRP shear reinforcement [35]. Figures 4.15 and 4.16 show setup and postmortem crack patterns for a large-sized beam that failed in shear.

Punching shear. ACI 440.1R-06 proposes the following equation to calculate the concentric punching shear capacity of FRP reinforced two-way concrete slabs that are supported by interior columns or subjected to concentrated loads, either square or circular in shape:

$$V_c = 10\sqrt{f_c'}b_o c \tag{4.95}$$

$$\left[\text{or } V_c = \frac{4}{5}\sqrt{f_c'}b_o c \text{ in SI units}\right]$$

Figure 4.15 Test setup for shear test. (A. K. Matta et al. *ACI Structures Journal* 110 (4): 617–628, 2013.)

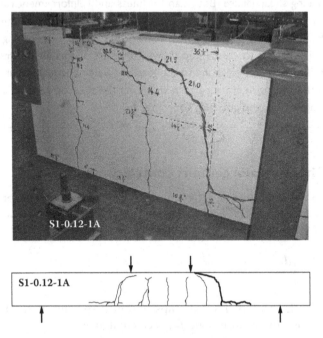

Figure 4.16 Crack pattern of large-sized GFRP-RC beam subject to shear-compression failure. (A. K. Matta et al. *ACI Structures Journal* 110 (4): 617–628, 2013.)

where b_o is the perimeter of the critical section for slabs and footings (computed at $d/2$ away from the column face), and c is the cracked transformed-section neutral axis depth.

Equation (4.95) corresponds to the basic ACI 318-11 concentric punching shear equation for steel reinforced slabs modified by the factor $5/2 \, k$ that accounts for the axial stiffness of the FRP reinforcement as shown:

$$V_c = \left(\frac{5}{2}k\right)4\sqrt{f_c'}b_o d \tag{4.96}$$

COMMENTARY

Ospina [36] demonstrated that the one-way shear design model proposed by Tureyen and Frosch [34], which accounts for reinforcement stiffness, can be modified to account for the shear transfer in two-way concrete slabs. Dulude et al. [37] investigated the punching shear behavior of full-scale, interior, GFRP RC two-way slabs. All slabs showed punching shear failure and similar crack patterns, regardless of the reinforcement ratio. The slabs with low reinforcement ratios showed some ductility and large deformation before the punching shear failure. Both slab thickness and reinforcement ratio significantly affected punching shear capacity. Figure 4.17 shows an example of a GFRP RC two-way slab tested to simulate the effects of punching shear. Figure 4.17a depicts the test setup showing the restraint at the perimeter of the slab that would correspond to the load in a field condition. Figure 4.17b displays the extensive crack pattern visible on the slab top surface, and the third one portrays the punching cone after failure.

4.10.2 Shear reinforcement contribution, V_f

The ACI 318-11 method used to compute the shear contribution of steel stirrups is adopted by ACI 440.1R-06 to compute the contribution to the shear capacity due to the FRP stirrups, V_f:

$$V_f = \frac{A_{fv}f_{fv}d_f}{s} \tag{4.97}$$

where A_{fv} is the area of FRP stirrups within a spacing of s. The tensile strength of FRP for shear design, f_{fv}, is calculated as

$$f_{fv} = 0.004E_f \le f_{fb} \tag{4.98}$$

(a) Test setup for 8 in. (200 mm) thick slab

(b) Crack pattern

Figure 4.17 Punching shear test of two-way GFRP-RC slab. (University of Sherbrooke, Sherbrooke, Quebec, Canada; courtesy of Prof. Brahim Benmokrane.)

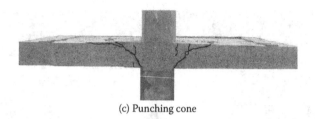

(c) Punching cone

Figure 4.17 (Continued)

where f_{fb} is the strength of the bent portion of the FRP stirrup. f_{fb} depends on the ratio of internal radius of bend in stirrups to their diameter, r_b/d_b, which may be assumed equal to 3. The tail length of an FRP stirrup should be not less than $12d_b$. The strength of the FRP bend recommended by the ACI 440.1R-06 is

$$f_{fb} = \left(0.05\frac{r_b}{d_b} + 0.3\right)f_{fu} \leq f_{fu} \qquad (4.99)$$

At this time, the shear contribution of bent longitudinal FRP bars has not been fully assessed. However, provisions similar to those adopted in ACI 318-11 for steel RC construction are envisioned to be applicable for the case of FRP reinforcement. An upper limit for V_f is discussed in Section 4.10.3.

Minimum shear reinforcement. The minimum amount of shear reinforcement, $A_{fv,min}$, defined in ACI 440.1R-06 is

$$A_{fv,min} = \frac{50b_w s}{f_{fv}} \qquad (4.100)$$

$$\left[\text{or } A_{fv,min} = \frac{0.35b_w s}{f_{fv}} \text{ in SI units}\right]$$

COMMENTARY

Equation (4.100) was derived from steel-reinforced members and is more conservative when used for their FRP reinforced counterparts.

The spacing of the stirrups cannot exceed the smaller half of the value of the effective depth of the flexural reinforcement, $d/2$, or 24 in. (610 mm).

Furthermore, to prevent shear failure due to crushing of the web, the following limitation is considered:

$$V_f \leq 8\sqrt{f_c'}b_w d \tag{4.101}$$

4.10.3 Strength-reduction factor for shear

The shear strength-reduction factor of 0.75 required by ACI 318-11 is also adopted by ACI 440.1R-06. However, in order to maintain the same level of reliability in FRP RC as expected for steel RC, it has been shown [14] that the following upper limit limitation on V_n must be imposed:

$$V_n = V_c + V_f \text{ if } V_f \leq 3V_c \tag{4.102}$$

$$V_n = 4V_c \text{ if } V_f > 3V_c \tag{4.103}$$

The same reliability analysis also demonstrates that for members with no shear reinforcement, such as slabs and footings, ACI 440.1R-06 tends to underpredict the shear strength. For these members, the contribution of concrete to the shear capacity need not be taken less than

$$V_c \geq 0.8\sqrt{f_c'}b_w d \quad \text{(for one-way shear)} \tag{4.104}$$

$$V_c \geq 1.6\sqrt{f_c'}b_w d \quad \text{(for two-way shear)} \tag{4.105}$$

provided that the flexural reinforcement satisfies the minimum requirement for temperature and shrinkage.

COMMENTARY

ACI 440.1R-06 states that "the strength-reduction factor of 0.75 given by ACI 318-05 for reducing nominal shear capacity of steel-reinforced concrete members should also be used for FRP reinforcement." This singles out this guideline for proposing a larger (i.e., less conservative) strength-reduction factor for shear as compared to flexure and makes the need for validation more deeply felt [14]. The process of calibrating the safety factors starts by obtaining the statistical parameters; thus, FRC RC beams with or without transverse FRP reinforcement (stirrups) were investigated independently as follows: (a) Test results compiled by Miano [38] in combination with results from Matta et al. [35] provided a statistical database for beams without stirrups, and (b) a similar database for beams with shear reinforcement (stirrups) was collected by Vitiello [39].

Based on the databases, statistical parameters of FRP RC beams without stirrups under shear failure are calculated as $\lambda_R = 1.93$ and $\delta_R = 0.238$ (Table 4.6, first row). Similarly, statistical parameters of FRP reinforced beams under shear failure are calculated as $\lambda_R = 1.64$ and $\delta_R = 0.353$ (Table 4.6, second row). For each of these two cases, the comparative reliability equation may be used to calculate the shear strength-reduction factor. Substituting the probabilistic parameters of shear failure of a steel RC beam from Table 4.6 (last row) ($\phi_1 = 0.75$, $\lambda_1 = 1.23$, and $\delta_1 = 0.109$) and an FRP RC beam from Table 4.6 (λ_2 and δ_2 from either of the first two rows), the strength-reduction factor for the latter can be calculated for the presumed value of target reliability of $\beta_T = 3.5$ (the first two rows of Table 4.7). Evidently, the current shear strength-reduction factor of 0.75 for FRC RC is relatively conservative for beams with no stirrups, while in presence of such reinforcement, a drastic modification (from existing $\phi = 0.75$ to no less than $\phi = 0.50$) appears to be necessary. However, two simple modifications to the limitations of the shear design equation (i.e., one for a minimum value of V_c and one for maximum amount of shear reinforcement) can reduce the likelihood of unnecessary overdesign or undesired underdesign, while the anticipated level of safety is maintained.

FRP RC beams without stirrups: With reference to ACI 318-11, the shear strength proposed by ACI 440.1R-06 may be rewritten as

$$V_{c,FRP} = \left(\frac{5}{2}k\right)V_{c,Steel} \tag{4.106}$$

In case of lightly FRP reinforced flexural members such as slabs and footings, for which the longitudinal reinforcement is normally next to the minimum allowable, the current formulation of shear resistance from ACI 440.1R-06 leads to results that might appear unrealistic. Here, an attempt is made to investigate whether a minimum can be imposed on the contribution of concrete.

From the database of beams with no stirrups, those specimens were extracted whose tensile stiffness ($\rho_f E_f$) is less than a steel reinforced slab with minimum reinforcement ($\rho_{s,min}$ = 0.0018, E_s = 29,000 ksi = 200 GPa), or $\rho_f E_f$ < 52.2 ksi (360 MPa). For a slab with such tensile stiffness, k is ≈0.16. For the subset of the database with $\rho_f E_f$ < 52.2 ksi (360 MPa), it is assumed that the shear strength can be calculated as

$$V_c \geq 0.80\sqrt{f_c'}b_w d \tag{4.107}$$

In other words, k, in the formulation proposed by ACI 440.1R-06, need not be taken less than 0.16. Table 4.6 (third row) presents the statistical parameters of this subset if V_c is calculated according to this equation. The strength-reduction factor of the subset can be calculated from the comparative reliability equation. This leads to ϕ_2 = 0.75 (Table 4.7, third row), which is equal to the current factor of ACI 440.1R-06, which in turn confirms that the proposed limit is of adequate safety:

$$V_c \geq 0.80\sqrt{f_c'}b_w d \tag{4.108}$$

This minimum would still guarantee safety and, at the same time, would prevent penalizing one-way shear in elements such as slabs and footings. Accordingly, for two-way shear (punching shear) calculated according to Equation (4.95), the corresponding minimum value for V_c can be expressed as

$$V_c \geq 1.6\sqrt{f_c'}b_w d \tag{4.109}$$

FRP RC beams with stirrups: To address the low strength-reduction factor for beams with FRP stirrups requires pinpointing the source of deviation in resistance, indicated by the large coefficient of variation of such elements (δ_R = 0.353). This variation represents the uncertainty associated with the element's strength that, subsequently, leads to a low strength-reduction factor for a target reliability of β_T = 3.5.

Grouping the beams based on the level of shear reinforcement (i.e., V_f/V_c or FRP to concrete shear contribution) reveals a direct relationship between this ratio and deviation of resistance, as the beams with $V_f \geq 3V_c$ were identified as the most inconsistent (δ_R = 0.614 from Table 4.7, fourth row). In line with ACI 318-11 and ACI 440.1R-06 approaches, but using a lower threshold, it is proposed to limit shear reinforcement contribution, V_f, to $3V_c$. By imposing a new ceiling on the combination of the two components of the shear resistance, the set of elements with $V_f \geq 3V_c$ is excluded from the sample population as expressed by Equations (4.102) and (4.103).

Equation (4.103) covers those cases in which the design is governed by nonstrength considerations, such as achieving higher ductility through a tight arrangement of the stirrups.

Leaving out the cases with $V_f \geq 3V_c$ from the sample population, the statistical parameters are calculated for the remainder population (Table 4.6, fifth row) with a considerable improvement in consistency of behavior (δ_R decreases from 0.353 to 0.226). Using these new parameters, the strength-reduction factors are recalculated (Table 4.7, last row) and show a strength-reduction factor of 0.77 for the target reliability of $\beta_T = 3.5$.

Combining these two limits, the shear strength equations of ACI 440.1R-06 may be reintroduced as

$$V_c = 5\sqrt{f_c'}b_w c \tag{4.110}$$

$$V_n = V_c + V_f \leq 4V_c \tag{4.111}$$

Table 4.6 Strength-reduction and statistical parameters of FRP reinforced beams subject to shear

FRP RC	ϕ[a]	β_T	Bias (λ)	CoV(δ)
$V_f = 0$ (no stirrups)	0.75	3.5	1.93	0.238
No limit on V_f	0.75	3.5	1.64	0.353
$V_f = 0$ ($k < 0.16$)	0.75	3.5	1.67	0.227
$V_f > 3V_c$	0.75	3.5	1.22	0.614
$V_f \leq 3V_c$	0.75	3.5	1.80	0.226
Steel RC	ϕ[b]	β_T[c]	Bias (λ)[c]	CoV(δ)[c]
Shear	0.75	3.5	1.23	0.109

[a] ACI 440.1R-06. ACI Committee 440. Guide for the design and construction of structural concrete reinforced with FRP bars.
[b] ACI 318-11. ACI Committee 318. Building code requirements for reinforced concrete, ACI 318-11. American Concrete Institute, Farmington Hills, MI (2011). ACI 318-11.
[c] A. S. Nowak and M. M. Szerszen. *ACI Structural Journal* 100 (3): 383–391 (2003).

Table 4.7 Calculated strength-reduction
factors for FRP RC beams subject
to shear for $\beta T = 3.5$

Limit state	ϕ
$V_f = 0$ (no stirrups)	0.84
No limit on V_f	0.49
$V_f = 0$ ($k < 0.16$)	0.75
$V_f \leq 3V_c$	0.77

4.10.4 Examples—One-way shear strength

The examples presented here (US customary only) are intended to show an application of the algorithms discussed in the preceding sections. More exhaustive design examples are given in Chapters 6, 7, and 8.

Example 4.13

Calculate the nominal and design shear strength of the following beam:

Concrete:	$f'_c = 4.0$ ksi
	E_c (ksi) $= 57\sqrt{f'_c}$(psi) $= 3600$ ksi
Longitudinal reinforcement:	$f_{fu} = 60$ ksi
	$E_f = 6000$ ksi
	$A_f = 4\#10 = 4.91$ in.^{2}
Shear reinforcement:	$f_{fu} = 60$ ksi
	$E_{fv} = 6000$ ksi
	$A_{fv} = \#3@10$ in.
	$r_b/d_b = (r_b/d_b)_{min} = 3.0$
Size:	$b_w = 16.0$ in. Width of the beam
	$h = 25.0$ in. Height of the beam
	$c_c = 3.0$ in. Concrete cover
	$d = h - c_c = 22.0$ in. Effective depth
Solution:	
Concrete contribution, V_c:	$n_f = E_f/E_c = (6000$ ksi$)/(3600$ ksi$) = 1.67$
	$\rho_f = A_f/(b_w d) = 0.0139$
	$n_f \rho_f = 0.0233$
	$k = 0.194$
	$c = kd = 4.26$ in.
	$V_c = 5\sqrt{f'_c} b_w c = 21.6$ kip

Continued

Shear reinforcement contribution, V_f:	A_{fv} = #3@10 in. = 0.22 in.²
	s = 10.0 in.
	$f_{fv} = 0.004E_{fv} \le f_{fb}$
	$f_{fb} = (0.05r_b/d_b + 0.3) f_{fu} \le f_{fu}$
	$f_{fb} = 0.45f_{fu} = 27.0$ ksi
	$f_{fv} = 24.0$ ksi
	$V_f = A_{fv} f_{fv}d/s = 11.6$ kip
The nominal shear strength:	$V_n = V_c + V_s = 33.2$ kip
The design shear strength:	$\phi V_n = 0.75V_n = 24.9$ kip

Example 4.14

Calculate the spacing of the stirrups for the beam in Example 4.13 so that a design shear strength of 30 kip can be provided.

Solution:
$\phi V_n = 0.75V_n = 30.0$ kip
$V_n = V_c + V_f = 40.0$ kip
$V_c = 21.6$ kip
$V_f = 18.4$ kip
$V_f = A_{fv} f_{fv}d/s = 18.4$ kip
$s = 6.3$ in.
Use: #3@6 in.
$V_n = 41.0$ kip
$\phi V_n = 0.75V_n = 30.8$ kip

Example 4.15

Compare the shear strength of the beams in Examples 4.13 and 4.14 with those of two similar beams with grade 60 steel flexural and shear reinforcement.

Solution:	Example	FRP RC	Steel RC
	4.13	$V_c = 21.6$ kip	$V_c = 44.5$ kip
		$V_f = 11.6$ kip	$V_s = 29.0$ kip
		$V_n = 33.2$ kip	$V_n = 73.5$ kip
		$\phi V_n = 24.9$ kip	$\phi V_n = 55.1$ kip
	4.14	$V_c = 21.6$ kip	$V_c = 44.5$ kip
		$V_f = 19.4$ kip	$V_s = 48.4$ kip
		$V_n = 41.0$ kip	$V_n = 92.9$ kip
		$\phi V_n = 30.8$ kip	$\phi V_n = 69.7$ kip

4.10.5 Examples—Two-way shear strength

The examples presented here (US customary only) are intended to show an application of the algorithms discussed in the preceding sections. More exhaustive design examples are given in Chapters 8 and 10.

Example 4.16

An 8 in. flat plate (h = 8 in., d = 6.5 in., f'_c = 5 ksi, E_c = 4030 ksi) is supported by 18 in. × 18 in. columns with tributary areas of A = 20 ft × 20 ft = 400 ft² and sustains a live load of 50 psf and an additional dead load of 15 psf (D = 115 psf, L = 50 psf). Check the adequacy of the slab if: (a) the flat plate is reinforced by GFRP bars with f_{fu} = 60 ksi and E_f = 6000 ksi, and (b) the flat plate is reinforced by steel (f_y = 60 ksi). Assume that the reinforcement ratio for both cases is equal to the shrinkage and temperature reinforcement for FRP bars or 0.0036.

$U = 1.2D + 1.6L = 218$ psf
$V_u = U.A = 87.2$ kip
$b_0 = 4(C + d) = 98$ in. C = 18 in.: column size

FRP RC	Steel RC
$n_f = E_f/E_c = (6000\ \text{ksi})/(4030\ \text{ksi})$ $= 1.49$	
$\rho_f = A_f/(b_w d) = (0.0036 b_w h)/(b_w d)$ $= 0.0044$	
$n_f \rho_f = 0.0066$	
$k = 0.108$	
$c = kd = 0.70$ in.	
$V_c = 10\sqrt{f'_c} b_0 c = 48.5$ kip	$V_c = 4\sqrt{f'_c} b_0 d = 180.2$ kip
$\phi V_n = (0.75) V_c = 36.4$ kip $< V_u$ $= 87.2$ kip	$\phi V_n = (0.75) V_c = 135.1$ kip $> V_u$
Inadequate	Adequate

The vast difference between the two cases in Example 4.13 reveals that for two-way flexural members such as slabs and footings, whose longitudinal reinforcement is normally next to the minimum allowable, the current formulations lead to results that might appear unworkable and unrealistic. The example provides evidence for a minimum to be imposed on the contribution of concrete as discussed in Section 4.10.3 and its commentary.

Example 4.17

Calculate the minimum thickness required for a footing based on the shear strength. The 10 × 10 ft single footing supports a 24 × 24 in.

column with an ultimate axial load of $P_u = 500$ kip. The net soil pressure under the footing is $q_u = 5$ ksf. Assume $f'_c = 5000$ psi and a concrete cover (from the bar center) of 3 in. The flexural reinforcement properties are not known, but it is assumed that the design minima are met. The ultimate one-way shear at distance d from the face of the column can be calculated as

$$V_{1u} \text{ (kip)} = (5 \text{ ksf})(4 \text{ ft}-d)(10 \text{ ft}) = 200 - 50d \text{ (with } d \text{ in ft)}$$

$$\phi V_{1c} \text{ (kip)} \geq (0.75)(0.8)\sqrt{f'_c}b_w d = 61.1d \text{ (with } d \text{ in ft)}$$

$$V_u = \phi V_c$$

$$d = 1.80 \text{ ft} \approx 22 \text{ in.}$$

Assuming $d = 22$ in., the critical section for punching is

$$b_1 = b_2 = 22 \text{ in.} + 24 \text{ in.} = 46 \text{ in.} = 3.83 \text{ ft}$$

$$b_0 = 2(b_1 + b_2) = 184 \text{ in.}$$

And the ultimate two-way shear acting outside the critical section is

$$V_{2u} \text{ (kip)} = (5 \text{ ksf})(10^2 \text{ ft}^2 - 3.83^2 \text{ ft}^2) = 426.7 \text{ kip}$$

$$\phi V_{2c} \text{ (kip)} \geq (0.75)(1.6)\sqrt{f'_c}b_0 d = 343.5 \text{ kip} < V_{2u} \text{ (not adequate)}$$

Try $d = 25$ in.

$$b_1 = b_2 = 25 \text{ in.} + 24 \text{ in.} = 49 \text{ in.} = 4.08 \text{ ft}$$

$$b_0 = 2(b_1 + b_2) = 196 \text{ in.}$$

And the ultimate two-way shear acting outside the critical section is

$$V_{2u} \text{ (kip)} = (5 \text{ ksf})(10^2 \text{ ft}^2 - 4.08^2 \text{ ft}^2) = 416.7 \text{ kip}$$

$$\phi V_{2c} \text{ (kip)} \geq (0.75)(1.6)\sqrt{f'_c}b_0 d = 415.8 \text{ kip} \approx V_{2u} \text{ (OK)}$$

Use a 30 in. footing with $d = 27$ in.

4.10.6 Shear friction

Shear-friction design is applicable where shear is transferred directly across a given plane such as the interface between concretes cast at different times and connections of precast constructions. The nominal shear strength, V_n, provided by reinforcement is computed as

$$V_n = A_{vf} f_{vf} \mu \tag{4.112}$$

where A_{vf} is the area of shear friction reinforcement perpendicular to the plane of shear and f_{vf} is defined as

$$f_{vf} = E_f \varepsilon_{vf} \leq f_{fu} \tag{4.113}$$

Unless tests demonstrate otherwise, the recommended value for ε_{fv} is 0.003.

According to ACI 318-11, μ is the coefficient of friction. For the common case of normal weight concrete placed against hardened concrete surfaces:

$\mu = 0.6$ (if the surface is not intentionally roughened)

$\mu = 1.0$ (if the surface is intentionally roughened)

This formulation of shear friction provides a resistance considerably lower than what steel reinforcement can deliver. Hence, other parameters, such as compressive axial force transferred across the plane, or other devices and mechanisms, such as shear keys, may be considered to enhance the strength. In this case, the recommended total nominal shear strength is:

$$V_n = \sqrt{\left[\left(A_{vf} f_{vf} \mu \right)^2 + \left(P_u \mu + V_n' \right)^2 \right]} \tag{4.114}$$

where $P_u \geq 0$ is the compressive axial force acting simultaneously with the transferred shear. V_n' is the shear strength provided by other mechanisms. If $P_u < 0$ (tensile force), then

$$V_n = \sqrt{\left[\left(A_{vf} f_{vf} \mu \right)^2 + V_n'^2 \right]} + P_u \mu \tag{4.115}$$

The reason that the contributions of reinforcement and other mechanisms are not considered to be directly additive is the relatively large deformations that are required to mobilize the normal force provided by reinforcement. Therefore, the presence of another mechanism may render the other partly ineffective. Finally, as always, Equation (4.86) has to be verified and other cases of shear design ($\phi = 0.75$). An example of the application of this design is presented in Chapter 5 for the case of a shear wall (Example 5.6).

4.10.7 Shear stresses due to torsion

The basic safety relationship at the ultimate limit state can be written as

$$\phi T_n \geq T_u \tag{4.116}$$

In Equation (4.116), ϕT_n is the factored torsional strength of the member and is a function of the member geometry and the mechanical properties of the materials; the term "factored" means that the nominal calculated torsional capacity has been reduced by a strength-reduction factor applicable to torsion. The second term of Equation (4.116), T_u, is the factored torsion force resulting from the analysis of the member for applied loads amplified by the safety factors related to the acting loads. Because of the very limited experimental evidence related to torsion, ACI 318-11 provisions are proposed here as applicable to FRP RC construction for the case limited to pure torsion.

For normal weight concrete, it is permitted to neglect torsion effects if the factored torsional moment T_u is

$$\phi\sqrt{f_c'}\left(\frac{A_{cp}^2}{p_{cp}}\right) \geq T_u \tag{4.117}$$

where the strength-reduction factor $\phi = 0.75$, A_{cp}, is the area enclosed by the outside perimeter, p_{cp}, of the full concrete cross section. The value expressed by Equation (4.117) is known as threshold torsion.

When T_u exceeds the threshold torsion, in addition to the longitudinal reinforcement, closed stirrups have to be used such that Equation (4.118) is satisfied:

$$\phi T_n = \frac{2A_{oh}A_t f_{ft}}{s} \geq T_u \tag{4.118}$$

where A_{oh} is the area enclosed by centerline of the outermost closed transverse torsional reinforcement and A_t is area of one leg of a closed stirrup resisting torsion within spacing s.

At the conclusion of the example shown in Chapter 7, sample calculations for torsion capacity are presented.

COMMENTARY

Very limited research on the torsional resistance of FRP RC members has been found in the technical literature. El-Awady, Husain, and Mandour [40] presented experimental as well as analytical investigations on the torsional behavior of 18 FRP RC beams tested under combined torsion and flexure.

The authors reported that the torsional resistance of the beams increased as the reinforcement ratio of the FRP longitudinal bars increased and that longitudinal GFRP and CFRP bars had comparable effectiveness to steel in torsion reinforcements.

Figure 4.18 shows an example of a GFRP RC beam tested in torsion. Figure 4.18(a) depicts the test setup while Figure 4.18(b) portrays the failure mode with the typical crack pattern due to torsion.

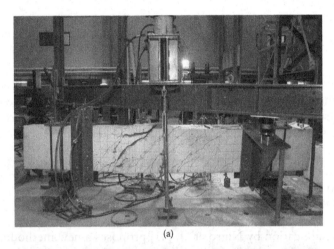

(a)

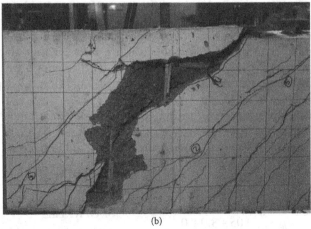

(b)

Figure 4.18 Torsion test of concrete beam with GFRP bars and stirrups. (University of Sherbrooke, Sherbrooke, Quebec, Canada; courtesy of Prof. Brahim Benmokrane.)

4.11 TEMPERATURE AND SHRINKAGE REINFORCEMENT

No experimental data are available in the technical literature to establish the minimum FRP reinforcement ratio for shrinkage and temperature, $\rho_{f,ts}$. Accordingly, the following equation adopted from ACI 318-11 is proposed by ACI 440.1R-06:

$$\rho_{f,ts} = 0.0018 \frac{60,000}{f_{fu}} \frac{E_s}{E_f} \tag{4.119}$$

[or $\rho_{f,ts} = 0.0018 \dfrac{414}{f_{fu}} \dfrac{E_s}{E_f}$ in SI units]

where E_s is the modulus of elasticity of steel. An upper bound equal to 0.0036 and a lower bound equal to 0.0014 are also set for $\rho_{f,ts}$. The spacing of the shrinkage and temperature reinforcement should not exceed three times the slab thickness, or 12 in., whichever is smaller.

4.12 SAFETY FIRE CHECKS FOR BENDING MOMENT CAPACITY

A recent publication by Nigro et al. [41] proposes a new methodology to perform fire safety checks for bending moment capacity of unprotected FRP RC flexural members exposed to fire on the side of the FRP bars under tension. The objective of this approach is to compute the bending moment capacity of the member taking into account the reduced FRP bar properties due to the exposure to high temperatures and the available anchorage developable in the zones not directly exposed to the fire.

Nigro et al. propose the following equations to estimate the "average" deterioration factors for tensile strength and modulus of elasticity of CFRP and GFRP bars at a specific temperature T expressed in degree Celsius as shown in equations (4.120) and (4.121) [1°C = 5(1°F–32)/9].

$$\rho_f(T) = \frac{f_f(T)}{f_{fu}} = \frac{0.05}{0.05 + 8.0 \cdot 10^{-11} \cdot T^{3.55}} \quad \text{for GFRP bars}$$

$$\rho_f(T) = \frac{f_f(T)}{f_{fu}} = \frac{0.06}{0.06 + 2.0 \cdot 10^{-10} \cdot T^{3.33}} \quad \text{for CFRP bars} \tag{4.120}$$

$$\rho_E = \frac{E_f(T)}{E_f} = \frac{0.28}{0.28 + 6.0 \cdot 10^{-12} \cdot T^{4.3}} \quad \text{for GFRP bars}$$

$$(4.121)$$

$$\rho_E = \frac{E_f(T)}{E_f} = \frac{2.4}{2.4 + 9.0 \cdot 10^{-12} \cdot T^{4.4}} \quad \text{for CFRP bars}$$

According to the authors, these equations can be used to estimate the reduced FRP tensile strength and elastic modulus at the temperature T reached by the bars after a fire exposure time t. This temperature can be estimated using the following equations where T is expressed in degree Celsius.

$$t \le 30\,\text{min} : T = A_1(c) \cdot t + 20$$

$$(4.122)$$

$$t > 30\,\text{min} : T = A_2(c) + A_3(c) \cdot t^{A_4(c)}$$

where c is the concrete cover in in., and the coefficients $A_i(c)$ are defined as shown in Table 4.8.

In computing the bending moment capacity, Nigro et al. recommend adopting the concrete constitutive law suggested by EN1992-1-2 [42], which differs from the constitutive law typically used at normal temperature because it includes a strong softening branch.

$$\sigma(\varepsilon_c) = f_c' \cdot \frac{3 \cdot \left(\dfrac{\varepsilon_c}{\varepsilon_{c1}} \right)}{2 + \left(\dfrac{\varepsilon_c}{\varepsilon_{c1}} \right)^3} \quad \text{for } \varepsilon_c \le \varepsilon_{c1}$$

$$(4.123)$$

$$\sigma(\varepsilon_c) = f_c' \cdot \frac{\varepsilon_{cu1} - \varepsilon_c}{\varepsilon_{cu1} - \varepsilon_{c1}} \quad \text{for } \varepsilon_{c1} \le \varepsilon_c \le \varepsilon_{cu1}$$

Table 4.8 Coefficients A_i

c (in.)	A_1	A_2	A_3	A_4
0.787	11.538	−4586.1	4221.2	0.0470
1.18	8.032	−2326.8	1935.7	0.0854
1.58	5.685	−892.3	592.2	0.1774
2.00	3.997	−509.4	271.7	0.2561
2.36	2.792	−312.0	130.8	0.3400

Note: 1 in. = 24.5 mm

Table 4.9 Coefficients B_i

c (in.)	B_1	B_2	B_3
0.787	−23.43	15.38	0.4660
1.18	−104.68	58.45	0.2821
1.58	−159.79	87.28	0.2437
2.00	−2159.93	1995.44	0.02800
2.36	−13582.44	13347.56	0.0055

Note: I in. = 24.5 mm

where $f'c$ is the concrete compressive strength, ε_{c1} is the strain corresponding to $f'c$ and equal to 0.0025, and ε_{cu1} is the ultimate strain equal to 0.002.

The typical elastic-brittle stress-strain relationship is, instead, assumed for the FRP. The ultimate usable FRP tensile strength is the smaller of the reduced strength due to fire, computed per Equation (4.120), and the maximum stress developable by the end anchorage in zones not directly exposed to fire. Based on the work by Katz and Berman [43], it is assumed that for temperatures exceeding 122°F (50°C) the bond between concrete and FRP reinforcement becomes negligible. For this reason, in computing the maximum developable tensile stress, Nigro et al. recommend considering only the length of the end anchorage whose temperature does not exceed 122°F (50°C). The following equation provided by the Italian CNR Guidelines [44] is proposed.

$$f_{fd,t} = \frac{l_d - l_{d,fi,t,T>Tcr}}{0.1 \cdot d} \tag{4.124}$$

where l_d is the design development length, $l_{d,fi,t,T>Tcr}$ is the embedment length of a bar with a temperature exceeding 122°F (50°C), d is the bar diameter.

The development length $l_{d,fi,t,T>50C}$ is proposed to be computed as follows:

$$l_{d,fi,t,T>Tcr} = B_1(c) + B_2(c) \cdot t^{B_3(c)} \tag{4.125}$$

where t is the fire exposure time, c is the concrete cover in in., and the coefficients $B_i(c)$ are computed based on thermal analysis results and reported in Table 4.9.

An application of this method is further discussed in Chapter 6 (Step 10).

REFERENCES

1. J. K. Wight and J. G. MacGregor. *Reinforced concrete mechanics and design*, 6th ed., Englewood Cliffs, NJ: Prentice Hall (2012).
2. ACI Committee 318. Building code requirements for reinforced concrete, ACI 318-11. American Concrete Institute, Farmington Hills, MI. (2011).

3. C. Todeschini, A. Bianchini, and C. Kesler. *Behavior of concrete columns reinforced with high-strength steels*. ACI Journal Proceedings 61 (6): 701–716 (1964).

4. ACI Committee 440. Guide for the design and construction of structural concrete reinforced with FRP bars. 440.1R-06. American Concrete Institute, Farmington Hills, MI (2006).

5. C. E. Ospina, S. Alexander, and J. J. Cheng. Behavior of concrete slabs with fiber-reinforced polymer reinforcement. Structural engineering report no. 242, Department of Civil and Environmental Engineering, University of Alberta, AB, Canada, p. 355 (2001).

6. C. E. Ospina and S. P. Gross. Rationale for the ACI 440.1R-06 indirect deflection control design provisions. *Proceedings of the 7th International Symposium on Fiber-Reinforced Polymer (FRP) Reinforcement for Concrete Structures (FRPRCS-7)*, SP-230, ed. C. Shield, J. Busel, S. Walkup, and D. Gremel, American Concrete Institute, Farmington Hills, MI, 651–670 (2005).

7. A. Belarbi, K. Chandrashkhara, and S. E. Watkins. Performance evaluation of FRP rebar featuring ductility and health monitoring capability. *Proceedings 4th International Symposium on Fiber Reinforced Polymer for Reinforced Concrete Structures (FRPRCS-4)*, American Concrete Institute, special publication SP-188, 1–12 (1999).

8. H. G. Harris, W. Somboonsong, and F. K. Ko. A new ductile hybrid fiber reinforced polymer (FRP) reinforcing bar for concrete structures. *ASCE Journal of Composites for Construction* 2 (1): 28–37 (1998).

9. H. Wang and A. Belarbi. Flexural behavior of fiber-reinforced-concrete beams reinforced with FRP rebars. *Proceedings Seventh International Symposium on Fiber Reinforced Polymer for Reinforced Concrete Structures (FRP7RCS)*, American Concrete Institute special publication SP-230, 895–914 (2005).

10. S. H. Alsayed and A. Alhozaimy. Ductility of concrete beams reinforced with FRP bars and steel fibers. *Journal of Composite Materials* 33 (4): 1792–1806 (1999).

11. B. Wambeke and C. Shield. Development length of glass fiber reinforced polymer 21 bars in concrete. *ACI Structural Journal* 103 (1): Jan.-Feb., 11–17(2006).

12. P. Gulbrandsen. Reliability analysis of the flexural capacity of fiber reinforced polymer bars in concrete beams. Master thesis, University of Minnesota, MN, p. 80 (2005).

13. A. S. Nowak and M. M. Szerszen. Calibration of design code for buildings (ACI 318): Part 2—Reliability analysis and resistance factors. *ACI Structural Journal* 100 (3): 383–391 (2003).

14. H. Jawaheri Zadeh and A. Nanni. Reliability analysis of concrete beams internally reinforced with FRP bars. *ACI Structural Journal* 110 (6): 1023–1032 (2013).

15. K. Bondy. Moment redistribution: Principles and practice using ACI 318-02. *PTI Journal* 1 (1): 3–21 (2003).

16. I. F. Kara and A. F. Ashour. Moment redistribution in continuous FRP reinforced concrete beams. *Construction and Building Materials,* http://dx.doi.org/10.1016/j.conbuildm at.2013.03.094 (2013).

17. G. W. Washa and P. G. Fluck. Plastic flow (creep) of reinforced concrete continuous beams. *Journal of the American Concrete Institute* 52: 549–561 (1956).

18. G. McClure and Y. Mohammadi. Compression creep of pultruded E-glass-reinforced-plastic angles. *J. Mater. Civ. Eng.*, 7(4): 269–276 (1995).

19. M. Z. Cohn, S. K. Ghosh, and S. R. Parimi. Unified approach to theory of plastic structures. *Journal of the Engineering Mechanics Division of ASCE* 98:1133–1155 (1972).

20. D. Gao, B. Benmokrane, and R. Masmoudi. A calculating method of flexural properties of FRP-reinforced concrete beam: Part 1: Crack width and deflection. Technical report, Department of Civil Engineering, University of Sherbrooke, Sherbrooke, QC, Canada, p. 24 (1998).

21. B. Tighiouart, B. Benmokrane, and D. Gao. Investigation of bond in concrete member with fiber reinforced polymer (FRP) bars. *Construction and Building Materials Journal* 12:453–462 (1998).

22. A. Nanni. Flexural behavior and design of reinforced concrete using FRP rods. *Journal of Structural Engineering* 119 (11): 3344–3359 (1993).

23. R. J. Frosch. Another look at cracking and crack control in reinforced concrete. *ACI Structural Journal* 96 (3): 437–442 (1999).

24. B. Broms. Crack width and crack spacing in reinforced concrete members. *Journal of the American Concrete Institute* 62 (10): 1237–1256 (1965).

25. C. E. Ospina and C. E. Bakis. Indirect flexural crack control of concrete beams and one-way slabs reinforced with FRP bars. *Proceedings of the 8th International Symposium on Fiber Reinforced Polymer Reinforcement for Concrete Structures, FRPRCS-8,* ed. T. C. Triantafillou, University of Patras, Greece, CD ROM (2007).

26. D. E. Branson. Instantaneous and time-dependent deflections of simple and continuous reinforced concrete beams. Alabama Highway Department, Bureau of Public Roads, Montgomery, AL, HPR report no. 7, part 1, p. 78 (1965).

27. ACI Committee 435. Control of deflection in concrete structures, ACI 435R-95. American Concrete Institute, Farmington Hills, MI, Reapproved in 2000 and 2003, pp. 89 (1995).

28. P. H. Bischoff. Reevaluation of deflection prediction for concrete beams reinforced with steel and fiber reinforced polymer bars. *Journal of Structural Engineering* 131 (5): 752–767 (2005).

29. P. Bischoff and S. Gross. Design approach for calculating deflection of FRP-reinforced concrete. *Journal of Composites for Construction* 15 (4): 490–499 (2011).

30. P. H. Bischoff and A. Scanlon. Effective moment of inertia for calculating deflections of concrete members containing steel reinforcement and FRP reinforcement. *ACI Structural Journal* 104 (1): 68–75 (2007).

31. E. A. Ahmed, E. F. El-Salakawy, and B. Benmokrane. Performance evaluation of GFRP shear reinforcement in concrete beams. *ACI Structural Journal* 107 (1): 53–62 (2010).

32. A. K. El-Sayed, E. F. El-Salakawy, and B. Benmokrane. Shear strength of concrete beams reinforced with FRP bars: Design method. *Proceedings, 7th International Symposium on Fiber Reinforced Polymer Reinforcement for Concrete Structures, FRPRCS-7,* Kansas City, MO, SP 230-54, 955–974 (2003).

33. A. Nanni. North American design guidelines for concrete reinforcement and strengthening using FRP: Principles, applications, and unresolved issues. *Construction and Building Materials* 17 (6/7): 439–446 (2003).

34. A. K. Tureyen and R. J. Frosch. Concrete shear strength: Another perspective. *ACI Structural Journal* 100 (5): 609–615 (2003).
35. F. Matta, A. K. El-Sayed, A. Nanni, and B. Benmokrane. Size effect on concrete shear strength in beams reinforced with FRP bars. *ACI Structures Journal* 110 (4): 617–628 (2013).
36. C. E. Ospina. Alternative model for concentric punching capacity evaluation of reinforced concrete two-way slabs. *Concrete International* 27 (9): 53–57 (2005).
37. C. Dulude, M. Hassan, E. Ahmed, and B. Benmokrane. Punching shear behaviour of two-way flat concrete slabs reinforced with GFRP bars. *ACI Structural Journal* 110 (5): 723–734 (2013).
38. A. Miano. Capacità a taglio di solette in C.A. con barre in FRP. Master thesis, Universitá di Napoli—Federico II, Napoli, Italy (2011).
39. U. Vitiello. Capacità a taglio di travi in C.A. con barre e staffe in FRP. Master thesis, Universitá di Napoli—Federico II, Napoli, Italy (2011).
40. E. El-Awady, M. Husain, and S. Mandour. FRP-reinforced concrete beams under combined torsion and flexure. *International Journal of Engineering Science and Innovative Technology* (IJESIT) 2 (1): 384–393 (2013).
41. E. Nigro, G. Cefarelli, A. Bilotta, G. Manfredi, and E. Cosenza, Guidelines for flexural resistance of FRP reinforced concrete slabs and beams in fire. *Composites Part B: Engineering,* 58: 103–112 (2014).
42. European Committee for Standardization, EN1992-1-2, Eurocode 2 – Design of concrete structures – Part 1-1: General rules and rules for building. Brussels, Belgium (2004).
43. A. Katz, and N. Berman, Modeling the effect of high temperature on the bond of FRP reinforcing bars to concrete. *Cement Concrete Composites,* 22(6): 433–443 (2000).
44. National Research Council CNR –DT203/2006, Guide for the design and construction of concrete structures reinforced with fiber reinforced polymer bars. Rome, Italy (2006).

Chapter 5

Members subjected to combined axial load and bending moment

NOTATION

A_1, A_2 = area of fiber-reinforced polymer (FRP) reinforcement along sides 1 and 2 of a rectangular section, in.^{2} (mm^2)

A_f = area of FRP reinforcement, in.^{2} (mm^2)

$A_{f,bar}$ = area of one FRP bar, in.^{2} (mm^2)

A_g = gross cross sectional area of column, in.^{2} (mm^2)

A_{fv} = amount of FRP shear reinforcement with spacing s, in.^{2} (mm^2)

$A_{fv,min}$ = minimum amount of FRP shear reinforcement with spacing s, in.^{2} (mm^2)

A_s = area of tension steel reinforcement, in.^{2} (mm^2)

A_{vf} = area of shear friction reinforcement perpendicular to the plane of shear

a = depth of equivalent rectangular stress block, in. (mm)

b = width of rectangular cross section, in. (mm)

C = compressive force, lb (N)

C_E = environmental reduction factor

c = distance from extreme compression fiber to the neutral axis of a fully cracked section, in. (mm)

D = diameter of circular cross section, in. (mm)

d = distance from extreme compression fiber to centroid of tension reinforcement, in. (mm)

d_b = diameter of tie, in. (mm)

d_f = effective depth of the FRP reinforcement

E_c = modulus of elasticity of concrete, psi (MPa)

E_s = modulus of elasticity of steel, psi (MPa)

E_f = design or guaranteed modulus of elasticity of FRP defined as mean modulus of sample population ($E_f = E_{f,ave}$), psi (MPa)

e = ratio of ε_{fu} over ε_{cu}

f'_c = specified compressive strength of concrete, psi (MPa)

f_f = stress in FRP reinforcement in tension, psi (MPa)

f_{fd} = design tensile strength (f_{fd} = smaller of f_{fu} or $0.01E_f$), psi (MPa)

f_{sf}	=	shear friction stress in reinforcement
f_{fu}	=	design tensile strength of FRP, considering reductions for service environment ($f_{fu} = C_E f^*_{fu}$), psi (MPa)
f^*_{fu}	=	guaranteed tensile strength of population FRP bar, defined as mean tensile strength of sample minus three standard deviations ($f^*_{fu} = f_{fu,ave} - 3\sigma$), psi (MPa)
f_{fv}	=	tensile strength of FRP for shear design, taken as smallest of design tensile strength f_{fu}, strength of bent portion of FRP stirrups f_{fb}, or stress corresponding to $0.004E_f$, psi (MPa)
$f_{u,ave}$	=	mean tensile strength of sample population, psi (MPa)
h	=	overall height of rectangular member, in. (mm)
I_g	=	moment of inertia of gross concrete section, neglecting reinforcement, in.^{4} (mm.^{4})
k	=	Effective length factor
L	=	distance between joints in a slab on grade, ft (m)
L_1, L_2, L_C	=	arm in moment computation of columns, in. (mm)
l_u	=	clear height of the column, in. (mm)
l_w	=	length of shear wall, in. (mm)
M_C	=	contribution of compressive force to nominal moment capacity for circular section, lb-in. (N-mm)
M_{nox}	=	nominal uniaxial moment strength about x-axis, lb-in. (N-mm)
M_{noy}	=	nominal uniaxial moment strength about y-axis, lb-in. (N-mm)
M_{nx}	=	nominal biaxial moment strength about x-axis, lb-in. (N-mm)
M_{ny}	=	nominal biaxial moment strength about y-axis, lb-in. (N-mm)
M_T	=	contribution of tensile force to nominal moment capacity for circular section, lb-in. (N-mm)
M_n	=	nominal moment capacity, lb-in. (N-mm)
M_u	=	factored moment, lb-in. (N-mm)
n_f	=	ratio of modulus of elasticity of FRP bars to modulus of elasticity of concrete
P_c	=	axial force carried by concrete, lb (N)
P_n	=	nominal axial capacity for non-zero eccentricity, lb (N)
P_{n0}, P_o	=	nominal axial load strength at zero eccentricity, lb (N)
P_{nx}	=	nominal axial load strength at given eccentricity along x-axis, lb (N)
P_{ny}	=	nominal axial load strength at given eccentricity along y-axis, lb (N)
P_o	=	nominal axial capacity for zero eccentricity, lb (N)
P_s	=	axial force carried by steel, lb (N)
P_u	=	ultimate axial force, lb (N)
r	=	radius of gyration of the column, in. (mm)
r_b	=	radius of bend, in. (mm)
s	=	stirrup spacing or pitch of continuous spirals, and longitudinal FRP bar spacing, in. (mm)

T_1, T_2	=	tensile force corresponding to A_1 and A_2, lb (N)
T, T_{max}	=	tensile force and maximum tensile force, lb (N)
V_c	=	nominal shear strength provided by concrete, lb (N)
V_f	=	shear resistance provided by FRP stirrups, lb (N)
V_n	=	nominal shear strength, lb (N)
V_u	=	factored shear force, lb (N)
x	=	distance of N.A. from compression edge, in. (mm)
x_b	=	distance of N.A. from compression edge at balanced condition, in. (mm)
α	=	interaction contour parameter
α	=	ratio of x over d
α_1	=	ratio of average stress of equivalent rectangular stress block to f'_c
β	=	ratio of distance from neutral axis to extreme tension fiber to distance from neutral axis to center of tensile reinforcement
β_1	=	factor relating depth of equivalent stress block to neutral axis depth
ε_c	=	strain in concrete
ε_{cu}	=	ultimate strain in concrete
ε_f	=	strain in FRP reinforcement
ε_{fd}	=	design strain for FRP reinforcement (ε_{fd} = smaller of ε_{fu} and 0.01)
ε_{fu}	=	design rupture strain of FRP reinforcement ($\varepsilon_{fu} = C_E \varepsilon^*_{fu}$)
ε^*_{fu}	=	guaranteed rupture strain of FRP reinforcement defined as the mean tensile strain at failure of sample population minus three standard deviations ($\varepsilon^*_{fu} = \varepsilon_{u,ave} - 3\sigma$)
ε_v	=	shear friction strain in reinforcement
η	=	ratio of distance from extreme compression fiber to centroid of tension reinforcement (d) to overall height of flexural member (h)
γ	=	ratio of d over h, or d over D
θ_C, θ_T	=	angles defining the compressive and tensile regions in a circular column (rad.)
ρ_f	=	FRP reinforcement ratio
ϕ	=	strength reduction factor
μ	=	coefficient of friction

5.1 INTRODUCTION

According to the current ACI 440.1R-06 [1] guide, reinforced concrete columns cannot be designed with FRP longitudinal bars and ties. Based on the work by Jawaheri and Nanni [2], this chapter discusses the theoretical approach at the basis of the behavior of FRP RC members subject to combined flexural and axial loads. A new rationale to develop strength-reduction factors for simultaneous flexural and axial resistance that is consistent with

ACI 318-11 [3] is also presented. An example of the design of an FRP RC square column for a medical facility is illustrated in Chapter 9.

5.2 FRP BARS AS COMPRESSION REINFORCEMENT

The behavior of FRP bars as longitudinal reinforcement in compression members is still a relevant issue to be addressed and not yet covered by ACI 440.1R-06 [1]. Research studies, including the recent ones reported by Lofty [4] and Tobbi et al. [5], investigated the effect of the compressive behavior of longitudinal FRP bars by testing full-scale reinforced concrete columns subjected to pure compressive load. Different modes of failure (transverse tensile failure, fiber microbuckling, or shear failure) may characterize the response of FRP bars in compression, depending on the type of fiber, fiber volume fraction, and type of resin. Testing of FRP bars in compression is typically complicated by the anisotropic and nonhomogeneous nature of the FRP material that can lead to inaccurate measurements [6]. A standard test method for FRP bars to be used as compression reinforcement in concrete has not yet been established.

COMMENTARY

Glass FRP (GFRP): This is the most commonly used FRP material system for internal reinforcement of FRP-reinforced concrete members. Experimental research studies [7–14] have investigated the effect of the compressive behavior of longitudinal GFRP bars by testing reinforced concrete (RC) columns. Alsayed et al. [9] investigated the effect of replacing longitudinal steel bars (reinforcement ratio of 1.07%) and ties with an equal amount of GFRP bars and ties. Based on the results of tests performed on 17.7 × 9.8 × 47.2 in. (450 × 250 × 1200 mm) columns under concentric loads, it was reported that replacing longitudinal steel bars with GFRP bars reduced the capacity by 13%, irrespective of the type of ties (steel or GFRP). Replacing steel ties with GFRP ties reduced the capacity by 10%, with no influence on the load-deformation response up to approximately 80% of the ultimate capacity. Mirmiran [8] conducted a parametric study for the analysis of slender GFRP RC columns. It was shown that even though GFRP RC columns are more susceptible to instability failure than steel RC columns, the design practice of using moment magnification factors is also applicable to GFRP RC columns. In other research by Mirmiran, Yuan, and Chen [15], it was concluded that the slenderness limits should be lowered when using longitudinal GFRP reinforcement and a minimum reinforcement ratio of 1% should be maintained. The overall conclusion of these studies is that GFRP RC columns are a doable application.

Aramid FRP (AFRP): Kawaguchi [16] tested 12 concrete members reinforced with AFRP bars and subjected to eccentric tension or compression. He reported that AFRP RC columns can be analyzed using the same procedure as that for steel RC columns. Similarly, Fukuyama et al. [17] tested a half-scale three-story AFRP RC frame under quasi-static loading. AFRP bars were used for columns, beams, and slabs. The frame remained elastic up to a drift angle of 1/50 rad, and no substantial decrease in strength took place after rupture of some AFRP bars in a main beam due to the high degree of indeterminacy of the frame.

Carbon FRP (CFRP): Kobayashi and Fujisaki [18] tested columns reinforced with CFRP grids and determined that strain compatibility was maintained up to the crushing of concrete. CFRP grids were later used by Grira and Saatcioglu [19] as transverse reinforcement for columns tested under cyclic loading and it was concluded that their performance was comparable to that of columns reinforced with steel ties.

5.3 OVERALL DESIGN LIMITATIONS FOR FRP RC COLUMNS

The design recommendations provided in this chapter are based on the outcomes of experimental studies that ultimately provide a convincing case to allow for the limited use of FRP bars in columns, particularly when corrosion resistance or electromagnetic transparency is sought. Given the novelty of the FRP RC technology for column design and construction, two limitations in addition to the ones derived from ACI 318-11 [3] are applicable:

(a) The recommendations proposed for design and illustrated in this chapter apply to buildings with five or fewer stories aboveground and no more than one basement level.

(b) Although some experimental and analytical work has been performed in FRP RC beam–column joint performance [14,20], due to the limited availability of data in this field, the design provisions proposed herein are not applicable to structures in seismic zones. This is a limitation that may be removed in the future based on the outcomes of more experimental evidence.

5.4 REINFORCED CONCRETE COLUMNS SUBJECTED TO AXIAL LOAD

5.4.1 Steel RC columns

The American Concrete Institute (ACI) Building Code [3] bases the axial load capacity equation and the tie requirements for steel RC columns on research carried out at Lehigh University and the University of Illinois in the early 1930s (see reference 10). The maximum compressive stress that the concrete can develop at a strain beyond the yield point of the reinforcing steel, f_y, is taken equal to 85% of the compressive strength of a 6 by 12 in. (152 by 308 mm) concrete cylinder. The nominal capacity of an axially loaded steel RC column, P_o, is defined as the sum of the forces carried by the concrete, P_c, and the steel, P_s, as given by the following equation:

$$P_o = P_c + P_s = 0.85f_c' \cdot (A_g - A_s) + f_y \cdot A_s \qquad (5.1)$$

where A_g is the gross cross-sectional area of the column, A_s is the area of the longitudinal steel reinforcement, and f_c' is the nominal compressive strength of the concrete. ACI 318-11 also requires that the vertical spacing of ties not exceed 16 longitudinal bar diameters (to prevent bar buckling), 48 tie diameters (to ensure sufficient tie area to restrain the lateral displacement of the longitudinal bars), or the least lateral dimension of the column (to develop the maximum strength of the concrete core).

COMMENTARY

Experimental studies performed between the late 1950s and early 1960s (referenced in De Luca, Matta, and Nanni [10]) showed that steel ties provide transverse constraint to the concrete core, allowing the column to fail in a more gradual manner than without ties. It was also found that ties offered sufficient restraint against buckling of the longitudinal bars up to compressive failure of the concrete, with negligible influence on the ultimate load.

Few experimental studies have attempted to characterize the influence of the size of reinforced concrete columns on their structural behavior; however, the current ACI design specifications for RC columns neglect any size effect on the nominal axial strength. Bazant and Kwon [21] tested a total of 26 scaled RC columns of different sizes under eccentric axial load. The existence of a size effect on the ultimate capacity was observed and it was consistent with the fracture mechanics-based mathematical formulation derived by Bazant [22]. Sener, Barr, and Abusiaf [23] tested a total of 27 square RC columns with different scales and slenderness ratios under concentric axial load. The largest cross section had dimensions of 7.9 × 7.9 in. (200 × 200 mm) and reinforcement ratio of 4.91%. It was found that a reduction in strength occurred at increasing size and slenderness, which was in good agreement with Bazant's size-effect law. Nemecek and Bittnar [24] tested square RC columns of three different scales, with maximum size of 11.8 × 11.8 × 78.7 in. (300 × 300 × 2000 mm) and reinforcement ratio of 2.18%, under eccentric axial load. No significant size effect was observed in the ultimate capacity.

5.4.2 FRP RC columns

As confirmed by other research programs [4,5], a recent study by De Luca et al. [10] on full-scale GFRP RC columns loaded in pure compression provided a compelling case to justify the applicability of ACI 318-based design recommendations to FRP RC columns and, as a consequence, to modify existing design guidelines to allow for limited use of FRP bars in members subjected to axial load. The main conclusions of this work may be summarized as follows:

1. The behavior of RC columns internally reinforced with GFRP bars was found to be similar to that of conventional steel RC columns when the longitudinal reinforcement ratio is equal to 1.0% and no appreciable difference could be determined in terms of ultimate capacity.
2. The failure of the steel RC specimen appeared to happen due to buckling of the longitudinal reinforcement when still in the elastic range, whereas the GFRP RC specimens failed due to the crushing of the concrete core at axial strains higher than those measured in the steel RC counterpart.
3. For a reinforcement ratio of 1%, the contribution of the GFRP bars to the column capacity was found to be less than 5% of the ultimate load, which is significantly lower than that of about 12% of the steel bars in the steel RC counterpart.
4. The spacing of the GFRP ties did not contribute to increasing the ultimate capacity, but strongly influenced the failure mode by delaying the buckling of the longitudinal bars, initiation and propagation of unstable cracks, and crushing of the concrete core.

Transverse reinforcement for FRP reinforced members is typically in the form of coupled C-shaped bars and, less commonly, continuous spirals. Given the infancy of the technology and the lack of experimental evidence related to the potentially added confinement of spiral reinforcement, FRP spirals and ties are treated equally. Thus, the geometrical provisions related to ties are extended to spirals where the spiral pitch would coincide with the tie spacing.

COMMENTARY

The experimental work by De Luca et al. [10] included a total of five specimens: one steel RC benchmark, and four GFRP RC columns. The specimens had a square cross section with 24 in. (610 mm) sides and length of 10 ft (3.0 m). The GFRP RC columns were subdivided into two sets of two; each set was identical to the other, but bars from different manufacturers were used.

The steel RC benchmark column was designed using the minimum amount of longitudinal reinforcement and the minimum tie cross-sectional area at maximum spacing as mandated by ACI 318-11. In particular, the total area of longitudinal bars was taken as 1.0% of the gross section area, A_g, choosing eight no. 8 (25 mm) bars. No. 4 (12 mm) ties were used as transverse reinforcement at a spacing of 16 in. (406 mm) on center. For each set of two GFRP RC columns, bar size and total area of longitudinal reinforcement were adopted as for the steel benchmark. For the GFRP ties, the same bar size as their steel counterpart was used, but the spacing was reduced to 12 in. (304 mm) and 3 in. (76 mm) when compared to the steel case. The 12 in. (304 mm) spacing was selected to prevent longitudinal bars buckling, while the 3 in. (76 mm) spacing was chosen as the minimum practical spacing for GFRP ties.

The GFRP RC specimens with the large tie spacing behaved similarly to the benchmark steel specimen. Failure typically initiated with vertical cracks—followed, first, by lateral deflection of the longitudinal bars contributing to the splitting of the concrete cover and, then finally, by crushing of the concrete core and buckling of the longitudinal bars. In the case of the GFRP RC specimens with the smaller tie spacing, energy absorption and deformability were greatly increased as capacity decreased steadily after the peak load until the test was intentionally interrupted.

In all columns, the concrete compressive stress at peak was close to 0.85 f'_c, which is the value defined in ACI 318-11 as the average concrete compressive stress when an adequately tied column reaches its axial strength. The average load carried by the longitudinal GFRP reinforcement (assuming an equal modulus of elasticity for GFRP in tension and compression) ranged between about 2.9% and 4.5% of the peak load, whereas the average load carried by the longitudinal grade 60 (413 MPa) steel reinforcement was about 11.6% of the peak load.

The failure of the steel RC specimen appeared to be ultimately caused by the buckling of the longitudinal bars preceding crushing of the concrete core. Conversely, in the case of the GFRP RC specimens with smaller tie spacing, failure could be attributed to the crushing of the concrete core, while for all the GFRP RC specimens the relatively low contribution of the GFRP bars to the load-carrying capacity resulted in higher strains compared with the steel RC counterpart. Photographs of failed specimens are shown in Figure 5.1 (rectangular cross section) and Figure 5.2 (circular cross section).

(a)

(b)

(c)

Figure 5.1 Concrete columns of rectangular cross section reinforced with GFRP longitudinal bars and ties. (a) overall view of failed column with 12-inch tie spacing; (b) detail of longitudinal bar buckling for column with 12-inch tie spacing; and (c) detail of longitudinal bar and tie rupture for column with 3-inch tie spacing; (De Luca, Matta, and Nanni. *9th International Symposium on Fiber Reinforced Polymer Reinforcement for Concrete Structures* (FRPRCS-9), D. Oehlers, M. Griffith, and R. Seracino, eds., July 13–15, 2009, Sydney, Australia, CD-ROM, 4, 2010.)

(a) MTS test frame with data-acquisition system   (b) LVDTs and steel collars

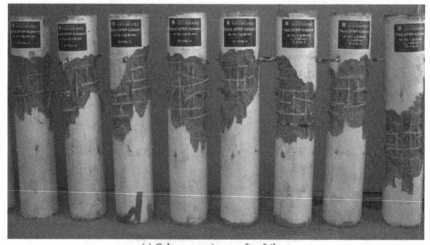

(c) Column specimens after failure

Figure 5.2 Circular concrete columns with GFRP longitudinal bars, hoops, or spirals under axial load. (University of Sherbrooke, Sherbrooke, Quebec, Canada; courtesy of Prof. Brahim Benmokrane.)

5.5 DESIGN RECOMMENDATIONS FOR FRP RC COLUMNS

For the purpose of this book, only GFRP bars are covered in detail. The same considerations can be extended to the other types of FRP systems.

5.5.1 Minimum longitudinal reinforcement

A minimum reinforcement ratio of 1% is recommended as the lower threshold for a longitudinal bar reinforcement area. Longitudinal bar detailing requirements (i.e., spacing, minimum bar number, and minimum bar diameter) set by ACI 318-11 for steel reinforcement are applicable to FRP bars.

COMMENTARY

Since 1936, the building code has required that the minimum reinforcement ratio be 0.01 of the gross area of concrete section. This minimum reinforcement area was intended to prevent "passive yielding" of the steel, which occurs when load is transferred gradually from concrete to the reinforcement as the concrete creeps under sustained axial load [25]. Even though it appears to have become an outdated restriction for modern concrete and steel [26] and notwithstanding the consideration that GFRP does not yield, this requirement has been retained also for the case of GFRP reinforcement for analogy.

5.5.2 Equivalency under compression between GFRP and concrete

Available test results indicate that the equivalency under compression between GFRP and concrete can be assumed.

COMMENTARY

With reference to the behavior of FRP bars in compression, it is known that their testing is complicated by the anisotropic and nonhomogeneous nature of the FRP material, which can lead to inaccurate measurements [27]. For the case of GFRP bars in particular, reductions in the compressive strength and elastic modulus by up to 45% and 20% with respect to the values in tension, respectively, have been reported [28]. Similar results were reported for GFRP bars by Deitz, Hark, and Gesund [29], who indicated that the compressive to tensile strength and modular ratios were approximately 50% and 100%, respectively. Accordingly, GFRP mechanical characteristics exceed those of concrete in compression and, therefore, the equivalency between the two materials when performing analysis and design is justifiable [6].

5.5.3 Limit on maximum tensile strain in GFRP

The tensile design strain of the longitudinal GFRP bars is limited to 0.01. This provision is made more general by defining the ultimate design strain, ε_{fd} and corresponding design strength, f_{fd}, as

$$\varepsilon_{fd} = Min(\varepsilon_{fu}, 0.010) \tag{5.2}$$

$$f_{fd} = Min(f_{fu}, 0.010E_f) \tag{5.3}$$

COMMENTARY

The tensile rupture strain of GFRP bars exceeds 2%. Such an ultimate strain would lead to unacceptably large deformations, if the full tensile capacity of the bars were to be achieved. To avoid this, it is recommended [2] that, for design purposes, the ultimate design strain not exceed a fixed limit of 1%.

5.5.4 Limit on maximum spacing of transverse reinforcement

As a result of the different characteristics between GFRP and steel, the spacing (or pitch) of the transverse reinforcement, s_{max}, should be limited to the least of the following quantities:

- Least dimension of the column (same as ACI 318-11)
- Twelve longitudinal bar diameters (75% of the limit in ACI 318-11)
- Twenty-four tie bar diameters (50% of the limit in ACI 318-11)

COMMENTARY

The spacing between ties can be related to the diameter of the longitudinal bars by a simplified model that assumes that the bar is a compressive member simply supported between two adjacent ties. The lateral support provided by the concrete cover is neglected as, at the point of failure, the loss of cover is very probable. For such a member to reach a strain level of ε without buckling, this condition must be upheld:

$$\frac{\pi^2 EI}{s^2} \leq AE\varepsilon \tag{5.4}$$

where s is the spacing of the ties, I and A are the moment of inertia and the area of the longitudinal bars, and E is their modulus of elasticity. For a solid round bar of diameter d_b, this can be rewritten as

$$s_{max} = \frac{\pi d_b}{4\sqrt{\varepsilon}} \tag{5.5}$$

For steel bars to yield before buckling, $\varepsilon = \varepsilon_y \approx 0.002$, which leads to $s_{max} \approx$ $17.5d_b$, a value in good agreement with the ACI 318-11 provision ($s_{max} \leq 16d_b$). For GFRP bars to avoid buckling before concrete crushing, $\varepsilon = \varepsilon_{cu} = 0.003$ and, therefore, $s_{max} \approx 14d_b$. This lower value justifies the more stringent provision adopted for GFRP bars ($s_{max} \leq 12d_b$).

The spacing is also related to the diameter of tie bars to achieve a desired level of concrete confinement at the core of the column. From this viewpoint, De Luca et al. [10] suggest that for GFRP RC columns, the tie spacing of ACI 318-11 as controlled by tie diameter must be halved from 48 to 24 tie bar diameters in consideration of the lower stiffness of GFRP.

5.5.5 Modified column stiffness

Simplified expressions like the ones adopted by ACI 318-11 for steel RC members are herein proposed for GFRP RC members. In the analysis stage of ordinary steel RC frames, the flexural stiffness of the members is modified to account for cross-section cracking. When a GFRP RC frame is analyzed, owing to the different mechanical properties of GFRP and steel, these modifications need to be adjusted to incorporate GFRP reinforcement effects. This section presents modification factors for the moment of inertia of GFRP RC members that are styled after those of ACI 318-11. ACI 318-11 recommends that the internal forces and the lateral deflections of RC frames resulting from factored loads be computed by linear analysis with modified moments of inertia, I, of the members, as follows:

Steel RC flexural members:

$$I_{beam} = 0.35I_g \tag{5.6}$$

$$I_{slab} = 0.25I_g \tag{5.7}$$

For any flexural member, a more accurate formulation is

$$I_{flexure} = (0.10 + 25\rho)\,(1.20 - 0.20b_w/d)\,I_g \leq 0.50I_g \tag{5.8}$$

Steel RC columns:

$$I_{column} = 0.7I_g \tag{5.9}$$

A more detailed approach is

$$I_{column} = (0.80 + 25\rho_{st})\left[1 - \frac{M_u}{P_u h} - 0.5\frac{P_u}{P_0}\right]I_g \leq 0.875I_g \tag{5.10}$$

Based on the current provisions of ACI 318-11 shown previously, similar provisions are proposed for the case of GFRP RC members as follows:

GFRP RC flexural members: It can be reasoned that for GFRP RC flexural members, Equation (5.8) may be written as

$$I_{\text{flexure}} = \left(0.10 + 25\rho_f \frac{E_f}{E_s}\right)\left(1.20 - 0.20\frac{b_w}{d}\right)I_g \leq 0.50I_g \tag{5.11}$$

Therefore, from the comparison of a GFRP RC flexural member to a geometrically similar steel RC member, Equations (5.8) and (5.11), it can be concluded that

$$\left(\frac{I_{\text{FRP}}}{I_{\text{steel}}}\right)_{\text{flexure}} = \frac{0.10 + 25\rho_f \dfrac{E_f}{E_s}}{0.10 + 25\rho_f} \tag{5.12}$$

The ratio of the moments of inertia decreases as a function of the reinforcement ratio ρ_f. Thus, to obtain a conservative estimate independent of the reinforcement ratio, a heavily reinforced beam or slab may be considered. Substituting the following into Equation (5.12):

- Equation (5.6) and an average of $\rho_f = 1.5\%$ for beams
- Equation (5.7) and an average of $\rho_f = 0.5\%$ for slabs

and rounding the results, the modified moments of inertia for GFRP RC flexural members can be expressed as

$$I_{\text{beam}} = \left[0.075 + 0.275\frac{E_f}{E_s}\right]I_g \leq 0.35I_g \tag{5.13}$$

$$I_{\text{slab}} = \left[0.10 + 0.15\frac{E_f}{E_s}\right]I_g \leq 0.25I_g \tag{5.14}$$

GFRP RC columns: The generalized form of Equation (5.10) for GFRP RC columns is

$$I_{\text{column}} = \left(0.80 + 25\rho_f \frac{E_f}{E_s}\right)\left[1 - \frac{M_u}{P_u h} - 0.5\frac{P_u}{P_0}\right]I_g \leq 0.875I_g \tag{5.15}$$

and, from Equations (5.10) and (5.15), if two columns are compared that only differ in the type of reinforcement,

$$\left(\frac{I_{FRP\,RC}}{I_{steel\,RC}}\right)_{column} = \left(0.80 + 25\rho_f\,\frac{E_f}{E_s}\right)\bigg/\left(0.80 + 25\rho_{st}\right) \tag{5.16}$$

Substituting Equation (5.9) and an average of $\rho_f = \rho_{st} = 2.5\%$, after rounding, Equation (5.16) can be simplified as

$$I_{column} = \left[0.40 + 0.30\frac{E_f}{E_s}\right]I_g \leq 0.70I_g \tag{5.17}$$

Similarly to ACI 318-11, lateral deflections resulting from service lateral loads may be computed by a linear analysis using 1.4 times the flexural stiffness defined in Equation (5.17).

5.5.6 Slenderness effects

The definitions of non-sway and sway frames adopted here for GFRP RC conform to ACI 318-11. In fact, a frame can be considered non-sway (i.e., braced) when the column end-moments due to second order effects do not exceed 5% of the first-order end-moments.

Sway column: According to ACI 318-11 [3], the effects of slenderness for compression members in a sway frame may be neglected when the slenderness ratio (kl_u/r) is less than 22. Mirmiran et al. [15] showed that for GFRP RC columns not braced against side sway, this limit should be reduced to 17. This value is therefore recommended as the new threshold for neglecting slenderness effects for a GFRP RC column free to sway.

Non-sway column. The effects of slenderness in a steel RC column in a non-sway frame may be neglected if

$$\frac{kl_u}{r} \leq \left(\frac{kl_u}{r}\right)_{max} = 34 - 12\left(\frac{M_1}{M_2}\right) \leq 40 \tag{5.18}$$

where M_2 is the larger end-moment and M_1 is the smaller end-moment. M_1 and M_2 are factored end-moments obtained by an elastic frame analysis and the ratio M_1/M_2 is positive if the column is bent in single curvature and negative if bent in double curvature.

Taking into account the conclusion by Mirmiran et al. [15], Equation (5.18) can be modified so that the effects of slenderness in GFRP RC columns may be neglected when

$$\frac{kl_u}{r} \leq \left(\frac{kl_u}{r}\right)_{max} = 29 - 12\left(\frac{M_1}{M_2}\right) \leq 35 \tag{5.19}$$

5.6 BENDING MOMENT AND AXIAL FORCE

Based on the existing knowledge, the design of concrete columns with rectangular or circular cross sections using FRP longitudinal bars and ties appears doable and a design methodology is presented as follows. As the basis of this methodology, the following considerations and assumptions are made:

- The strength of a GFRP RC cross section under combined flexure and axial load can be calculated by satisfying strain compatibility.
- GFRP longitudinal reinforcement is considered effective only in tension. The maximum design tensile strain of longitudinal bars must be less than 0.01 to limit lateral deflections.
- The area of the FRP reinforcement subject to compression is replaced with an equivalent area of concrete as if the FRP bars in compression were not present in the cross section.
- A modified and unified formulation of the strength-reduction factor for the interaction diagram is derived using comparative target reliability indices to meet appropriate safety requirements.
- Based on ACI 440.1R-06, the contributions of concrete, V_c, and ties, V_f, to the total shear strength, V_n, are reformulated to accommodate the case of column cross sections.

The combined nominal moment and axial force (M_n, P_n) are multiplied by the appropriate strength-reduction factor, ϕ, to obtain the design strength $(\phi M_n, \phi P_n)$ of the cross section. The design strength must be equal to or greater than the factored ultimate moment and axial load:

$$(\phi M_n, \phi P_n) \geq (M_u, P_u) \tag{5.20}$$

The factored pair of ultimate moment and axial load (M_u, P_u) denotes the effects of the various combinations of loads to which a structure is subjected.

Similarly to steel RC columns, an "interaction diagram" can be generated by plotting the nominal axial force strength, P_n, against the corresponding nominal moment strength, M_n. This diagram defines the strength of a cross section at different eccentricities of the load, which must encompass all the points associated with (M_u, P_u). With such assumptions, typical design interaction diagrams, $(\phi M_n, \phi P_n)$, may be formulated as shown in the following section.

5.6.1 Interaction diagram for rectangular cross section

This section summarizes a procedure to build the interaction diagram of a rectangular cross section in a way that is similar to what is typically

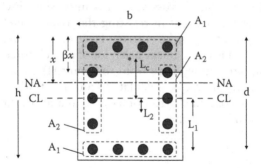

Figure 5.3 Cross section of a rectangular column with N.A. within the cross section.

done for steel-reinforced cross sections. Differently from the steel case, the contribution of the FRP reinforcement placed in the compression zone is replaced by that of concrete. The assumed parameters and the geometry of the generic cross section are displayed in Figure 5.3. For the computation of the combined nominal axial and moment capacities (P_n, M_n), the compressive force is assumed to be positive and the moment is assumed to act around the center line (C.L.) of the cross section. For the analysis, it is also assumed that the GFRP reinforcement is symmetrical with respect to both the horizontal and vertical cross-sectional axes and can be approximated by a thin rectangular tube. With reference to Figure 5.3, A_1 is the GFRP area of the top or bottom side; A_2 is the area of one vertical side; L_c is distance from C.L. of the centroid of the concrete compressive zone; L_1 is the distance from C.L. of the centroid of the tensile (bottom) reinforcement; and L_2 is the distance from C.L. of the centroid of the tensile side reinforcement neglecting the portion above the neutral axis.

The interaction diagram can be constructed by locating a few critical points as enumerated next. The succession of points is according to the level of axial load: from maximum tensile load to maximum compressive load or, equally, according to x, the location of the neutral axis (N.A.). Failure is either initiated by the FRP tensile limit $(\varepsilon_f = \varepsilon_{fd})$—called tensile rupture from here on—or by concrete crushing $(\varepsilon_c = \varepsilon_{cu})$, which are referred to, respectively, as tension controlled and compression controlled modes of failure.

1. $x = -\infty$. The maximum tensile force constitutes the lowermost point of any interaction diagram and is one of the two points corresponding to zero eccentricity. If $\varepsilon_f = \varepsilon_{fd}$:

$$P_n = T_{\max} = -2(A_1 + A_2) f_{fd}; \quad M_n = 0 \tag{5.21}$$

2. $x = 0$. Since the entire cross section is in tension, FRP bars are the only component engaged in resisting the load. Therefore, failure is triggered by rupture in the extreme layer of reinforcement ($\varepsilon_f = \varepsilon_{fd}$). Furthermore, due to the linear behavior of the reinforcement, the portion of the interaction curve from the point (1) to (2) is linear. The load and moment can be calculated from the traditional equations regarding the combined effect of axial load and bending moment on a section with linear behavior:

$$P_n = -\frac{A_1 + A_2}{\gamma} f_{fd} \tag{5.22}$$

$$M_n = \frac{(2\gamma - 1)^2}{2\gamma}(A_1 + \frac{A_2}{3})f_{fd}h \tag{5.23}$$

where $\gamma = d/h$.

3. $x_b \leq x \leq d$. The two modes of failure (i.e., tension and compression controlled modes) are separated by the "balanced failure," when FRP ruptures and concrete crushes, simultaneously. x_b marks the location of the neutral axis for such a balanced condition and can be calculated as

$$x_b = \frac{d}{1+e}; e = \frac{\varepsilon_{fd}}{\varepsilon_{cu}}; \varepsilon_{cu} = 0.003 \tag{5.24}$$

If the neutral axis is situated beyond the balanced location ($x \geq x_b$), the failure mode shifts from tension to compression and starts with concrete crushing. Assuming that x_b lies within the reinforced area (i.e., $x_b \geq h-d$ = concrete cover), which is normally the case, a range that covers both conditions can be defined in terms of dimensionless parameters:

$$Max\left(\frac{1}{\gamma} - 1, \frac{1}{1+e}\right) \leq \alpha = \frac{x}{d} \leq 1 \tag{5.25}$$

For any arbitrary α within this range, the forces in concrete and bars and their corresponding lever arms may be calculated as

$$C = 0.85\alpha\beta_1\gamma f_c'bh; L_c = (1-\alpha\beta_1\gamma)\frac{h}{2} \tag{5.26}$$

$$T_1 = \frac{1-\alpha}{\alpha}\varepsilon_{cu}E_f A_1; L_1 = (2\gamma-1)\frac{h}{2} \tag{5.27}$$

$$T_2 = \frac{\gamma}{2\gamma-1}\frac{(1-\alpha)^2}{\alpha}\varepsilon_{cu}E_fA_2; L_2 = \left[\frac{2}{3}(2+\alpha)\gamma - 1\right] \tag{5.28}$$

$$P_n = C - T_1 - T_2 \tag{5.29}$$

$$M_n = CL_c + T_1L_1 + T_2L_2 \tag{5.30}$$

4. $d < x \le h$. If the neutral axis lies within the concrete cover on the tension side, then the contribution of the reinforcement vanishes from the equations presented in (3). For example, if $x = h$ or $\alpha = 1/\gamma$:

$$C = 0.85\,\beta_1 f_c' bh; L_c = (1-\beta_1)\frac{h}{2} \tag{5.31}$$

$$T_1 = T_2 = 0 \tag{5.32}$$

$$P_n = C; M_n = CL_c \tag{5.33}$$

5. $x = +\infty$. The maximum compressive force is the uppermost point of any interaction diagram and is the other point corresponding to zero eccentricity:

$$P_n = P_0 = 0.85\,f_c' bh; M_n = 0 \tag{5.34}$$

These points provide a set of nominal moment-axial force couples, (M_n, P_n), that allows for the generation of a sufficiently smooth and accurate interaction diagram.

5.6.2 Interaction diagram for circular cross section

A procedure similar to the one discussed for rectangular cross sections is presented for the circular ones. Figure 5.4 displays the assumed parameters and the geometry of a generic section. For the analysis, the area of FRP reinforcement is uniformly distributed along the circumference as if it were a continuous thin tube.

Similarly to a rectangular cross section, the interaction curve can be constructed based on selected M_n–P_n pairs calculated at certain positions of the neutral axis as presented next.

1. $x = -\infty$. Maximum tensile force:

$$P_n = T_{max} = -A_f f_{fd}; M_n = 0 \tag{5.35}$$

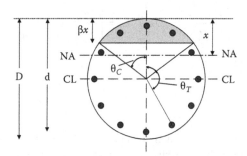

Figure 5.4 Cross section of a circular column with N.A. within the cross section.

2. $x = 0$. When failure is initiated by rupture in the extreme layer of reinforcement ($\varepsilon_f = \varepsilon_{fd}$):

$$P_n = -\frac{A_f f_{fd}}{2\gamma} \qquad (5.36)$$

$$M_n = \frac{(2\gamma - 1)^2}{8\gamma} A_f f_{fd} D \qquad (5.37)$$

where $\gamma = d/D$.

3. $x_b \leq x \leq d$. Again, x_b marks the location of the neutral axis for balanced failure:

$$x_b = \frac{d}{1+e}; e = \frac{\varepsilon_{fd}}{\varepsilon_{cu}}; \varepsilon_{cu} = 0.003 \qquad (5.38)$$

Also, for the neutral axis to be located in the reinforced region,

$$Max\left(\frac{1}{\gamma} - 1, \frac{1}{1+e}\right) \leq \alpha = \frac{x}{d} \leq 1 \qquad (5.39)$$

For any arbitrary α within this range, the forces and moments may be calculated as

$$C = 0.85 f_c' \frac{(\theta_C - \sin\theta_C \cos\theta_C) D^2}{4} \qquad (5.40)$$

$$M_C = 0.85 f_c' \frac{\sin^3\theta_C D^3}{12} \qquad (5.41)$$

where $\cos\theta_C = 1 - 2\alpha\beta_1\gamma$ and $0 \leq \theta_C \leq \pi$.

$$T = \frac{\pi\cos\theta_T - \theta_T\cos\theta_T + \sin\theta_T}{\pi(1+\cos\theta_T)} \frac{1-\alpha}{\alpha}\varepsilon_{cu}E_f A_f \qquad (5.42)$$

$$M_T = \frac{\pi - \theta_T + \sin\theta_T \cos\theta_T}{2\pi(1+\cos\theta_T)}\frac{1-\alpha}{\alpha}(\gamma - \frac{1}{2})\varepsilon_{cu}E_f A_f D \qquad (5.43)$$

where $\cos\theta_T = (1-2\alpha\gamma)/(2\gamma-1)$ and $0 \le \theta_T \le \pi$.

$$P_n = C - T \qquad (5.44)$$

$$M_n = M_c + M_T \qquad (5.45)$$

4. $d < x \le D$. If the neutral axis lies within the concrete section, but beyond the reinforced area, then the contribution of the reinforcement vanishes from the equations presented in (3). For example, if $x = D$ or $\alpha = 1/\gamma$:

$$P_n = C = 0.85f'_c\frac{(\theta_c - \sin\theta_c\cos\theta_c)D^2}{4} \qquad (5.46)$$

$$M_n = M_c = 0.85f'_c\frac{\sin^3\theta_c D^3}{12} \qquad (5.47)$$

where $\cos\theta_C = 1-2\beta_1$ and $0 \le \theta_C \le \pi$.

5. $x = +\infty$. Maximum compressive force:

$$P_n = P_0 = 0.85f'_c\frac{\pi D^2}{4}; M_n = 0 \qquad (5.48)$$

5.6.3 Example—P–M diagram

Example 5.1

Calculate the nominal interaction diagrams of a square column and a circular one with the following geometry and materials:

Square column: $b \times h$ =	24 × 24 in. (610 × 610 mm)
Circular column: D =	24 in. (610 mm)
Cover to the center of bars:	2.5 in. (63.5 mm)

Continued

Concrete:	$f'_c = 4000$ psi (27.6 MPa); $\beta_1 = 0.85$
Reinforcement:	$f_{fu} = 60$ ksi (413 MPa); $E_f = 6000$ ksi (41.3 GPa); $\varepsilon_{fu} = f_{fu}/E_f = 0.01$ therefore: $f_{fd} = 60$ ksi (413 MPa); $\varepsilon_{fd} = 0.01$
Square column: A_1; A_2; =	4#8 = 3.14 in.^{2} (2026 mm^2); 2#8 = 1.57 in.^{2} (1012 mm^2)
Square column: A_f =	$2(A_1 + A_2) = 12$#8 = 9.42 in.^{2} (6077 mm^2)
Circular column: A_f =	12#8 = 9.42 in.^{2} (6077 mm^2)

The design strength of GFRP, f_{fd}, is deliberately selected equal to the yield strength of ordinary grade 60 steel bars, $f_y = 60$ ksi (413 MPa), so that the effects of the two reinforcing materials (i.e., GFRP and steel) can be compared.

Figure 5.5 shows the construction of the P–M diagram based on six critical points (a) to (f) for the GFRP RC rectangular cross section and compares it to that of a similar section with grade 60 steel bars. Understandably, the compressive strength of steel and its yielding ability allow its interaction diagram to surround the diagram associated with GFRP bars.

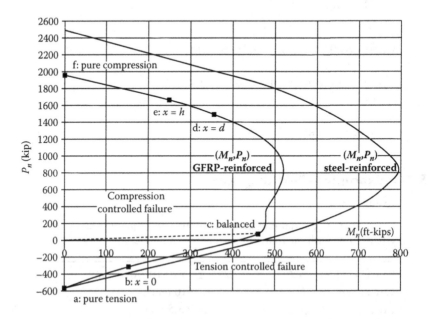

Figure 5.5 Interaction diagrams of GFRP and steel RC columns (square cross section).

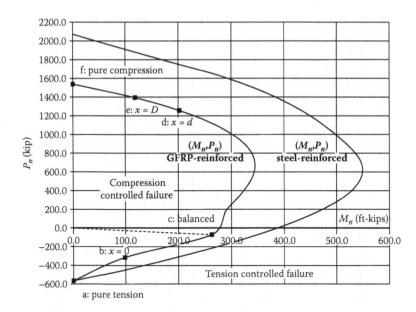

Figure 5.6 Interaction diagrams of GFRP and steel RC columns (circular cross section).

Similarly, Figure 5.6 shows the interaction diagrams for the circular cross section reinforced with the same number of GFRP and steel bars. The considerations made for the rectangular cross section hold here. It should be noted that the balanced condition for the GFRP RC column occurs when the axial force produces tension. This is generally the case for large values of the reinforcement ratio ($\rho_f \geq 2\%$).

5.7 STRENGTH-REDUCTION FACTOR FOR COMBINED BENDING MOMENT AND AXIAL FORCE

As ACI 440.1R-06 [1] is silent about columns, the method presented herein to calculate the strength-reduction factor for columns relies on a reliability analysis study as discussed in detail in Chapter 4. This method aims to unify the strength-reduction factors of columns and flexural members as a function of the maximum tensile strain in the reinforcement. The proposed formulation to compute the ϕ-factor is the following:

$$0.65 \leq \phi = 1.15 - \frac{\varepsilon_f}{2\varepsilon_{fd}} \leq 0.75 \tag{5.49}$$

ε_f in Equation (5.49) is the tensile strain taken as positive.

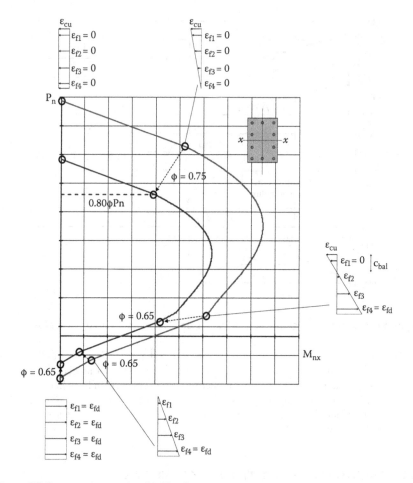

Figure 5.7 Rectangular section P_n–M_{nx} diagram.

When multiple levels of reinforcement are considered, the ϕ-factor is computed based on the reinforcement level subjected to the largest tensile strain.

Per ACI 318-11 steel tied columns, the design compressive strength of a GFRP RC column should be limited to $0.8\phi P_n$. For the case of GFRP RC, this limit is extended to include spirally reinforced columns.

Typical P–M diagrams are shown in Figure 5.7 through Figure 5.9. The interaction diagrams in Figures 5.7 and 5.8 are built for the same rectangular section with bending moment around the strong axis (x-axis) and weak axis (y-axis), respectively. The interaction diagram in Figure 5.9 is relative to a circular section having the same concrete area and amount of GFRP reinforcement of the rectangular one.

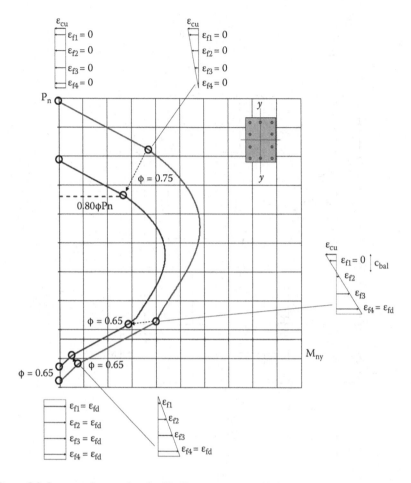

Figure 5.8 Rectangular section P_n–M_{ny} diagram.

5.8 COLUMNS SUBJECTED TO AXIAL LOAD AND BIAXIAL BENDING

A column is subjected to biaxial bending when equally significant bending moments act in two orthogonal directions. For the design of an FRP RC rectangular cross section subjected to moments about two axes, the same approach valid for steel RC appears to be applicable. Two analysis methods are herein proposed. They are the reciprocal load method and the load contour method. Both methods refer to a three-dimensional (3-D) failure surface to describe the interaction of the nominal axial load and the nominal biaxial bending moments. For example, the interaction surface for compression combined to biaxial bending is illustrated in Figure 5.10.

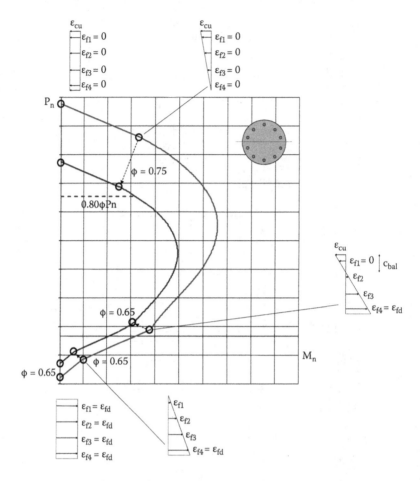

Figure 5.9 Circular section P_n–M_n diagram.

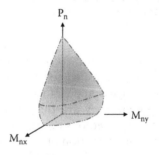

Figure 5.10 Interaction surface for the case of compression and biaxial bending.

Bresler's reciprocal load method. ACI 318-11 references the following equation originally developed by Bresler [30] to compute the column capacity under axial load and biaxial bending:

$$\frac{1}{P_n} = \frac{1}{P_{nx}} + \frac{1}{P_{ny}} - \frac{1}{P_{n0}} \tag{5.50}$$

where P_n is the approximation of the nominal axial capacity applied for eccentricities e_x and e_y; P_{nx} and P_{ny} are the nominal axial capacities for eccentricities e_y along the y-axis and e_x along the x-axis, respectively; and P_{n0} is the nominal axial capacity for zero eccentricity. This procedure is acceptably accurate for design purposes provided that P_n is larger than $0.10\, P_{n0}$.

Load contour method. The load contour method uses the load contour defined by a plane at a constant value of P_n intersecting the 3-D (P_n, M_{nx}, M_{ny}) interaction diagram. The adimensional expression of the load contour M_{nx}, M_{ny} at a generic value of P_n is the following:

$$\left(\frac{M_{nx}}{M_{nox}}\right)^\alpha + \left(\frac{M_{ny}}{M_{noy}}\right)^\alpha = 1 \tag{5.51}$$

with a suggested value of $\alpha \leq 1.0$ for the case of FRP RC. This equation is still valid if the ϕ-factors are applied.

In the case of $\alpha \neq 1.0$, it is deemed justifiable for practical purposes that Equation (5.51) be simplified by a bi-linear curve as proposed by Parme et al. [31] based on the assumption illustrated in Equation (5.53):

$$\left(\frac{M_{nx}}{M_{nox}}\right)^{\log 0.5/\log \beta} + \left(\frac{M_{ny}}{M_{noy}}\right)^{\log 0.5/\log \beta} = 1 \tag{5.52}$$

$$M_{nx} = \beta M_{nox} \text{ if } M_{ny} = \beta M_{noy} \tag{5.53}$$

where β is the constant portion of the uniaxial moment capacities that may act simultaneously on the column section. β usually ranges between 0.55 and 0.70, with a value of 0.50 recommended for design of FRP RC members. The load contour expression in Equation (5.52) is illustrated in Figure 5.11. For design, the load contour can be approximated by straight lines.

5.9 SHEAR STRENGTH, V_n

The nominal shear strength of a column, V_n, is the sum of the contributions of concrete, V_c, and FRP ties (or spiral), V_f:

$$V_n = V_c + V_f \tag{5.54}$$

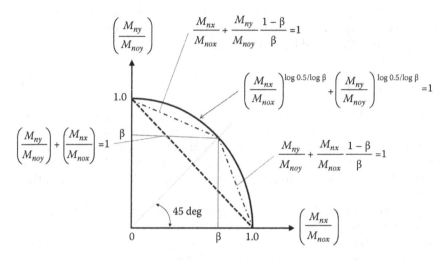

Figure 5.11 Adimensional load contour at a constant P_n.

5.9.1 Concrete contribution, V_c, for rectangular sections

According to ACI 440.1R-06, the concrete shear capacity, V_c, of flexural members using FRP as main reinforcement can be evaluated as

$$V_c = 5k\sqrt{f_c'}b_w d \qquad (5.55)$$

[or $V_c = \dfrac{2}{5}k\sqrt{f_c'}b_w d$ for SI Units]

where k is the ratio of depth of neutral axis to reinforcement depth, d. For members with only one layer of tensile reinforcement, k is calculated according to ACI 440.1R-06 as

$$k = \sqrt{2(\rho_f n_f) + (\rho_f n_f)^2} - (\rho_f n_f) \qquad (5.56)$$

where ρ_f is the ratio of the longitudinal FRP bars and n_f is the modular ratio: $n_f = E_f/E_c$.

For columns, however, multiple layers of longitudinal reinforcement render Equation (5.56) inaccurate. To generalize the V_c formulation for rectangular columns, the location of the neutral axis in a cracked section can be calculated by equalizing the first moments of area of the compressive and transformed tensile portions of the cross section around a separating neutral axis located at $x = kd$. This equation can be written as

$$k^2 = 2(\rho_{f1}n_f)(1-k) + 2(\rho_{f2}n_f)\frac{\gamma}{2\gamma-1}(1-k)^2 \tag{5.57}$$

where $\rho_{f1} = A_{f1}/bd$ and $\rho_{f2} = A_{f2}/bd$ (See Figure 5.3 for the definitions of A_{f1} and A_{f2}.) and $\gamma = d/h$.

Note that the last term of Equation (5.57) accounts for the presence of the two side layers of reinforcement. Although this equation can be solved for the exact value of k, the final closed-form solution is rather unwieldy. Since the exact value of k is of little practical importance, a simpler iterative solution is suggested as an alternative. Defining the "effective" tensile reinforcement ratio, ρ_f, as

$$\rho_f = \rho_{f1} + \frac{\gamma(1-k)}{2\gamma-1}\rho_{f2} \tag{5.58}$$

and starting with an initial guess for k ($k = 0.2$ is recommended), Equation (5.56) can be used repetitively with Equation (5.58) until convergence to calculate k.

Equation (5.55) provides a conservative estimate of the concrete contribution, V_c, if the column is subjected to a compressive axial force. In the uncommon case of tensile axial force, the use of this equation remains valid with the appropriate value of kd. To attain this, Equation (5.59) must be employed, which, with negligible approximation, predicts the location of the neutral axis, c, under the combined effects of axial load and flexure based on the value of k calculated from Equation (5.57):

$$0 \le \frac{c}{d} = k + \frac{k^2\left(1-\dfrac{k}{3}\right)}{2}\left(\frac{P_u d}{M_u}\right) \le 0.4 \tag{5.59}$$

where P_u and M_u are the ultimate axial load and moment corresponding to V_u. Compressive axial force is positive and tensile axial force is negative while the moment is always positive. With c from Equation (5.59) and noting that $c = kd$, Equation (5.55) may be used to achieve an accurate assessment of the shear strength provided by concrete when the section is subjected to a tensile axial force.

5.9.2 Shear reinforcement contribution, V_f, for rectangular sections

Once V_c is calculated, the shear strength provided by ties (or spiral) of the column, V_f, can be calculated as discussed in Chapter 4 and shown in Equation (5.60):

$$V_f = \frac{A_{fv}f_{fv}d}{s} \tag{5.60}$$

A_{fv} is the area of FRP ties with a spacing of s (or spiral with a pitch of s). For FRP RC columns, the spacing is limited to the least dimension of the column, 12 longitudinal bar diameters or 24 tie bar diameters. Other parameters are only briefly discussed herein as they are similar to flexural members. The tensile strength of the FRP tie for shear design, f_{fv}, is calculated as

$$f_{fv} = 0.004E_{fv} \le f_{fb} \tag{5.61}$$

where E_{fv} is the modulus of elasticity of the FRP ties. To prevent loss of aggregate interlock in the concrete as the result of a wide crack, the maximum usable tensile strain of the FRP ties is limited to 0.004. f_{fb}, strength of the bent portion of FRP tie, depends on r_b/d_b, the ratio of the internal radius of the bend to the diameter of the tie, which has to be at least equal to 3, the minimum recommended by the ACI 440.1R-06 guidelines. f_{fb} is given by the following expression:

$$f_{fb} = \left(0.05\frac{r_b}{d_b} + 0.3\right)f_{fu} \le f_{fu} \tag{5.62}$$

where f_{fu} is the tensile strength of the straight portion of the GFRP tie.

5.9.3 Shear strength-reduction factor

Jawaheri Zadeh and Nanni [32] showed that the current shear strength-reduction factor of 0.75 may be maintained so long as the maximum effective level of shear resistance of a member with FRP shear reinforcement does not exceed four times the strength provided by concrete ($V_n \le 4V_c$ or, equally, $V_f \le 3V_c$). Accordingly, the shear strength-reduction factor for columns is 0.75.

5.9.4 Examples—Shear strength for square columns

Example 5.2

Calculate the design shear strength of the following square column with no P_u effect:

Concrete:	f'_c = 4.0 ksi
	E_c (ksi) = 57 $\sqrt{f'_c}$ (psi) = 3600 ksi
Longitudinal GFRP reinf.:	$f_{fu} = f_{fd}$ = 60 ksi
	E_f = 6000 ksi

Continued

$A_f = 12\#8 = 9.42$ in.^{2}

$A_{f1} = 4\#8 = 3.14$ in.^{2}

$A_{f2} = 2\#8 = 1.57$ in.^{2}

GFRP shear reinforcement: $f_{fu} = 60$ ksi

$E_{fv} = 6000$ ksi

$A_{fv} = \#3@8$ in.

$r_b/d_b = (r_b/d_b)_{min} = 3.0$

Size: $b = 24.0$ in.   Width of the column

$h = 24.0$ in.   Height of the column

$c_c = 2.5$ in.   Concrete cover from bar center

$d = h - c_c = 21.5$ in.   Effective depth

Solution:

Concrete contribution, V_c: $\rho_{f1} = A_{f1}/(bd) = 0.00609$

$\rho_{f2} = A_{f2}/(bd) = 0.00304$

$\gamma = d/h = 0.896$

Assuming $k = 0.2$ and using

Equation (5.58): $\rho_f = 0.00884$

$$n_f = E_f/E_c = (6000 \text{ ksi})/(3600 \text{ ksi}) = 1.67$$

$$n_f\,\rho_f = 0.01476$$

and k can be calculated according to Equation (5.56) as

$k = 0.158$

The new value of k is of sufficient accuracy, but to demonstrate the convergence trend, another step is performed:

Assuming $k = 0.158$ and using Equation (5.58):

$\rho_f = 0.00899$

$n_f\,\rho_f = 0.01501$

$k = 0.159$

$c = 3.42$ in.

$V_c = 26.0$ kip

Shear reinforcement contribution, V_f:

$A_{fv} = \#3@8$ in. $= 0.22$ in.^{2}

$s = 8.0$ in.

Continued

$$f_{fv} = 0.004 E_{fv} = 24 \text{ ksi} \leq f_{fb}$$
$$f_{fb} = (0.05 r_b/d_b + 0.3) f_{fu} \leq f_{fu}$$
$$f_{fb} = 0.45 f_{fu} = 27.0 \text{ ksi}$$
$$f_{fv} = 24.0 \text{ ksi}$$
$$V_f = A_{fv} f_{fv} d/s = 14.2 \text{ kip}$$

And the nominal shear strength:

$$V_n = V_c + V_s = 40.2 \text{ kip}$$
$$\phi V_n = 30.2 \text{ kip}$$

Example 5.3

Calculate the concrete shear strength V_c of the square column in Example 5.2 for various values of P_u, M_u when accounting for the effect of the axial load. In Example 5.2, the shear strength of the column is calculated disregarding the axial load. To account for the effect of the axial load, Equation (5.59) is used to evaluate c/d that replaces k = 0.159, calculated in Example 5.2. Therefore:

$$0 \leq c/d = 0.159 + 0.01197 \, (P_u d/M_u) \leq 0.4$$

The contribution of concrete to the shear strength is calculated for the loading cases that follow:

1. No axial force	$c/d = 0.159$	$c = 3.42$ in.	$V_c = 26.0$ kip
2. $M_u = 0$, tensile force	$c/d = 0$	$c = 0$	$V_c = 0$
3. $M_u = 0$, compressive force	$c/d = 0.4$	$c = 8.6$ in.	$V_c = 65.3$ kip
4. $P_u = 500$ kip, $M_u = 100$ ft-kip	$c/d = 0.266$	$c = 5.72$ in.	$V_c = 43.4$ kip
5. $P_u = 500$ kip, $M_u = 200$ ft-kip	$c/d = 0.213$	$c = 4.58$ in.	$V_c = 34.8$ kip
6. $P_u = 500$ kip, $M_u = 300$ ft-kip	$c/d = 0.195$	$c = 4.19$ in.	$V_c = 31.8$ kip

5.9.5 Circular sections

For circular sections, the same equations introduced in rectangular sections to compute shear contributions of concrete and transverse reinforcement can be used, if an equivalent rectangular section is defined as

$$b_w = D; \; d = 0.8D; \; A_1 = A_2 = A_f/4 \qquad (5.63)$$

The equivalent dimensions replicate the provisions of ACI 318-11 for shear strength of circular sections. Similarly, the strength-reduction factor of rectangular sections equal to 0.75 applies to circular ones under the same limitations for V_c and V_f.

5.9.6 Example—Shear strength for circular columns

Example 5.4

Repeat Example 5.2 for a circular column with the same concrete area ($D = 27$ in.) and reinforcement with no P_u effect:

Concrete:	$f'_c = 4.0$ ksi
	E_c (ksi) $= 57\sqrt{f'_c}$ (psi) $= 3600$ ksi
Longitudinal reinf.:	$f_{fu} = f_{fd} = 60$ ksi
	$E_f = 6000$ ksi
	$A_f = 12\#8 = 9.42$ in.^{2}
	$A_{f1} = A_{f2} = 3\#8 = 2.36$ in.^{2}
Shear reinforcement:	$f_{fu} = 60$ ksi
	$E_{fv} = 6000$ ksi
	$A_{fv} = \#3@8$ in.
Size:	$D = 27.0$ in. Diameter of the column
	$c_c = 2.5$ in. Concrete cover to the center of #8

Solution:

Concrete contribution, V_c:

$$d = 0.8D = 21.6 \text{ in.}$$
$$b = D = 27 \text{ in.}$$
$$\rho_{f1} = \rho_{f2} = A_{f1}/(bd) = 0.00405$$
$$\gamma = (D-c_c)/D = 0.907$$

Assuming $k = 0.2$

$$\rho_f = 0.00766$$
$$n_f = E_f/E_c = (6000 \text{ ksi})/(3600 \text{ ksi}) = 1.67$$
$$n_f \rho_f = 0.01280$$

And k can be calculated according to Equation (5.56) as:

$$k = 0.148$$

Assuming $k = 0.148$ and using Equation (5.58):

$$\rho_f = 0.00789$$
$$n_f \rho_f = 0.01317$$
$$k = 0.150$$
$$c = k_c d = 3.24 \text{ in.}$$
$$V_c = 27.7 \text{ kip}$$

Shear reinforcement contribution, V_f:

$$A_{fv} = \#3@8 \text{ in.} = 0.22 \text{ in.}^2$$
$$s = 8.0 \text{ in.}$$
$$f_{fv} = 0.004 E_{fv} \leq f_{fb}$$
$$f_{fb} = (0.05 r_b/d_b + 0.3) f_{fu} \leq f_{fu}$$

Continued

r_b can be calculated as:

$$r_b = (D-2c_c + 1 \text{ in.})/2 = 11.5 \text{ in.}$$

Where c_c = concrete cover to the center of #8

$$r_b/d_b = 30.7$$
$$f_{fb} = f_{fu} = 60.0 \text{ ksi}$$
$$f_{fv} = 24.0 \text{ ksi}$$
$$V_f = A_{fv} f_{fv} d/s = 16.0 \text{ kip}$$

And the nominal shear strength:

$$V_n = V_c + V_s = 43.7 \text{ kip}$$
$$\phi V_n = 32.8 \text{ kip}$$

5.9.7 Shear walls

The in-plane shear strength of walls can be calculated similarly to columns; however, for most cases Equation (5.58) can be simplified to obtain a more straightforward solution. For a shear wall, the in-plane horizontal dimension, l_w, is normally long enough to justify the approximation of $\gamma = (l_w - c_c)/l_w \approx 1$. Furthermore, if the total vertical reinforcement, A_f, is uniformly distributed throughout l_w, the parameters in Equation (5.58) can be evaluated as $\rho_{f1} = 0$ and $\rho_{f2} = A_f/(2bl_w) = \rho_f/2$, where b is the thickness of the wall. Substituting these values into Equation (5.58), k can be calculated as

$$k = \frac{\sqrt{n_f \rho_f}}{1 + \sqrt{n_f \rho_f}} \tag{5.64}$$

The contribution of horizontal (shear reinforcement) can be evaluated, assuming that

$$d = 0.8 l_w \tag{5.65}$$

COMMENTARY

Mohamed et al. [33] tested GFRP RC shear walls to attain strength and drift data. Four large-scale shear walls were constructed and failed under quasi-static reversed cyclic lateral loading. The GFRP RC walls had different aspect ratios covering the range of medium-rise walls. Experimental results show that properly designed and detailed GFRP RC walls can attain their flexural capacities with no strength degradation and that shear, sliding shear, and anchorage failures can be effectively controlled. Figure 5.12 shows a GFRP RC shear wall tested and failed under lateral cyclic loads.

(a) Test setup of the wall specimens

(b) Concrete crushing causing failure

Figure 5.12 GFRP RC shear walls reinforced with GFRP bars under lateral cyclic loading. (University of Sherbrooke, Sherbrooke, Quebec, Canada; courtesy of Prof. Brahim Benmokrane.)

5.9.8 Examples—Shear wall strength and shear friction

Example 5.5

Calculate the design shear strength of the shear wall detailed as follows:

Concrete:	$f'_c = 4.0$ ksi
	E_c (ksi) $= 57\sqrt{f'_c}$ (psi) $= 3600$ ksi
Vertical reinforcement:	$f_{fu} = f_{fd} = 60$ ksi
	$E_f = 6000$ ksi
	$A_{f1} = $ #5@12 in. $= 0.3$ in.^{2}/ft
	Layer 1
	$A_{f2} = $ #5@8 in. $= 0.45$ in.^{2}/ft Layer 2
Horizontal reinforcement:	$f_{fu} = 60$ ksi
	$E_{fv} = 6000$ ksi
	$A_{fv} = $ #3@12 in. Two layers
Size:	$b = 12.0$ in. Wall thickness
	$l_w = 16$ ft Wall length

Solution:
Concrete contribution, V_c:

$$\rho_f = (A_{f1} + A_{f2})/b = 0.0625 \text{ in.}/\text{ft} = 0.00521$$

$$n_f = E_f/E_c = (6000 \text{ ksi})/(3600 \text{ ksi}) = 1.67$$

$$n_f \rho_f = 0.00868$$

$$k = 0.0852 \quad \text{Equation (5.64)}$$

$$c = k l_w = 16.4 \text{ in.}$$

$$V_c = 62.2 \text{ kip}$$

Shear reinforcement contribution, V_f:

$$A_{fv} = 2\#3@12 \text{ in.} = 0.22 \text{ in.}^2$$

$$s = 12.0 \text{ in.}$$

$$f_{fv} = 0.004 E_{fv} \le f_{fb}$$

$$f_{fb} = f_{fu} = 60 \text{ ksi} \quad \text{No bends}$$

$$f_{fv} = 24.0 \text{ ksi}$$

$$d = 0.8 l_w = 153.6 \text{ in.}$$

$$V_f = A_{fv} f_{fv} d/s = 67.6 \text{ kip}$$

The nominal shear strength:

$$V_n = V_c + V_s = 129.8 \text{ kip}$$

And the design shear strength:

$$\phi V_n = 97.4 \text{ kip}$$

Example 5.6

Investigate the necessity of additional dowels to achieve the full strength of the shear wall in Example 5.5 (ϕV_n = 97.4 kip), if the wall is placed on a foundation with a surface not intentionally roughened. The axial compressive load is P_u = 100 kip.

Size:	b = 12.0 in. Thickness of the wall
	l_w = 16 ft Horizontal length of the wall

Vertical reinforcement:

$$f_{fu} = f_{fd} = 60 \text{ ksi}$$
$$E_f = 6000 \text{ ksi}$$
$$\rho_f = 0.00521$$
$$A_{vf} = \rho_f \,(bl_w) = 12 \text{ in.}^2$$

Solution:

$$f_{vf} = E_f \varepsilon_{vf} = 18 \text{ ksi} < f_{fu} \text{ (Equation (4.113))}$$
$$\mu = 0.6 \quad \text{The surface is not intentionally roughened}$$
$$A_{vf} f_{vf} \mu = 129.6 \text{ kip}$$
$$P_u \mu = 60 \text{ kip}$$
$$V_n = \sqrt{(129.6)^2 + (60)^2} = 142.8 \text{ kip (Equation (4.114))}$$
$$\phi V_n = (0.75)(142.8) = 107.1 \text{ kip} > 97.4 \text{ kip}$$

No additional dowel is required.

REFERENCES

1. ACI Committee 440. Guide for the design and construction of structural concrete reinforced with FRP bars, 440.1R-06. American Concrete Institute, Farmington Hills, MI (2006).
2. H. Z. Jawaheri and A. Nanni. Design of RC columns using glass FRP reinforcement. *ASCE Journal of Composites for Construction* 17 (3): 294–304 (2013).
3. ACI Committee 318 (2011). Building code requirements for reinforced concrete (ACI 318-11). American Concrete Institute, Farmington Hills, MI (2011).
4. E. M. Lofty. Behavior of reinforced concrete short columns with fiber reinforced polymer bars. *International Journal of Civil and Environmental Engineering* 1 (3): 545–557 (2010).
5. H. Tobbi, A. S. Farghali, and B. Benmokrane. Concrete columns reinforced longitudinally and transversally with glass FRP bars. *ACI Structural Journal* 109 (4): 1–8 (2012).
6. C. C. Choo, I. E. Harik, and H. Gesund. Minimum reinforcement ratio for fiber-reinforced polymer reinforced concrete rectangular columns. *ACI Structural Journal* 103 (3): 460–466 (2006).
7. N. S. Paramanantham. Investigation of the behavior of concrete columns reinforced with fiber-reinforced plastic rebars. MS Thesis, Lamar University, Beaumont, TX (1993).

8. A. Mirmiran. Length effects on FRP-reinforced concrete columns. *Proceedings, 2nd International Conference on Composites in Infrastructure,* Tucson, AZ, 518–532 (1998).

9. S. H. Alsayed, Y. A. Al-Salloum, T. H. Almusallam, and M. A. Amjad. Concrete columns reinforced by GFRP rods. *Fourth International Symposium on Fiber-Reinforced Polymer Reinforcement for Reinforced Concrete Structures,* SP-188, ed. C. W. Dolan, S. H. Rizkalla, and A. Nanni, American Concrete Institute, Farmington Hills, MI, 103–112 (1999).

10. A. De Luca, F. Matta, and A. Nanni. Structural response of full-scale reinforced concrete columns with internal FRP reinforcement under compressive load. *9th International Symposium on Fiber Reinforced Polymer Reinforcement for Concrete Structures (FRPRCS-9),* ed. D. Oehlers, M. Griffith, and R. Seracino, July 13–15, 2009, Sydney, Australia, CD-ROM, 4 (2010).

11. A. De Luca, F. Matta, and A. Nanni. Behavior of full-scale GFRP-reinforced concrete columns under axial load. *ACI Structural Journal* 107 (5): 589–596 (2009).

12. M. S. Issa, M. M. Metwally, and S. M. Elzeiny. Influence of fibers on flexural behavior and ductility of concrete columns reinforced with GFRP rebars. *Engineering Structures* 33 (5): 1754–1763 (2011).

13. A. Deiveegan and G. Kumaran. Reliability analysis of concrete columns reinforced internally with glass fiber reinforced polymer reinforcements. *ICFAI University Journal of Structural Engineering* 2 (2): 49–59 (2009), IC-FAI University Press, India.

14. A. Deiveegan and G. Kumaran. Joint shear strength of FRP-reinforced concrete beam-column joints. *Central European Journal of Engineering* 1 (1): 89–102 (2011).

15. A. Mirmiran, W. Yuan, and X. Chen. Design for slenderness in concrete columns internally reinforced with fiber-reinforced polymer bars. *ACI Structural Journal* 98 (1): 116–125 (2001).

16. N. Kawaguchi. Ultimate strength and deformation characteristics of concrete members reinforced with AFRP rods under combined axial tension or compression and bending. *Fiber-Reinforced-Plastic Reinforcement for Concrete Structures,* SP-138, ed. A. Nanni and C. W. Dolan, American Concrete Institute, Farmington Hills, MI, 671–685 (1993).

17. H. Fukuyama, Y. Masuda, Y. Sonobe, and M. Tanigaki. Structural performances of concrete frame reinforced with FRP reinforcement. Non-metallic (FRP) reinforcement for concrete structures. *Second International Symposium on Non-Metallic (FRP) Reinforcement for Structures, FRPRCS-2,* Ghent, Belgium, 275–286 (1995).

18. K. Kobayashi and T. Fujisaki. Compressive behavior of FRP reinforcement in non-prestressed concrete members. *Proceedings, 2nd International RILEM Symposium on Non-Metallic (FRP) Reinforcement for Concrete Structures (FRPRCS-2),* Ghent, Belgium, 267–274 (1995).

19. M. Grira and M. Saatcioglu. Reinforced concrete columns confined with steel or FRP grids. *Proceedings, 8th Canadian Conference on Earthquake Engineering,* Vancouver, Canada, 445–450 (1999).

20. A. M. Said and M. L. Nehdi. Use of FRP for RC frames in seismic zones: Part II. Performance of steel-free GFRP-reinforced beam-column joints. *Journal of Applied Composite Materials* 11 (4): 227–245 (2004).

21. Z. P. Bazant and Y. K. Kwon. Failure of slender and stocky reinforced concrete columns: Tests of size effect. *Journal of Materials and Structures* 27:79–90 (1994).
22. Z. P. Bazant. Size effect in blunt fracture: Concrete, rock, metal. *ASCE Journal of Engineering Mechanics* 110 (4): 518–535 (1984).
23. S. Sener, B. I. G. Barr, and H. F. Abusiaf. Size effect in axially loaded reinforced concrete columns. *ASCE Journal of Structural Engineering* 130 (4): 662–670 (2004).
24. J. Nemecek and Z. Bittnar. Experimental investigation and numerical simulation of post-peak behavior and size effect of reinforced concrete columns. *Journal of Materials and Structures* 37:161–169 (2004).
25. C. H. Lin and R. W. Furlong. Longitudinal steel limits for concrete columns. *ACI Structural Journal* 92 (3): 282–287 (1995).
26. P. H. Ziehl, J. E. Cloyd, and M. E. Kreger. Evaluation of minimum longitudinal reinforcement requirements for reinforced concrete columns. Texas Department of Transportation, Austin. TX, FHWA/TX-02/1473-S, 128 (October 1998).
27. P. K. Mallick. *Fiber reinforced composites materials manufacturing and design*, 469. New York: Marcell Dekker, Inc. (1988).
28. W. P. Wu. Thermomechanical properties of fiber reinforced plastic (FRP) bars. PhD dissertation, West Virginia University, Morgantown, WV, 292 (1990).
29. D. H. Deitz, I. E. Hark, and H. Gesund. Physical properties of glass fiber reinforced polymer rebars in compression. *Journal of Composites for Construction* 7 (4): 363–366 (2003).
30. B. Bresler. Design criteria for reinforced concrete columns under axial load and biaxial bending. *ACI Journal Proceedings* 57 (5): 481–490 (1960).
31. A. L. Parme, J. M. Nieves, and A. Gouwens. Capacity of reinforced rectangular columns subject to biaxial bending. *ACI Journal Proceedings* 63:911–923 (1966).
32. H. Jawaheri Zadeh and A. Nanni. Reliability analysis of concrete beams internally reinforced with FRP bars. *ACI Structural Journal* 110 (6): 1023–1032 Nov.-Dec. (2013).
33. N. Mohamed, A. Farghaly, B. Benmokrane, and K. W. Neale. Evaluation of a concrete shear wall reinforced with GFRP bars subjected to lateral cyclic loading. *Proceedings, 3rd Asia Pacific Conference on FRP Structures,* Sapporo, Japan, Feb. 2–4, 10 (2012).

Part III

Design examples

Chapter 6

Design of a one-way slab

6.1 INTRODUCTION

The floor plan of a two-story medical facility building is shown in Figure 6.1. The column spacing was dictated by the size of the equipment that occupies the ground floor. The second floor system is a one-way RC slab spanning along the east–west plan direction. The building is located in a region of low seismicity. Loading of each floor consists of the self-weight, a superimposed dead load of 2.5 psf, and a live load of 100 psf.

This example describes the procedure to design a 1-foot slab strip of the second floor. The design is presented as a sequence of ten steps as summarized here:

Step 1   Define slab geometry and concrete properties
Step 2   Compute factored loads
Step 3   Compute ultimate and bending moments and shear forces
Step 4   Design FRP primary reinforcement for bending moment capacity
Step 5   Check creep rupture stress
Step 6   Check crack width
Step 7   Check maximum midspan deflection
Step 8   Check shear capacity
Step 9   Design FRP secondary reinforcement for temperature and shrinkage
Step 10 Fire safety check

The results of the slab strip design are summarized next to facilitate understanding of the nine sequential steps devoted to calculations.

6.2 DESIGN SUMMARY

Based on a slab thickness of 8 in., the following loads are considered.

Slab self-weight	96.7 psf
Superimposed dead load	2.5 psf

Live load	100 psf
Total factored load	279 psf (0.279 kip/ft)
Service load	199 psf (0.199 kip/ft)

Bending moments and shear forces for the 1-foot slab strip are computed at the support and midspan sections of the interior and exterior bays, as summarized in Table 6.1.

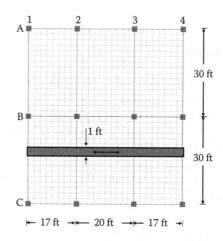

Figure 6.1 Second-floor plan.

Table 6.1 Bending moments and shear forces

		Bending moment			Shear force	
Section		Moment coefficient	Ultimate moment (kip-ft)	Service moment (kip-ft)	Shear coefficient	Ultimate shear (kip)
Ext. bay	Exterior support	1/24	2.91	2.08	1.0	2.21
	Midspan	1/14	5.00	3.57	0.15	0.331
	Interior support	1/10	6.99	4.99	1.15	2.54
Int. bay	First support	1/11	6.36	4.54	1.15	3.02
	Midspan	1/16	4.37	3.12	1.15	0.394
	Second support	1/11	6.36	4.54	1.15	3.02

Table 6.2 Slab geometry and reinforcement

Section	Slab thickness	Primary reinforcement	Secondary reinforcement
1		No. 4 @ 6 in. (top and bottom)	No. 4 @ 12 in. (top and bottom)
2	8 in.		
3			

Table 6.3 Slab design summary

Limit state		Section	Demand/computed	Capacity/limit
Ultimate	Flexural strength	1	2.91 kip-ft	4.06 kip-ft
		2	5.00 kip-ft	11 kip-ft
		3	6.99 kip-ft	11 kip-ft
	Shear strength[a]	4	3.02 kip	4.8 kip
Serviceability	Creep rupture	1	5.6 ksi	16 ksi
		2	9.7 ksi	
		3	13.5 ksi	
	Crack width	1	0.011 in.	0.028 in.
		2	0.019 in.	
		3	0.026 in.	
	Maximum midspan deflection		0.27 in.	0.425 in.

[a] When the shear capacity of the concrete slab is checked, the maximum shear force at the supports of the interior bay (identified as Section 4) is used.

The design for bending moment is limited to the exterior bay because it represents the worst condition. The following sections are considered:

Section 1   Exterior support
Section 2   Midspan
Section 3   Interior support

The design of FRP reinforcement for a slab thickness of 8 in. is summarized in Table 6.2. Bar sizes and layout are selected in order to optimize bar production time and construction effort.

The design step results are summarized in Table 6.3.

Bar layout and typical details are shown in Figures 6.2(a) and 6.2(b).

Table 6.4 is provided to convert US customary units to the SI system.

6.3 STEP 1—DEFINE SLAB GEOMETRY AND CONCRETE PROPERTIES

6.3.1 Geometry

The one-way slab has three spans of length l_1, l_2, and l_3, respectively:

$l_1 := 17$ ft

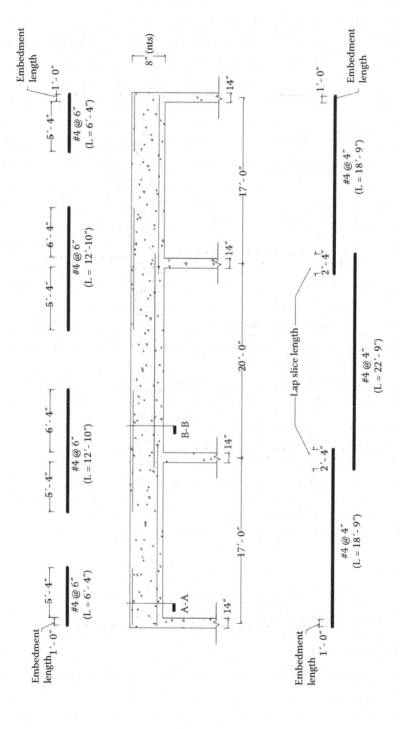

Figure 6.2 (a) Reinforcement layout.

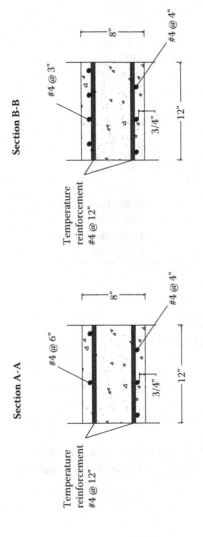

Section A-A

Section B-B

Temperature reinforcement #4 @ 12"

#4 @ 6"

8"

3/4"

12"

#4 @ 4"

Temperature reinforcement #4 @ 12"

#4 @ 3"

8"

3/4"

12"

#4 @ 4"

Notes: 1. 3/4-inch clear cover for top and bottom primary reinforcement.

Figure 6.2 (Continued) **(b) typical details.**

Table 6.4 Conversion table

US customary	SI units
Lengths, areas, section properties	
I in.	25.4 mm
	0.025 m
I ft	304.8 mm
	0.305 m
I in.^{2}	645 mm^2
I ft^2	0.093 m^2
I in.^{3}	16,387 mm^3
I in.^{4}	416,231 mm^4
Forces, pressures, strengths	
I lbf	4.448 N
I kip	4.448 kN
I lbf-ft	1.356 N·m
I kip-ft	1.356 kN·m
I psi	6.895 kPa
I psf	47.88 N/m^2
I ksi	6.895 MPa
I ksf	47.88 kN/m^2
I lbf/ft^3	157.1 N/m^3

$l_2 := 20 \text{ ft}$
$l_3 := 17 \text{ ft}$

A clear top and bottom concrete cover, c_c, is of 0.75 in.:

$c_c := 0.75 \text{ in.}$

Slab width used for computation is

$b := 1 \text{ ft}$

Define slab thickness: Table 8-2 of ACI 440.1R-06 guides the selection by recommending minimum values for the slab thickness. For an end bay, the recommended thickness is

$$t_{end} := \text{round}\left(\frac{l_1}{17 \cdot \text{in}}\right) \text{in} = 12 \cdot \text{in.}$$

For the interior bays, the recommended thickness is

$$t_{int} := \text{round}\left(\frac{l_2}{22 \cdot \text{in}}\right) \text{in} = 11 \cdot \text{in.}$$

The recommended slab thickness can be selected as $t_{slabACI} := \max(t_{end}, t_{int})$ = 12·in. It has to be noted that these values are only a starting point for the design. In this example, a different thickness is selected:

$$t_{slab} := 8 \text{ in.}$$

6.3.2 Concrete properties

The following concrete properties are considered for the design:

$f'_c := 5000$ psi   Compressive strength

$\varepsilon_{cu} := 0.003$ Ultimate compressive strain

$\rho_c := 145 \dfrac{lbf}{ft^3}$ Density

$E_c := 33 \, psi^{0.5} \left(\dfrac{\rho_c}{lbf \cdot ft^{-3}} \right)^{1.5} \cdot \sqrt{f'_c}$ Compressive modulus of elasticity

$E_c = 4074 \cdot ksi$ Computed as indicated in ACI 318-11

$f_r := 7.5 \cdot \sqrt{f'_c \cdot psi}$ Concrete tensile strength

$f_r = 530 \cdot psi$ Computed as indicated in ACI 318-11

The stress-block factor, β_1, is computed as indicated in ACI 318-11:

$$\beta_1 := \begin{vmatrix} 0.85 & \text{if } f'_c = 4000 \, psi & = 0.8 \\ 1.05 - 0.05 \cdot \dfrac{f'_c}{1000 \, psi} & \text{if } 4000 \, psi < f'_c < 8000 \, psi \\ 0.65 & \text{otherwise} \end{vmatrix}$$

6.3.3 Analytical approximations of concrete compressive stress–strain curve—Todeschini's model

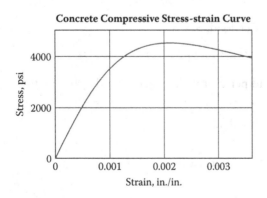

Concrete Compressive Stress-strain Curve

Compressive strain at peak is

$$\varepsilon_{c0} := \frac{1.71 \cdot f_c'}{E_c} = 0.0021$$

Compressive stress at peak is

$$\sigma''_c := 0.9 \cdot \frac{f_c'}{psi}$$

Stress–strain curve equation is

$$\sigma_c(\varepsilon_c) := \frac{2 \cdot \sigma''_c \cdot \left(\dfrac{\varepsilon_c}{\varepsilon_{c0}}\right)}{1 + \left(\dfrac{\varepsilon_c}{\varepsilon_{c0}}\right)^2}$$

6.4 STEP 2—COMPUTE THE FACTORED LOADS

The self-weight is computed considering a concrete density of 150 psf. Other dead loads such as floor cover (0.5 psf) and ceiling (2 psf) are considered. A live load of 100 psf was requested by the owner. The following unfactored uniform loads are considered:

$SW := t_{slab} \cdot \rho_c = 96.7 \cdot psf$ Slab self-weight
$OD := 2.5 \ psf$ Other dead loads
$DL := SW + OD = 99.2 \cdot psf$ Total dead load
$LL := 100 \ psf$ Live load

The governing load combination for computing the total factored load, TFL, is load combination (9-2) defined in ACI 318-11:

$$TFL := 1.2 \cdot DL + 1.6 \cdot LL = 279 \cdot psf$$

The total service load, SL, is

$$SL := DL + LL = 199.2 \cdot psf$$

The dead load per unit width (including the slab's self-weight) is

$$w_D := DL \cdot b = 99.2 \cdot \frac{lbf}{ft}$$

The live load per unit width is

$$w_L := LL \cdot b = 100 \cdot \frac{lbf}{ft}$$

The total factored load per unit width is

$$w_{TFL} := 1.2 w_D + 1.6 \cdot w_L = 279 \cdot \frac{lbf}{ft}$$

The total service load per unit width is

$$w_S := w_D + w_L = 199 \cdot \frac{lbf}{ft}$$

6.5 STEP 3—COMPUTE BENDING MOMENTS AND SHEAR FORCES

Bending moments and shear forces are determined as indicated in ACI 318-11. The moment coefficients can be used as the slab satisfies the specified geometry requirements. In fact, there are three spans; the ratio of the longer clear span to the shorter clear span is less than 1.2, the loads are uniformly distributed, the unfactored live load does not exceed three times the unfactored dead load, and, the members are prismatic.

Clear span values, l_n, are computed considering a constant beam width of 14 in.:

$$b_{beam} := 14 \text{ in.}$$

$$l_{n1} := l_1 - b_{beam} = 15.8 \cdot ft$$

$$l_{n2} := l_2 - b_{beam} = 18.8 \cdot ft$$

Bending moments (exterior bay)

Exterior support

$$C_{mNeg1} := \frac{1}{24}$$ Moment coefficient

$$M_{uNeg1} := C_{mNeg1} \cdot w_{TFL} \cdot l_{n1}^2 = 2.9 \cdot ft \cdot kip$$ Ultimate bending moment

$$M_{S1} := C_{mNeg1} \cdot w_S \cdot l_{n1}^2 = 2.1 \cdot ft \cdot kip$$ Service bending moment

Midspan

$$C_{mPos2} := \frac{1}{14}$$ Moment coefficient

$$M_{uPos2} := C_{mPos2} \cdot w_{TFL} \cdot l_{n1}^2 = 5 \cdot ft \cdot kip$$ Ultimate bending moment

$$M_{S2} := C_{mPos2} \cdot w_S \cdot l_{n1}^2 = 3.6 \cdot ft \cdot kip$$ Service bending moment

Interior support

$$C_{mNeg3} := \frac{1}{10}$$ Moment coefficient

$$M_{mNeg3} := C_{mNeg3} \cdot w_{TFL} \cdot l_{n1}^2 = 7 \cdot ft \cdot kip$$ Ultimate bending moment

$$M_{S3} := C_{mNeg3} \cdot w_S \cdot l_{n1}^2 = 5 \cdot ft \cdot kip$$ Service bending moment

Bending moments (interior bay)

First support

$$C_{mNeg4} := \frac{1}{11}$$ Moment coefficient

$$M_{uNeg4} := C_{mNeg4} \cdot w_{TFL} \cdot l_{n1}^2 = 6.4 \cdot ft \cdot kip$$ Ultimate bending moment

$$M_{S4} := C_{mNeg4} \cdot w_S \cdot l_{n1}^2 = 4.5 \cdot ft \cdot kip$$ Service bending moment

Midspan

$$C_{mPos5} := \frac{1}{16}$$ Moment coefficient

$$M_{uPos5} := C_{mPos5} \cdot w_{TFL} \cdot l_{n1}^2 = 4.4 \cdot ft \cdot kip$$ Ultimate bending moment

$$M_{S5} := C_{mPos5} \cdot w_S \cdot l_{n1}^2 = 3.1 \cdot ft \cdot kip$$ Service bending moment

Second support

$$C_{mNeg6} := \frac{1}{11}$$ Moment coefficient

$$M_{uNeg6} := C_{mNeg6} \cdot w_{TFL} \cdot l_{n1}^2 = 6.4 \cdot ft \cdot kip$$ Ultimate bending moment

$$M_{S6} := C_{mNeg6} \cdot w_S \cdot l_{n1}^2 = 4.5 \cdot ft \cdot kip$$ Service bending moment

Shear forces (exterior bay)

Exterior support

$$C_{v1} := 1$$ Shear coefficient

$$V_{u1} := C_{v1} \cdot w_{TFL} \cdot \frac{l_{n1}}{2} = 2.2 \cdot kip$$ Ultimate shear force

Midspan

$$C_{v2} := 0.15$$ Shear coefficient

$$V_{u2} := C_{v2} \cdot w_{TFL} \cdot \frac{l_{n1}}{2} = 0.33 \cdot kip$$ Ultimate shear force

Interior support

$$C_{v3} := 1.15 \qquad \text{Shear coefficient}$$

$$V_{u3} := C_{v3} \cdot w_{TFL} \cdot \frac{l_{n1}}{2} = 2.5 \cdot kip \qquad \text{Ultimate shear force}$$

Shear forces (interior bay)

First support

$$C_{v4} := 1.15 \qquad \text{Shear coefficient}$$

$$V_{u4} := C_{v4} \cdot w_{TFL} \cdot \frac{l_{n2}}{2} = 3 \cdot kip \qquad \text{Ultimate shear force}$$

Midspan

$$C_{v5} := 0.15 \qquad \text{Shear coefficient}$$

$$V_{u5} := C_{v5} \cdot w_{TFL} \cdot \frac{l_{n2}}{2} = 0.39 \cdot kip \qquad \text{Ultimate shear force}$$

Second support

$$C_{v6} := 1.15 \qquad \text{Shear coefficient}$$

$$V_{u6} := C_{v6} \cdot w_{TFL} \cdot \frac{l_{n2}}{2} = 3 \cdot kip \qquad \text{Ultimate shear force}$$

6.6 STEP 4—DESIGN FRP PRIMARY REINFORCEMENT

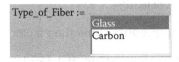

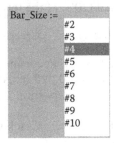

Select the FRP reinforcement: For the purpose of this design example, it is assumed that GFRP bars of the same size are used everywhere in the slab.

The ACI 440.6.1R-06 minimum manufacturer's guaranteed mechanical properties of the selected bars are the following:

$f_{fuu} = 100 \cdot ksi$ Ultimate guaranteed tensile strength of the FRP
$\varepsilon_{fuu} = 0.018$ Ultimate guaranteed rupture strain of the FRP
$E_f = 5700 \cdot ksi$ Guaranteed tensile modulus of elasticity of the FRP

$n_f := \dfrac{E_f}{E_c} = 1.399$ Ratio of modulus of elasticity of bars to modulus of elasticity of concrete

The geometrical properties of the selected bars are the following:

$\phi_{f\_bar} = 0.5 \cdot in.$ Bar diameter
$A_{f\_bar} = 0.196 \cdot in.^2$ Bar area

FRP reduction factors: Table 7-1 of ACI 440.1R-06 is used to define the environmental reduction factor, C_E. The type of exposure has to be selected:

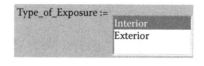

$C_E = 0.8$ Environmental reduction factor for GFRP

Table 8-3 of ACI 440.1R-06 is used to define the reduction factor to take into account the FRP creep rupture stress. Creep stress in the FRP is evaluated considering the total unfactored dead loads and the sustained portion of the live load (20% of the total live load):

$k_{creep-R} = 0.2$ Creep-rupture stress limitation factor

Crack width is checked using Equation (8-9) of ACI 440.1R-06. A crack width limit, w_{lim}, of 0.028 in. is used for interior exposure, while 0.020 in. is used for exterior exposure:

$w_{lim} = 0.028 \cdot in$ Crack width limit

FRP ultimate design properties: The ultimate design properties are calculated per Section 7.2 of ACI 440.1R-06:

$f_{fu} := C_E \cdot f_{fuu} = 80 \cdot ksi$ Design tensile strength
$\varepsilon_{fu} := C_E \cdot \varepsilon_{fuu} = 0.014$ Design rupture strain

FRP creep-rupture limit stress: The FRP creep-rupture limit stress is calculated per Section 8.4 of ACI 440.1R-06:

$$f_{f-creep} := k_{creep-R} \cdot f_{fu} = 16 \cdot ksi$$

6.6.1 Case I—Exterior support

Reinforcement required to resist bending moments: As discussed in Chapter 4, the following design condition shall be satisfied: $\phi M_n > M_u$. When the failure is due to FRP rupture: $\phi M_n = \phi \, A_f \, f_{fu} \, (d_f - \beta_1/2c_u)$. Considering the lower bound condition ($\phi M_n = M_u$) and solving for A_f, the following can be written:

$$A_{f\_req\_bend} = M_u \, / \, (\phi \, A_f \, f_{fu} \, (d_f - \beta_1/2c_u)) \Rightarrow$$

$$\Rightarrow \rho_{f\_req\_bend} = A_{f\_req\_bend}/bd_f = M_u \, / \, (\phi \, A_f \, f_{fu} \, (d_f - \beta_1/2c_u)bd_f)$$

Assuming a ϕ-factor of 0.65 and a neutral axis depth equal to 15% of the effective reinforcement depth, the longitudinal reinforcement ratio required for bending, ρ_{f-req_bend}, can be estimated:

ϕ-factor $\phi_{b-trial} := 0.65$

The effective reinforcement depth is

$$d_{f1} := t_{slab} - c_c - \frac{\phi_{f\_bar}}{2} = 7.0 \cdot in.$$

The longitudinal reinforcement ratio required for bending is

$$\rho_{f\_req\_bend1} := \frac{M_{u1}}{\phi_{b\_trial} \cdot f_{fu} \cdot (d_{f1} - \beta_1/2 \cdot 0.15 d_{f1})} \cdot \frac{1}{b \cdot d_{f1}}$$

$$= 0.00122 \qquad (M_{u1} = 2.9 \cdot kip \cdot ft)$$

The minimum reinforcement requirement has to be verified. Equation (8-8) of ACI 440.1R-06 is used. If the failure is not governed by FRP rupture, this requirement is automatically achieved:

$$A_{f\_min1} := min\left[\frac{4.9 \cdot \sqrt{f_c'} \cdot psi}{f_{fu}} b \cdot d_{f1}, \frac{300 psi}{f_{fu}} (b \cdot d_{f1})\right] = 0.315 \cdot in^2$$

$$\rho_{f\_min1} := \frac{A_{f\_min1}}{b \cdot d_{f1}} = 0.00375$$

The design reinforcement ratio required for bending is taken as

$$\rho_{f-bend1} := \max(\rho_{f-req-bend1}, \rho_{f-min1}) = 0.00375$$

Reinforcement required for concrete shear strength: As discussed in Chapter 4, for FRP reinforced members the concrete contribution to shear strength is also dependent upon the longitudinal reinforcement ratio per Equation (9-1) of ACI 440.1R-06:

$$V_c = 5\sqrt{f'_c}\ b\ k_f d_f \text{ with } k_f = \sqrt{2 \cdot \rho_f \cdot n_f + (\rho_f \cdot n_f)^2} - \rho_f \cdot n_f$$

As discussed in Chapter 4, the lower bound value of V_c is $0.8\sqrt{f'_c}b \cdot d_f$. Therefore, the detailed expression of V_c should only be computed if $V_u > \phi \cdot 0.8\sqrt{f'_c}b \cdot d_f$.

The lower-bound concrete shear strength is

$$V_{c1\_lower} := 0.8\sqrt{f'_c \cdot psib} \cdot d_{f1} = 4.8 \cdot kip$$

The ultimate shear force is

$$V_{u1} = 2.209 \cdot kip$$

$$\text{Check\_ShearLongRenf1} := \begin{vmatrix} \text{"Compute detailed Vc" if } V_{u1} > V_{c1\_lower} \\ \text{"Minimum concrete shear strength" otherwise} \end{vmatrix}$$

$$\text{Check\_ShearLongRenf1} := \text{"Minimum concrete shear strength"}$$

FRP longitudinal reinforcement design: The design reinforcement ratio can be selected equal to ρ_{f-bend} because the design for shear is not required:

$$\rho_{f-des1} := \rho_{f-bend1} = 0.00375$$

This reinforcement ratio corresponds to an area of

$$A_{f-des1} := \rho_{f-des1} \cdot b \cdot d_{f1} = 0.315 \cdot in.^2$$

The required number of no. 4 bars is

$$N_{f\_des1} := \frac{A_{f\_des1}}{A_{f\_bar}} = 1.604$$

The corresponding required spacing is

$$S_{f\_des1} := \frac{b - N_{f\_des1} \cdot \phi_{f\_bar}}{N_{f\_des1}} = 6.98 \cdot in.$$

The following bar spacing is selected:

$$S_{f-bar1} := 6 \ in.$$

The minimum required clear bar spacing is

$$S_{f-min1} := max \ (1 \cdot in., \ \phi_{f-bar}) = 1 \cdot in.$$

The bar clear spacing is

$$S_{f-bar1} - \phi_{f-bar} = 5.5 \cdot in.$$

$$Check\_BarSpacing1 := \left| \begin{array}{ll} \text{"OK"} & if \ S_{f\_bar1} - \phi_{f\_bar} \geq S_{f\_min1} \\ \text{"Too many bars"} & otherwise \end{array} \right.$$

$$Check\_BarSpacing1 = \text{"OK"}$$

The area of FRP reinforcement is

$$A_{f1} := \frac{b}{S_{f\_bar1}} \cdot A_{f\_bar} = 0.39 \cdot in^2$$

The FRP reinforcement ratio (Equation 8-2 of ACT 440.1R-06) is

$$\rho_{f1} := \frac{A_{f1}}{b \cdot d_{f1}} = 0.00467$$

Design flexural strength: The failure mode depends on the amount of FRP reinforcement. If ρ_f is larger than the balanced reinforcement ratio, ρ_{fb}, then concrete crushing is the failure mode. If ρ_f is smaller than the balanced reinforcement ratio, ρ_{fb}, then FRP rupture is the failure mode (Equation 8-3 of ACI 440.1R-06):

$$\rho_{fb1} := 0.85\beta_1 \cdot \frac{f'_c}{f_{fu}} \cdot \frac{E_f \cdot \varepsilon_{cu}}{E_f \cdot \varepsilon_{cu} + f_{fu}} = 0.00748$$

Based on cross-section compatibility, the effective concrete compressive strain distribution at failure can be computed as a function of the neutral axis depth, x:

$$\varepsilon_{c1}(x,y) := \left| \begin{array}{l} \dfrac{\varepsilon_{cu}}{x} y \ if \ \rho_{f1} \geq \rho_{fb1} \\ \\ \dfrac{\varepsilon_{fu}}{d_{f1} - x} \cdot y \ if \ \rho_{f1} < \rho_{fb1} \end{array} \right.$$

Similarly, the effective tensile strain in the FRP reinforcement can be computed as a function of the neutral axis depth, x:

$$\varepsilon_{f1}(x) := \left| \begin{array}{l} \varepsilon_{fu} \ \text{if} \ \rho_{f1} < \rho_{fb1} \\[2mm] \min\left[\dfrac{\varepsilon_{cu}}{x}\cdot\left(d_{f1} - x\right), \varepsilon_{fu}\right] \quad \text{if} \ \rho_{f1} \geq \rho_{fb1} \end{array}\right.$$

The compressive force in the concrete as a function of the neutral axis depth, x, is

$$C_{c1}(x) := b\cdot \int_{0in}^{x} \frac{2\cdot\sigma''_{c}\cdot\left(\dfrac{\varepsilon_{c1}(x,y)}{\varepsilon_{c0}}\right)}{1+\left(\dfrac{\varepsilon_{c1}(x,y)}{\varepsilon_{c0}}\right)^2}\text{psi dy}$$

The tensile force in the FRP reinforcement as a function of the neutral axis depth, x, is

$$T_{f1}(x) := A_{f1}\cdot E_f\cdot\varepsilon_{f1}(x)$$

The neutral axis depth, c_u, can be computed by solving the equation of equilibrium $C_c - T_f = 0$:
First guess:

$$x_{01} := 0.1d_{f1}$$

Given:

$$f_1(x) := C_{c1}(x) - T_{f1}(x)$$

$$c_{u1} := \text{root}\ (f_1(x_{01}),x_{01})$$

The neutral axis depth is

$$c_{u1} = 0.86\cdot\text{in.}$$

The nominal bending moment capacity can be computed as follows:

$$M_{n1} := b\cdot \int_{0}^{c_{u1}} y\cdot\frac{\left(2\cdot\sigma''_{c}\cdot\dfrac{\varepsilon_{c1}(c_{u1},y)}{\varepsilon_{c0}}\right)}{\left[1+\left(\dfrac{\varepsilon_{c1}(c_{u1},y)}{\varepsilon_{c0}}\right)^2\right]}\text{psi dy} + T_{f1}(c_{u1})\cdot(d_{f1} - c_{u1}) = 17\cdot\text{ft}\cdot\text{kip}$$

The strain distribution over the cross section is shown next.

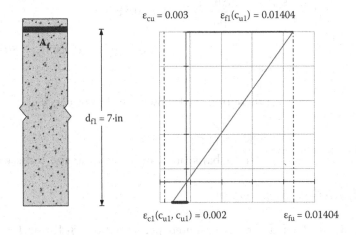

The concrete crushing failure mode is less brittle than the one due to FRP rupture. The ϕ-factor is calculated according to Jawahery and Nanni [1].

$$\phi_{b1} := \begin{vmatrix} 0.65 & \text{if } 1.15 - \dfrac{\varepsilon_{f1}(c_{u1})}{2\varepsilon_{fu}} \leq 0.65 \\[4mm] 0.75 & \text{if } 1.15 - \dfrac{\varepsilon_{f1}(c_{u1})}{2\varepsilon_{fu}} \geq 0.75 \\[4mm] 1.15 - \dfrac{\varepsilon_{f1}(c_{u1})}{2\varepsilon_{fu}} & \text{otherwise} \end{vmatrix}$$

$$\phi_{b1} = 0.65$$

The design flexural strength equation is computed per Equation (8-1) of ACI 440.1R-06:

$$\text{Check\_Flexure1} := \begin{vmatrix} \text{"OK"} & \text{if } \phi_{b1} \cdot M_{n1} \geq M_{u1} \;\; (\phi_{b1} \cdot M_{n1} = 11 \cdot \text{kip} \cdot \text{ft}) \\[2mm] \text{"Not good" otherwise} & \quad (M_{u1} = 2.9 \cdot \text{kip} \cdot \text{ft}) \end{vmatrix}$$

$$\text{Check\_Flexure1} = \text{"OK"}$$

Flexural strength computed per ACI 440.1R-06: The tensile stress in the GFRP is computed, per Equation (8-4c) of ACI 440.1R-06, when $\rho_f > \rho_{fb}$, or is f_{fu} if $\rho_f < \rho_{fb}$.

$$f_{f1} := \left| \begin{array}{l} \sqrt{\dfrac{(E_f \cdot \varepsilon_{cu})^2}{4} + \dfrac{0.85\beta_1 \cdot f'_c}{\rho_{f1}}} E_f \cdot \varepsilon_{cu} - 0.5 E_f \cdot \varepsilon_{cu} \\ \qquad\qquad\qquad\qquad \text{if } \rho_{f1} \geq \rho_{fb1} = 80 \cdot \text{ksi } (f_{fu} = 80 \cdot \text{ksi}) \\ f_{fu} \text{ otherwise} \end{array} \right.$$

f_f cannot exceed f_{fu}; therefore, the following has to be checked:

$$\text{CheckMaxStress1} := \left| \begin{array}{l} \text{"OK" if } f_{f1} \leq f_{fu} \\ \text{"Reduce bar spacing or increase bar size" otherwise} \end{array} \right.$$

$$\text{CheckMaxStress1} = \text{"OK"}$$

The stress-block depth is computed per Equation (8-4b) or Equation (8-6c) depending on whether $\rho_f > \rho_{fb}$, or $\rho_f < \rho_{fb}$, respectively:

$$a_{f1} := \left| \begin{array}{ll} \dfrac{A_{f1} \cdot f_{f1}}{0.85 \cdot f'_c \cdot b} & \text{if } \rho_{f1} \geq \rho_{fb1} \quad \text{Equation (8-4b) of ACI 440.1R-06} \\[4mm] \left[\beta_1 \cdot \left(\dfrac{\varepsilon_{cu}}{\varepsilon_{cu} + \varepsilon_{fu}} \right) d_{f1} \right] & \text{otherwise Equation (8-6c) of ACI 440.1R-06} \end{array} \right.$$

$$a_{f1} = 0.99 \cdot \text{in.}$$

$$c_{f1} := \frac{a_{f1}}{\beta_1} = 1.233 \cdot \text{in.} \quad \text{(neutral axis depth)}$$

The nominal moment capacity is

$$M_{nACI\_1} := A_{f1} \cdot f_{f1} \cdot \left(d_{f1} - \frac{a_{f1}}{2} \right) = 17 \cdot \text{ft} \cdot \text{kip}$$

The concrete crushing failure mode is less brittle than the one due to FRP rupture. The ϕ-factor is computed according to Equation (8-7) of ACI 440.1R-06:

$$\phi_{bACI\_1} := \left| \begin{array}{ll} 0.55 \text{ if } \rho_{f1} \leq \rho_{fb1} & \phi_{bACI\_1} = 0.55 \\[3mm] 0.30 + 0.25 \cdot \dfrac{\rho_{f1}}{\rho_{fb1}} & \text{if } \rho_{fb1} < \rho_{f1} < 1.4 \cdot \rho_{fb1} \\[3mm] 0.65 \text{ otherwise} \end{array} \right.$$

The design flexural strength equation is computed per Equation (8-1) of ACI 440.1R-06:

$$\phi_{bACI-1}\, M_{nACI-1} = 9 \cdot kip \cdot ft$$

$$Check\_FlexureACI\_1 := \left| \begin{array}{l} \text{``OK'' if } \phi_{bACI\_1} \cdot M_{nACI\_1} \geq M_{u1}(M_{u1} = 2.9 \cdot kip \cdot ft) \\ \text{``Not good'' otherwise} \end{array} \right.$$

$$Check\_FlexureACI\_1 = \text{``OK''}$$

Embedment length: Because this is a case of negative reinforcement, it has to be checked if adequate moment capacity can be achieved at the end of the embedment length. The available length for embedment is

$$l_{emb1} := 12 \text{ in.}$$

The developable tensile stress is calculated per Equation (11-3) of ACI 440.1R-06. Minimum between cover to bar center and half of the center-to-center bar spacing is

$$C_{b1} := \left(c_c + \frac{\phi_{f\_bar}}{2}, \frac{s_{f\_bar1}}{2} \right) = 1 \cdot in.$$

Bar location modification factor for top reinforcement but less than 12 in. of concrete below it is

$$\alpha_{Neg1} := 1.0$$

Required stress in the FRP is

$$f_{f1} := E_f \cdot \varepsilon_{f1}(c_{u1}) = 80 \cdot ksi$$

The developable tensile stress (ACI 440.1R-06 Equation 11-3) is

$$f_{fd11} := \left| \begin{array}{l} f_{fu} \text{ if } \dfrac{\sqrt{f'_c \cdot psi}}{\alpha_{Neg1}} \cdot \left(13.6 \cdot \dfrac{l_{emb1}}{\phi_{f\_bar}} + \dfrac{C_{b1}}{\phi_{f\_bar}} \cdot \dfrac{l_{emb1}}{\phi_{f\_bar}} + 340 \right) \geq f_{fu} \\[3mm] \left[\dfrac{\sqrt{f'_c \cdot psi}}{\alpha_{Neg1}} \cdot \left(13.6 \cdot \dfrac{l_{emb1}}{\phi_{f\_bar}} + \dfrac{C_{b1}}{\phi_{f\_bar}} \cdot \dfrac{l_{emb1}}{\phi_{f\_bar}} + 340 \right) \right] \text{otherwise} \end{array} \right.$$

$$= 33.7 \cdot ksi$$

$$\text{CheckFailure1} := \begin{vmatrix} \text{"Bar ultimate strength" if } f_{fd11} \geq f_{fr1} \\ \text{"Bond strength" otherwise} \end{vmatrix}$$

CheckFailure1 = "Bond strength"

The cross section of interest is a bond critical section. The nominal moment capacity, therefore, has to be computed per ACI 440.1R-06 Equation (8-5) or Equation (8-6b) when the failure mode is concrete crushing or bond, respectively.

$$M_{n1b1} := \begin{vmatrix} \left[A_{f1} \cdot f_{fd11} \cdot \left(d_{f1} - \frac{1}{2} \cdot \frac{A_{f1} \cdot f_{fd11}}{0.85 \cdot f'_c \cdot b} \right) \right] \text{ if CheckFailure1} = \text{"Bond strength"} \\ M_{n1} \text{ if CheckFailure1} = \text{"Bar ultimate strength"} \end{vmatrix}$$

$M_{n1b1} = 7.4 \cdot ft \cdot kip$

The ultimate moment is

$M_{u1} = 2.9 \cdot ft \cdot kip$

The strength-reduction factor when failure is controlled by bond is

$$\phi_{b1\_bond} := \begin{vmatrix} 0.55 \text{ if CheckFailure1} = \text{"Bond strength"} = 0.55 \\ \phi_{b1} \text{ if CheckFailure1} = \text{"Bond ultimate strength"} \end{vmatrix}$$

The design flexural strength is therefore:

$\phi_{b1\_bond} \cdot M_{n1b1} = 4.06 \cdot kip \cdot ft$

$$\text{Check\_FlexureNeg1b} := \begin{vmatrix} \text{"OK" if } \phi_{b1\_bond} \cdot M_{n1b1} \geq M_{u1} \left(M_{u1} = 2.9 \cdot kip \cdot ft \right) \\ \text{"Not good" otherwise} \end{vmatrix}$$

Check_FlexureNeg1b = "OK"

The embedment length is adequate. If the embedment length was not adequate, a bent bar could have been used as indicated in the following.

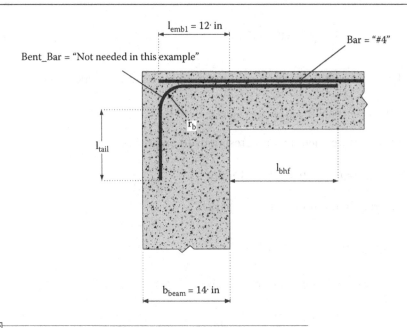

$l_{emb1} = 12 \cdot in$

Bent_Bar = "Not needed in this example"

Bar = "#4"

r_b

l_{tail}

l_{bhf}

$b_{beam} = 14 \cdot in$

6.6.2 Case 2—Midspan

Reinforcement required to resist bending moments: The same procedure discussed for Case 1 is followed.

The effective reinforcement depth is

$$d_{f2} := t_{slab} - c_c - \frac{\phi_{f\_bar}}{2} = 7 \cdot in.$$

ϕ-factor $\phi_{b\_trial} = 0.65$

The longitudinal reinforcement ratio required for bending is

$$\rho_{f\_req\_bend2} := \frac{M_{u2}}{\phi_{b\_trial} \cdot f_{fu} \cdot \left(d_{f2} - \frac{\beta_1}{2} \cdot 0.15 d_{f2}\right)} \cdot \frac{1}{b \cdot d_{f2}}$$

$$= 0.00209 \qquad (M_{u2} = 5 \cdot kip \cdot ft)$$

The minimum reinforcement requirement has to be verified. Equation (8-8) of ACI 440.1R-06 is used. If the failure is not governed by FRP rupture, this requirement is automatically achieved:

$$A_{f\_min2} := min\left[\frac{4.9 \cdot \sqrt{f_c' \cdot psi}}{f_{fu}} b \cdot d_{f2}, \frac{300 psi}{f_{fu}}(b \cdot d_{f2})\right] = 0.315 \cdot in^2$$

$$\rho_{f\_min2} := \frac{A_{f\_min2}}{b \cdot d_{f2}} = 0.00375$$

The design reinforcement ratio required for bending is taken as

$$\rho_{f\_des2} := \max\left(\rho_{f\_req\_bend2}, \rho_{f\_min2}\right) = 0.00375$$

This reinforcement ratio corresponds to an area of

$$A_{f\_des2} := \rho_{f\_des2} \cdot b \cdot d_{f2} = 0.315 \cdot in^2$$

The required number of no. 4 bars is

$$N_{f\_des2} := \frac{A_{f\_des2}}{A_{f\_bar}} = 1.604$$

The corresponding required spacing is

$$s_{f\_des2} := \frac{b - N_{f\_des2} \cdot \phi_{f\_bar}}{N_{f\_des2}} = 6.98 \cdot in$$

The following bar spacing is selected:

$$s_{f-bar2} := 6 \text{ in.}$$

The minimum required clear bar spacing is

$$s_{f-min2} := \max(1 \cdot in., \phi_{f-bar}) = 1 \cdot in.$$

The bar clear spacing is

$$s_{f-bar2} - \phi_{f-bar} = 5.5 \cdot in.$$

$$\text{Check\_BarSpacing2} := \left| \begin{array}{l} \text{``OK'' if } s_{f\_bar2} - \phi_{f\_bar} \geq s_{f\_min2} \\ \text{``Too many bars'' otherwise} \end{array} \right.$$

$$\text{Check\_BarSpacing2} = \text{``OK''}$$

The area of FRP reinforcement is

$$A_{f2} := \frac{b}{s_{f\_bar2}} \cdot A_{f\_bar} = 0.39 \cdot in^2$$

The FRP reinforcement ratio (Equation 8-2 of ACI 440.1R-06) is

$$\rho_{f2} := \frac{A_{f2}}{b \cdot d_{f2}} = 0.00467$$

Design flexural strength (Equation 8-3 of ACI 440.1R-06):

$$\rho_{fb2} := 0.85\beta_1 \cdot \frac{f'_c}{f_{fu}} \cdot \frac{E_f \cdot \varepsilon_{cu}}{E_f \cdot \varepsilon_{cu} + f_{fu}} = 0.00748$$

The effective concrete compressive strain at failure as a function of the neutral axis depth, x, is

$$\varepsilon_{c2}(x,y) := \begin{vmatrix} \dfrac{\varepsilon_{cu}}{x} \cdot y \text{ if } \rho_{f2} \geq \rho_{fb2} \\[2mm] \dfrac{\varepsilon_{fu}}{d_{f2} - x} \cdot y \text{ if } \rho_{f2} < \rho_{fb2} \end{vmatrix}$$

The effective tensile strain in the first layer of FRP reinforcement as a function of the neutral axis depth, x, is

$$\varepsilon_{f2}(x) := \begin{vmatrix} \varepsilon_{fu} \text{ if } \rho_{f2} < \rho_{fb2} \\[2mm] \min\left[\dfrac{\varepsilon_{cu}}{x} \cdot (d_{f2} - x), \varepsilon_{fu}\right] \text{if } \rho_{f2} \geq \rho_{fb2} \end{vmatrix}$$

The compressive force in the concrete as a function of the neutral axis depth, x, is

$$C_{c2}(x) := b \cdot \int_{0in}^{x} \frac{2 \cdot \sigma''_c \cdot \left(\dfrac{\varepsilon_{c2}(x,y)}{\varepsilon_{c0}}\right)}{1 + \left(\dfrac{\varepsilon_{c2}(x,y)}{\varepsilon_{c0}}\right)^2} \text{psi} \, dy$$

The tensile force in the first layer of FRP reinforcement as a function of the neutral axis depth, x, is

$$T_{f2}(x) := A_{f2} \cdot E_f \cdot \varepsilon_{f2}(x)$$

The neutral axis depth, cu, can be computed by solving the equation of equilibrium $C_c - T_f = 0$:

Given: $f_2(x) := C_{c2}(x) - T_{f2}(x)$

First guess: $x_{02} := 0.1_{df2}$

$cu_2 := \text{root}\,(f_2(x_{02}),\,x_{02})$

The neutral axis depth is

$cu_2 = 0.86\cdot\text{in.}$

The nominal bending moment capacity can be computed as follows:

$$M_{n2} := b\cdot\int_0^{cu_2} y\cdot\left[\dfrac{2\cdot\sigma"_c\cdot\dfrac{\varepsilon_{c2}(c_{u2},y)}{\varepsilon_{c0}}}{1+\left(\dfrac{\varepsilon_{c2}(c_{u2},y)}{\varepsilon_{c0}}\right)^2}\right]\text{psi}\,dy + T_{f2}(c_{u2})\cdot(d_{f2}-c_{u2}) = 17\cdot\text{ft}\cdot\text{kip}$$

The strain distribution over the cross section is shown next.

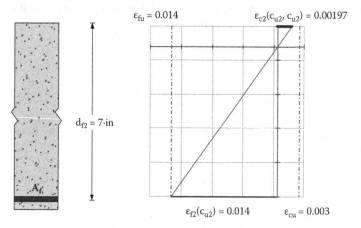

$d_{f2} = 7\cdot\text{in}$

$\varepsilon_{fu} = 0.014$ $\varepsilon_{c2}(c_{u2},\,c_{u2}) = 0.00197$

$\varepsilon_{f2}(c_{u2}) = 0.014$ $\varepsilon_{cu} = 0.003$

The ϕ-factor is calculated according to Jawahery and Nanni [1]:

$$\phi_{b2} := \begin{vmatrix} 0.65\,\text{if}\,1.15-\dfrac{\varepsilon_{f2}(c_{u2})}{2\varepsilon_{fu}}\le 0.65 \\[2mm] 0.75\,\text{if}\,1.15-\dfrac{\varepsilon_{f2}(c_{u2})}{2\varepsilon_{fu}}\ge 0.65 \\[2mm] 1.15-\dfrac{\varepsilon_{f2}(c_{u2})}{2\varepsilon_{fu}}\,\text{otherwise} \end{vmatrix}$$

$\phi_{b2} = 0.65$

The design flexural strength equation is computed per Equation (8-1) of ACI 440.1R-06:

$$\phi_{b2} \cdot M_{n2} = 11 \cdot \text{kip} \cdot \text{ft}$$

$$\text{Check\_Flexure2} := \begin{vmatrix} \text{"OK" if } \phi_{b2} \cdot M_{n2} \geq M_{u2} \quad \left(M_{u2} = 5 \cdot \text{kip} \cdot \text{ft} \right) \\ \text{"Not good" otherwise} \end{vmatrix}$$

$$\text{Check\_Flexure2} = \text{"OK"}$$

Flexural strength computed per ACI 440.1R-06: The tensile stress in the GFRP is computed per Equation (8-4c) when $\rho_f > \rho_{fb}$, or is f_{fu} if $\rho_f < \rho_{fb}$.

$$f_{f2} := \begin{vmatrix} \sqrt{\dfrac{\left(E_f \cdot \varepsilon_{cu} \right)^2}{4} + \dfrac{0.85 \beta_1 \cdot f'_c}{\rho_{f2}} E_f \cdot \varepsilon_{cu}} - 0.5 E_f \cdot \varepsilon_{cu} \text{ if } \rho_{f2} \geq \rho_{fb2} = 80 \cdot \text{ksi} \\ f_{fu} \text{ otherwise} \end{vmatrix}$$

f_f cannot exceed f_{fu}; therefore, the following has to be checked:

$$\text{CheckMaxStress2} := \begin{vmatrix} \text{"OK" if } f_{f2} \leq f_{fu} \qquad \left(f_{fu} = 80 \cdot \text{ksi} \right) \\ \text{"Reduce bar spacing or increase bar size" otherwise} \end{vmatrix}$$

$$\text{CheckMaxStress1} = \text{"OK"}$$

The stress-block depth is computed per Equation (8-4b) or Equation (8-6c) depending on whether $\rho_f > \rho_{fb}$ or $\rho_f < \rho_{fb}$, respectively:

$$a_{f2} := \begin{vmatrix} \dfrac{A_{f2} \cdot f_{f2}}{0.85 \cdot f'_c \cdot b} \text{ if } \rho_{f2} \geq \rho_{fb2} & \text{Equation} (8\text{-}4b) \text{ of ACI } 440.1R\text{-}06 \\ \left[\beta_1 \cdot \left(\dfrac{\varepsilon_{cu}}{\varepsilon_{cu} + \varepsilon_{fu}} \right) d_{f2} \right] \text{otherwise} & \text{Equation} (8\text{-}6c) \text{ of ACI } 440.1R\text{-}06 \end{vmatrix}$$

$$a_{f2} = 0.986 \cdot \text{in.}$$

Neutral axis depth is

$$c_{f2} := \frac{a_{f1}}{\beta_1} = 1.233 \cdot in.$$

The nominal moment capacity is

$$M_{nACI\_2} := A_{f2} \cdot f_{f2} \cdot \left(d_{f2} \frac{a_{f2}}{2}\right) = 17 \cdot ft \cdot kip$$

The concrete crushing failure mode is less brittle than the one due to FRP rupture. The ϕ-factor is computed according to Equation (8-7) of ACI 440.1R-06:

$$\phi_{bACI\_2} := \begin{vmatrix} 0.55 \, if \, \rho_{f2} \leq \rho_{fb2} & \phi_{bACI\_2} = 0.55 \\ 0.30 + 0.25 \cdot \dfrac{\rho_{f2}}{\rho_{fb2}} \, if \, \rho_{fb2} < \rho_{f2} < 1.4 \cdot \rho_{fb2} \\ 0.65 \, otherwise \end{vmatrix}$$

The design flexural strength equation is computed per Equation (8-1) of ACI 440.1R-06:

$$\phi_{bACI-2} \cdot M_{nACI-2} = 9 \cdot kip \cdot ft$$

$$Check\_FlexureACI\_2 := \begin{vmatrix} \text{"OK" if } \phi_{bACI\_2} \cdot M_{nACI\_2} \geq M_{u2} & (M_{u2} = 5 \cdot kip \cdot ft) \\ \text{"Not good" otherwise} \end{vmatrix}$$

$$Check\_FlexureACI\_2 = \text{"OK"}$$

Development of positive moment reinforcement: The development length, l_d, for straight bars can be calculated using Equation (11-3) of ACI 440.1R-06. Minimum between cover to bar center and half of the center-to-center bar spacing is

$$C_{b2} := min\left(c_c + \frac{\phi_{f\_bar}}{2}, \frac{s_{f\_bar2}}{2}\right) = 1 \cdot in.$$

The bar location modification factor for bottom reinforcement is

$$\alpha_{Pos} := 1$$

The minimum development length is computed according to ACI 440.1R-06 Equation (11-6):

$$1_{d\_min} := \frac{\alpha_{Pos} \cdot \dfrac{f_{fu}}{\sqrt{f'_c \cdot psi}} - 340}{13.6 + \dfrac{C_{b2}}{\phi_{f\_bar}}} \phi_{f\_bar} = 25.364 \cdot in.$$

The following development length is considered and is available to provide the required moment capacity:

$$l_d := 26 \text{ in.}$$

6.6.3 Case 3—Interior support

Reinforcement required to resist bending moments: The same procedure discussed for case 1 is followed.

The effective reinforcement depth is

$$d_{f3} := t_{slab} - c_c - \frac{\phi_{f\_bar}}{2} = 7 \cdot in.$$

ϕ-factor $\phi_{b-trial} = 0.65$

The longitudinal reinforcement ratio required for bending is

$$\rho_{f\_req\_bend3} := \frac{M_{u3}}{\phi_{b\_trial} \cdot f_{fu} \cdot \left(d_{f3} - \dfrac{\beta_1}{2} \cdot 0.15 d_{f3}\right)} \cdot \frac{1}{b \cdot d_{f3}} 0.00292 \quad (M_{u3} = 7 \cdot kip \cdot ft)$$

The minimum reinforcement requirement has to be verified. Equation (8-8) of ACI 440.1R-06 is used. If the failure is not governed by FRP rupture, this requirement is automatically achieved:

$$A_{f\_min3} := min\left[\frac{4.9 \cdot \sqrt{f'_c \cdot psi}}{f_{fu}} b \cdot d_{f3}, \frac{300 psi}{f_{fu}} (b \cdot d_{f3})\right] = 0.315 \cdot in^2$$

$$\rho_{f\_min3} := \frac{A_{f\_min3}}{b \cdot d_{f1}} = 0.00375$$

The design reinforcement ratio required for bending is taken as

$$\rho_{f\_bend3} := max(\rho_{f\_red\_bend3}, \rho_{f\_min3}) = 0.00375$$

Reinforcement required for concrete shear strength: The same procedure discussed for Case 1 is followed.

The lower bound concrete shear strength is

$$V_{c3\_lower} := 0.8\sqrt{f'_c \cdot psi}b \cdot d_{f3} = 4.8 \cdot kip$$

The ultimate shear force is

$$Max(V_{u3}, V_{u4}) = 3.021 \cdot kip$$

$$Check\_ShearLongRenf3 := \left| \begin{array}{l} \text{``Compute detailed } V_c\text{'' if } max\left(V_{u3}, V_{u4}\right) > V_{c3\_lower} \\ \text{``Minimum concrete shear strength'' otherwise} \end{array} \right.$$

$$Check\_ShearLongRenf3 := \text{``Minimum concrete shear strength''}$$

FRP longitudinal reinforcement design: The design reinforcement ratio can be selected as the maximum between ρ_{f-bend} and $\rho_{f-shear}$:

$$\rho_{f-des3} := \rho_{f-bend3} \; 0.00375$$

This reinforcement ratio corresponds to an area of

$$A_{f-des3} := \rho_{f\_des3} \cdot b \cdot d_{f3} = 0.315 \cdot in.^2$$

The required number of no. 4 bars is

$$N_{f\_des3} := \frac{A_{f\_des3}}{A_{f\_bar}} = 1.604$$

The corresponding required spacing is

$$s_{f\_des3} := \frac{b - N_{f\_des3}\phi_{f\_bar}}{N_{f\_des3}} = 6.98 \cdot in$$

The following bar spacing is selected:

$$s_{f-bar3} := 6 \; in.$$

The minimum required clear bar spacing is

$$s_{f-min3} := max(1 \cdot in., \phi_{f-bar}) = 1 \cdot in.$$

$$Check\_ShearLongRenf3 := \left| \begin{array}{l} \text{``OK'' if } s_{f\_bar3} - \phi_{f\_bar} \geq s_{f\_min3} \\ \text{``Too many bars'' otherwise} \end{array} \right.$$

Check_BarSpacing3 = "OK"

The area of FRP reinforcement is

$$A_{f3} := \frac{b}{s_{f\_bar3}} \cdot A_{f\_bar} = 0.39 \cdot in^2$$

The FRP reinforcement ratio (Equation 8-2 of ACI 440.1R-06) is

$$\rho_{f3} := \frac{A_{f3}}{b \cdot d_{f3}} = 0.00467$$

Design flexural strength: Equation (8-3) of ACI 440.1R-06 is

$$\rho_{fb3} := 0.85\beta_1 \cdot \frac{f'_c}{f_{fu}} \cdot \frac{E_f \cdot \varepsilon_{cu}}{E_f \cdot \varepsilon_{cu} + f_{fu}} = 0.00748$$

The effective concrete compressive strain at failure as a function of the neutral axis depth, x, is

$$\varepsilon_{c3}(x,y) := \begin{vmatrix} \dfrac{\varepsilon_{cu}}{x}y & \text{if } \rho_{f3} \geq \rho_{fb3} \\[4mm] \dfrac{\varepsilon_{fu}}{d_{f3} - x} \cdot x & \text{if } \rho_{f3} < \rho_{fb3} \end{vmatrix}$$

The effective tensile strain in the FRP reinforcement as a function of the neutral axis depth, x, is

$$\varepsilon_{f3}(x) := \begin{vmatrix} \varepsilon_{fu} \text{ if } \rho_{f3} < \rho_{fb3} \\[4mm] \min\left[\dfrac{\varepsilon_{cu}}{x} \cdot (d_{f3} - x), \varepsilon_{fu} \right] \text{if } \rho_{f3} \geq \rho_{fb3} \end{vmatrix}$$

The compressive force in the concrete as a function of the neutral axis depth, x, is

$$C_{c3}(x) := b \cdot \int_{0in}^{x} \frac{2.\sigma''_c \cdot \left(\dfrac{\varepsilon_{c3}(x,y)}{\varepsilon_{c0}} \right)}{1 + \left(\dfrac{\varepsilon_{c3}(x,y)}{\varepsilon_{c0}} \right)^2} psi\,dy$$

The tensile force in the FRP reinforcement as a function of the neutral axis depth, x, is

$$T_{f3}(x) := A_{f3} \cdot E_f \cdot \varepsilon_{f3}(x)$$

The neutral axis depth, cu, can be computed by solving the equation of equilibrium $C_c - T_f = 0$:

Given: $f_3(x) := C_{c3}(x) - T_{f3}(x)$

First guess: $x_{03} := 0.1 d_{f3}$

$c_{u3} := \text{root}(f_3(x_{03}), x_{03}))$

The neutral axis depth is

$c_{u3} = 0.63 \cdot \text{in.}$

The nominal bending moment capacity can be computed as follows:

$$M_{n3} := b \cdot \int_0^{c_{u3}} y \cdot \frac{\left(2.\sigma''_c \cdot \left(\frac{\varepsilon_{c3}(\varepsilon_{u3}, y)}{\varepsilon_{c0}} \right) \right)}{\left[1 + \left(\frac{\varepsilon_{c3}(\varepsilon_{u3}, y)}{\varepsilon_{c0}} \right)^2 \right]} \text{psi} \, dy + T_{f3}(c_{u3}) \cdot (d_{f3} - c_{u3}) = 17 \cdot \text{ft} \cdot \text{kip}$$

The strain distribution over the cross section is shown next.

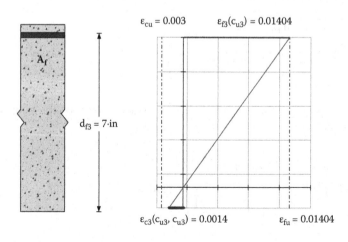

$\varepsilon_{cu} = 0.003$ $\varepsilon_{f3}(c_{u3}) = 0.01404$

A_f

$d_{f3} = 7 \cdot \text{in}$

$\varepsilon_{c3}(c_{u3}, c_{u3}) = 0.0014$ $\varepsilon_{fu} = 0.01404$

The ϕ-factor is calculated according to Jawahery and Nanni [1]:

$$\phi_{b3} := \left| \begin{array}{l} 0.65 \,\text{if}\, 1.15 - \dfrac{\varepsilon_{f3}\left(c_{u3}\right)}{2\varepsilon_{fu}} \leq 0.65 = 0.65 \\[3mm] 0.75 \,\text{if}\, 1.15 - \dfrac{\varepsilon_{f3}\left(c_{u3}\right)}{2\varepsilon_{fu}} \geq 0.75 \\[3mm] 1.15 - \dfrac{\varepsilon_{f3}\left(c_{u3}\right)}{2\varepsilon_{fu}} \,\text{otherwise} \end{array} \right.$$

The design flexural strength equation is computed per Equation (8-1) of ACI 440.1R-06:

$$\phi_{b3} \cdot M_{n3} = 11 \cdot \text{kip} \cdot \text{ft}$$

$$\text{Check\_Flexure3} := \left| \begin{array}{l} \text{``OK'' if}\, \phi_{b3} \cdot M_{n3} \geq M_{u3} \qquad \left(M_{u3} = 7 \cdot \text{kip} \cdot \text{ft}\right) \\[2mm] \text{``Not good'' otherwise} \end{array} \right.$$

$$\text{Check\_Flexure3} = \text{``OK''}$$

Flexural strength computed per ACI 440.1R-06: The tensile stress in the GFRP is computed per Equation (8-4c) when $\rho_f > \rho_{fb}$, or is f_{fu} if $\rho_f < \rho_{fb}$.

$$f_{f3} := \left| \begin{array}{l} \sqrt{\dfrac{\left(E_f \cdot \varepsilon_{cu}\right)^2}{4} + \dfrac{0.85\beta_1 \cdot f_c'}{\rho_{f3}} E_f \cdot \varepsilon_{cu}} - 0.5 E_f \cdot \varepsilon_{cu} \,\text{if}\, \rho_{f3} \geq \rho_{fb3} = 80 \cdot \text{ksi} \\[3mm] f_{fu} \,\text{otherwise} \end{array} \right.$$

f_f cannot exceed f_{fu}; therefore, the following has to be checked:

$$\text{CheckMaxStress3} := \left| \begin{array}{l} \text{``OK'' if}\, f_{f3} \leq f_{fu} \qquad\qquad \left(f_{fu} = 80 \cdot \text{ksi}\right) \\[2mm] \text{``Reduce bar spacing or increase bar size'' otherwise} \end{array} \right.$$

$$\text{CheckMaxStress3} = \text{``OK''}$$

The stress-block depth is computed per Equation (8-4b) or Equation (8-6c) depending on whether $\rho_f > \rho_{fb}$ or $\rho_f < \rho_{fb}$, respectively:

$$a_{f3} := \begin{vmatrix} \dfrac{A_{f3} \cdot f_{f3}}{0.85 \cdot f'_c \cdot b} \text{ if } \rho_{f3} \geq \rho_{fb3} & \text{Equation}(8\text{-}4b)\text{ of ACI440.1R-06} \\[4mm] \left[\beta_1 \cdot \left(\dfrac{\varepsilon_{cu}}{\varepsilon_{cu} + \varepsilon_{fu}} \right) d_{f3} \right] \text{otherwise} & \text{Equation}(8\text{-}6c)\text{ of ACI440.1R-06} \end{vmatrix}$$

$$a_{f3} = 0.986 \cdot \text{in.}$$

Neutral axis depth is

$$c_{f3} := \frac{a_{f3}}{\beta_1} = 1.23 \cdot \text{in.}$$

The nominal moment capacity is

$$M_{nACI\_3} := A_{f3} \cdot f_{f3} \cdot \left(d_{f3} - \frac{a_{f3}}{2} \right) = 17 \cdot \text{ft} \cdot \text{kip}$$

The concrete crushing failure mode is less brittle than the one due to GFRP rupture. The ϕ-factor is computed according to Equation (8-7) of ACI 440.1R-06:

$$\phi_{bACI\_3} := \begin{vmatrix} 0.55 \text{ if } \rho_{f3} \leq \rho_{fb3} \\[2mm] 0.30 + 0.25 \cdot \dfrac{\rho_{f3}}{\rho_{fb3}} \text{ if } \rho_{fb3} < \rho_{f3} < 1.4 \cdot \rho_{fb3} & = 0.55 \\[2mm] 0.65 \text{ otherwise} \end{vmatrix}$$

The design flexural strength equation is computed per Equation (8-1) of ACI 440.1R-06:

$$\phi_{bACI-3} \cdot M_{nACI-3} = 9 \cdot \text{kip} \cdot \text{ft}$$

$$\text{Check\_FlexureACI\_3} := \begin{vmatrix} \text{"OK" if } \phi_{bACI\_3} \cdot M_{nACI\_3} \geq M_{u3} & (M_{u3} = 7 \cdot \text{kip} \cdot \text{ft}) \\[2mm] \text{"Not good" otherwise} \end{vmatrix}$$

$$\text{Check\_FlexureACI\_3} = \text{"OK"}$$

Tension lap splice: The recommended development length of FRP tension lap splices is $1.3l_d$ (Section 11.4 of ACI 440.1R-06). The minimum recommended tension lap splice development length is

$$1.3l_d = 33.8 \cdot \text{in}.$$

A tension lap splice development length of 34 in. is considered and implemented.

Embedment length: Because this is a case of negative reinforcement, it has to be checked if adequate moment capacity can be achieved at the end of the embedment length. The maximum available length for embedment is equal to half the length of the adjacent span:

$$l_{emb3} := 0.25 \cdot l_2 = 60 \cdot \text{in}.$$

The developable tensile stress is calculated per Equation (11-3) of ACI 440.1R-06. Minimum between cover to bar center and half of the center-to-center bar spacing is

$$C_{b3} := \min\left(c_c + \frac{\phi_{f\_bar}}{2}, \frac{s_{f\_bar3}}{2}\right) = 1 \cdot \text{in}.$$

Bar location modification factor for top reinforcement but less than 12 in. of concrete below it is

$$\alpha_{Neg3} := 1.0$$

Required stress in the FRP is

$$f_{fr3} := E_f \cdot \varepsilon_{f3}(c_{u3}) = 80 \cdot \text{ksi}$$

The developable tensile stress (ACI 440.1R-06 Equation 11-3 is

$$f_{fd13} := \begin{vmatrix} f_{fu} \text{ if } \frac{\sqrt{f'_c \cdot \text{psi}}}{\alpha_{Neg3}} \cdot \left(13.6 \cdot \frac{l_{emb3}}{\phi_{f\_bar}} + \frac{C_{b3}}{\phi_{f\_bar}} \cdot \frac{l_{emb3}}{\phi_{f\_bar}} + 340\right) \geq f_{fu} \\ \left[\frac{\sqrt{f'_c \cdot \text{psi}}}{\alpha_{Neg3}} \cdot \left(13.6 \cdot \frac{l_{emb3}}{\phi_{f\_bar}} + \frac{C_{b3}}{\phi_{f\_bar}} \cdot \frac{l_{emb3}}{\phi_{f\_bar}} + 340\right)\right] \text{otherwise} \end{vmatrix}$$

$$f_{fd13} = 80 \cdot \text{ksi}$$

$$\text{CheckFailure3} := \begin{vmatrix} \text{"Bar ultimate strength" if } f_{fd11} \geq f_{fr1} \\ \text{"Bond strength" otherwise} \end{vmatrix}$$

$$\text{CheckFailure3} = \text{"Bond strength"}$$

The cross section of interest is not a bond critical section and adequate moment capacity can be achieved.

6.6.4 Ultimate bending moment diagram—Exterior bay

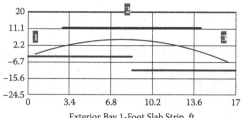

• Ultimate bending moment diagram - Exterior bay

Exterior Bay 1-Foot Slab Strip, ft

Section **1**	Section **2**	Section **3**
$M_{u1} = 2.9 \cdot \text{kip·ft}$	$M_{u2} = 5 \cdot \text{kip·ft}$	$M_{u3} = 7 \cdot \text{kip·ft}$
$\phi_{b1\_bond} \cdot M_{n1b1} = 4.1 \cdot \text{kip·ft}$	$\phi_{b2} \cdot M_{n2} = 11.4 \cdot \text{kip·ft}$	$\phi_{b3} \cdot M_{n3} = 11.4 \cdot \text{kip·ft}$

6.6.5 Ultimate bending moment diagram—Interior bay

This diagram shows that, for the interior bay, the same FRP reinforcement design adopted for the exterior bay can be used.

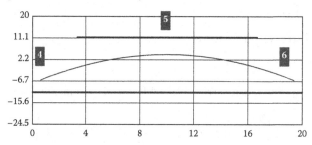

Interior bay 1-foot slab strip, ft

Section **4**	Section **5**	Section **6**
$M_{u4} = 6.4 \cdot \text{kip·ft}$	$M_{u5} = 4.4 \cdot \text{kip·ft}$	$M_{u6} = 6.4 \cdot \text{kip·ft}$
$\phi_{b3} \cdot M_{n3} = 11.4 \cdot \text{kip·ft}$	$\phi_{b2} \cdot M_{n2} = 11.4 \cdot \text{kip·ft}$	$\phi_{b3} \cdot M_{n3} = 11.4 \cdot \text{kip·ft}$

6.7 STEP 5—CHECK CREEP-RUPTURE STRESS

6.7.1 Case 1—Exterior support

The stress level in the FRP reinforcement for checking creep rupture failure is evaluated considering the total unfactored dead loads and the sustained portion of the live load (20% of the total live load).

Bending moment due to dead load plus 20% of live load is

$$M_{1\_creep} := M_{s1} \cdot \frac{w_D + 0.20 w_L}{w_S} = 1.2 \cdot ft \cdot kip$$

Ratio of modulus of elasticity of bars to modulus of elasticity of concrete is

$$n_f = 1.4$$

Ratio of depth of neutral axis to reinforcement depth, calculated per Equation (8-12) in ACI 440.1R-06, is

$$k_1 := \sqrt{2_{\rho_{f1}} \cdot n_f + \left(\rho_{f1} \cdot n_f\right)^2} - \rho_{f1} \cdot n_f = 0.108$$

The tensile stress in the FRP is

$$f_{f1\_creep} := \frac{M_{1\_creep}}{A_{f1} . d_{f1} . \left(1 - \frac{k_1}{3}\right)} = 5.6 \cdot ksi$$

$$Check\_Creep1 := \begin{vmatrix} \text{"OK" if } f_{f1\_creep} \le f_{f\_creep} & (f_{f\_creep} = 16 \cdot ksi) \\ \text{"Not good" otherwise} \end{vmatrix}$$

$$Check\_Creep1 = \text{"OK"}$$

The strain distribution is shown next.

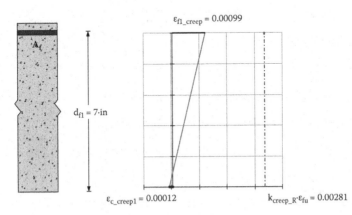

$$\varepsilon_{f1\_creep} = 0.00099$$

$$d_{f1} = 7 \cdot in$$

$$\varepsilon_{c\_creep1} = 0.00012 \qquad k_{creep\_R} \cdot \varepsilon_{fu} = 0.00281$$

Navier's equation is applicable because the maximum concrete stress is smaller than $0.45f'_c$:

$$E_c \cdot \varepsilon_{c\_creep1} = 488 \cdot psi < 0.45 f'_c = 2250 \cdot psi$$

6.7.2 Case 2—Midspan

Bending moment due to dead load plus 20% of live load is

$$M_{2\_creep} := M_{s2} \cdot \frac{w_D + 0.20 w_L}{w_S} = 2.1 \cdot ft \cdot kip$$

Ratio of modulus of elasticity of bars to modulus of elasticity of concrete is

$$n_f = 1.4$$

Ratio of depth of neutral axis to reinforcement depth, calculated per Equation (8-12), is

$$k_2 := \sqrt{2_{\rho_{f2}} \cdot n_f + \left(\rho_{f2} \cdot n_f\right)^2} - \rho_{f2} \cdot n_f = 0.108$$

The tensile stress in the FRP is computed, conservatively, assuming one single layer of reinforcement, is

$$f_{f2\_creep} := \frac{M_{2\_creep}}{A_{f2} \cdot d_{f2} \cdot \left(1 - \dfrac{k_2}{3}\right)} = 9.7.ksi$$

$$\text{Check\_Creep2} := \begin{vmatrix} \text{"OK" if } f_{f2\_creep} \le f_{f\_creep} & (f_{f\_creep} = 16 \cdot ksi) \\ \text{"Not good" otherwise} \end{vmatrix}$$

$$\text{Check\_Creep2} = \text{"OK"}$$

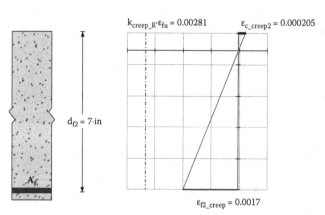

Navier's equation is applicable because the maximum concrete stress is smaller than $0.45f_c'$:

$$E_c \cdot \varepsilon_{c\_creep2} = 836 \cdot psi < 0.45 f_c' = 2250 \cdot psi$$

6.7.3 Case 3—Interior support

Bending moment due to dead load plus 20% of live load is

$$M_{3\_creep} := M_{s3} \cdot \frac{w_D + 0.20 w_L}{w_S} = 3 \cdot ft \cdot kip$$

Ratio of modulus of elasticity of bars to modulus of elasticity of concrete is

$$n_f = 1.4$$

Ratio of depth of neutral axis to reinforcement depth, calculated per Equation (8-12) of ACI 440.1R-06, is

$$k_3 := \sqrt{2\rho_{f3} \cdot n_f + (\rho_{f3} \cdot n_f)^2} - \rho_{f3} \cdot n_f = 0.108$$

The tensile stress in the FRP is

$$f_{f3\_creep} := \frac{M_{3\_creep}}{A_{f3} \cdot d_{f3} \cdot \left(1 - \dfrac{k_3}{3}\right)} = 13.5 \cdot ksi$$

$$Check\_Creep3 := \left| \begin{array}{l} \text{``OK'' if } f_{f3\_creep} \leq f_{f\_creep} \quad (f_{f\_creep} = 16 \cdot ksi) \\ \text{``Not good'' otherwise} \end{array} \right.$$

$$Check\_Creep3 = \text{``OK''}$$

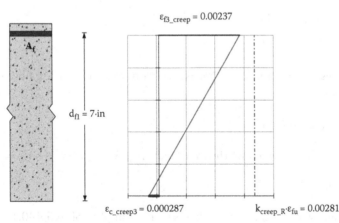

$\varepsilon_{f3\_creep} = 0.00237$

$d_{f1} = 7 \cdot in$

$\varepsilon_{c\_creep3} = 0.000287$ $k_{creep\_R} \cdot \varepsilon_{fu} = 0.00281$

Navier's equation is applicable because the maximum concrete stress is smaller than $0.45 f'_c$:

$$E_c \cdot \varepsilon_{c\_creep3} = 1171 \cdot psi < 0.45 f'_c = 2250 \cdot psi$$

6.8 STEP 6—CHECK CRACK WIDTH

The crack width is checked using Equation (8-9) of ACI 440.1R-06. A crack width limit, w_{lim}, of 0.028 in. is used for interior exposure. The following example is limited to the exterior bay of the slab strip.

6.8.1 Case I—Exterior support

Ratio of modulus of elasticity of bars to modulus of elasticity of concrete is

$n_f = 1.4$

Ratio of depth of neutral axis to reinforcement depth, calculated per Equation (8-12) is

$k_1 = 0.108$

Tensile stress in GFRP under service loads is

$$f_{fs1} := \frac{M_{s1}}{A_{f1} \cdot d_{f1} \cdot \left(1 - \dfrac{k_1}{3}\right)} = 9.4 \cdot ksi$$

Ratio of distance from neutral axis to extreme tension fiber to distance from neutral axis to center of tensile reinforcement is

$$\beta_{11} := \frac{t_{slab} - k_1 \cdot d_{f1}}{d_{f1} \cdot (1 - k_1)} = 1.16$$

Thickness of concrete cover measured from extreme tension fiber to center of bar is

$d_{c1} := t_{slab} - d_{f1} = 1 \cdot in.$

Bond factor (provided by the manufacturer) is

$k_b = 0.9$

The crack width under service loads (Equation (8-9) of ACI 440.1R-06) is

$$w_1 := 2 \frac{f_{fs1}}{E_f} \beta_{11} \cdot k_b \cdot \sqrt{d_{c1}^2 + \left(\frac{s_{f\_bar1}}{2}\right)^2} = 0.011 \cdot in.$$

The crack width limit for the selected exposure is

$w_{lim} = 0.028 \cdot in.$

$$\text{Check\_Crack1} := \begin{vmatrix} \text{"OK" if } w_1 \leq w_{lim} \\ \text{"Not good" otherwise} \end{vmatrix}$$

Check_Crack1 = "OK"

The maximum recommended bar spacing to limit cracking based on Ospina and Bakis [2] is

$$s_{Ospina\_1} := \min\left(1.2 \cdot \frac{E_f \cdot w_{lim}}{f_{fs1} \cdot k_b} - 2.5 \cdot c_c, \, 0.95 \cdot \frac{E_f \cdot w_{lim}}{f_{fs1} \cdot k_b}\right) = 17.9 \cdot in.$$

$$\text{Check\_SpacingOspina1} := \begin{vmatrix} \text{"OK" if } s_{f\_bar1} \leq s_{Ospina\_1} \\ \text{"Not good" otherwise} \end{vmatrix}$$

Check_SpacingOspina1 = "OK"

6.8.2 Case 2—Midspan

Ratio of modulus of elasticity of bars to modulus of elasticity of concrete is

$n_f = 1.4$

Ratio of depth of neutral axis to reinforcement depth, calculated per Equation (8-12), is

$k_2 = 0.108$

Tensile stress in GFRP under service loads is

$$f_{fs2} := \frac{M_{s2}}{A_{f2} \cdot d_{f2} \cdot \left(1 - \dfrac{k_2}{3}\right)} = 16.2 \cdot ksi$$

Ratio of distance from neutral axis to extreme tension fiber to distance from neutral axis to center of tensile reinforcement is

$$\beta_{12} := \frac{t_{slab} - k_2 \cdot d_{f2}}{d_{f2} \cdot (1 - k_2)} = 1.16$$

Thickness of concrete cover measured from extreme tension fiber to center of bar is

$$d_{c2} := t_{slab} - d_{f2} = 1 \cdot in.$$

Bond factor (provided by the manufacturer) is

$$k_b = 0.9$$

The crack width under service loads (Equation (8-9) of ACI 440.1R-06) is

$$w_2 := 2\frac{f_{fs2}}{E_f}\beta_{12} \cdot k_b \cdot \sqrt{d_{c2}^2 + \left(\frac{s_{f\_bar2}}{2}\right)^2} = 0.019 \cdot in.$$

The crack width limit for the selected exposure is

$$w_{lim} = 0.028 \cdot in.$$

$$\text{Check\_Crack2} := \begin{vmatrix} \text{"OK" if } w_2 \leq w_{lim} \\ \text{"Not good" otherwise} \end{vmatrix}$$

$$\text{Check\_Crack2} = \text{"OK"}$$

The maximum recommended bar spacing to limit cracking based on Ospina and Bakis [2] is

$$S_{Ospina\_2} := \min\left(1.2 \cdot \frac{E_f \cdot w_{lim}}{f_{fs2} \cdot k_b} - 2.5 \cdot c_c, 0.95 \cdot \frac{E_f \cdot w_{lim}}{f_{fs2} \cdot k_b}\right) = 10.4 \cdot in.$$

$$\text{Check\_SpacingOspina2} := \begin{vmatrix} \text{"OK" if } s_{f\_bar2} \leq s_{Ospina\_2} \\ \text{"Not good" otherwise} \end{vmatrix}$$

$$\text{Check\_SpacingOspina2} = \text{"OK"}$$

6.8.3 Case 3—Interior support

Ratio of modulus of elasticity of bars to modulus of elasticity of concrete is

$$n_f = 1.4$$

Ratio of depth of neutral axis to reinforcement depth, calculated per Equation (8-12), is

$$k_3 = 0.108$$

Tensile stress in GFRP under service loads is

$$f_{fs3} := \frac{M_{s3}}{A_{f3} \cdot d_{f3} \cdot \left(1 - \dfrac{k_3}{3}\right)} = 22.6 \cdot ksi$$

Ratio of distance from neutral axis to extreme tension fiber to distance from neutral axis to center of tensile reinforcement is

$$\beta_{13} := \frac{t_{slab} - k_3 \cdot d_{f3}}{d_{f3} \cdot (1 - k_3)} = 1.16$$

Thickness of concrete cover measured from extreme tension fiber to center of bar is

$$d_{c3} := t_{slab} - d_{f3} = 1 \cdot in.$$

Bond factor (provided by the manufacturer) is

$$k_b = 0.9$$

The crack width under service loads (Equation (8-9) of ACI 440.1R-06) is

$$w_3 := 2 \frac{f_{fs3}}{E_f} \beta_{13} \cdot k_b \cdot \sqrt{d_{c3}^2 + \left(\frac{s_{f\_bar3}}{2}\right)^2} = 0.026 \cdot in.$$

The crack width limit for the selected exposure is

$$w_{lim} = 0.028 \cdot in.$$

$$\text{Check\_Crack3} := \begin{vmatrix} \text{``OK'' if } w_3 \leq w_{lim} \\ \text{``Not good'' otherwise} \end{vmatrix}$$

$$\text{Check\_Crack3} = \text{``OK''}$$

The maximum recommended bar spacing to limit cracking based on Ospina and Bakis [2] is

$$s_{Ospina\_3} := \min\left(1.2 \cdot \frac{E_f \cdot w_{lim}}{f_{fs3} \cdot k_b} - 2.5 \cdot c_c, \; 0.95 \cdot \frac{E_f \cdot w_{lim}}{f_{fs3} \cdot k_b}\right) = 7.5 \cdot in.$$

$$\text{Check\_SpacingOspina3} := \begin{vmatrix} \text{``OK'' if } s_{f\_bar3} \leq s_{Ospina\_3} \\ \text{``Not good'' otherwise} \end{vmatrix}$$

$$\text{Check\_SpacingOspina3} = \text{``OK''}$$

6.9 STEP 7—CHECK MAXIMUM MIDSPAN DEFLECTION

The service bending moment diagram for the exterior bay is shown next.

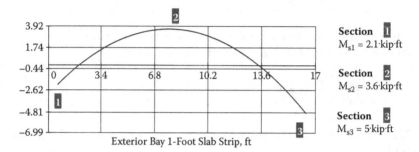

Section 1
$M_{s1} = 2.1 \cdot \text{kip·ft}$

Section 2
$M_{s2} = 3.6 \cdot \text{kip·ft}$

Section 3
$M_{s3} = 5 \cdot \text{kip·ft}$

Exterior Bay 1-Foot Slab Strip, ft

Exterior bay 1-foot slab strip:
The maximum allowable deflection is

$$\Delta_{\lim} := \frac{l_1}{480} = 0.425 \cdot \text{in.}$$

Preliminary calculations: The depth of the neutral axis of the gross section is

$$d_g := \frac{t_{slab}}{2} = 4 \cdot \text{in.}$$

The gross moment of inertia is

$$I_g := \frac{b \cdot t_{slab}^3}{12} = 512 \cdot \text{in}^4$$

The negative cracking moment is

$$M_{crNeg} := \frac{f_r \cdot I_g}{d_g} = 6 \cdot \text{kip} \cdot \text{ft}$$

The positive cracking moment is

$$M_{crPos} := \frac{f_r \cdot I_g}{t_{slab} - d_g} = 6 \cdot \text{kip} \cdot \text{ft}$$

Cracked moment of inertia: The cracked moment of inertia, I_{cr}, is computed per Equation (8-11) of ACI 440.1R-06 at the following locations:

6.9.1 Case 1—Exterior support

$$I_{cr1} := \frac{b \cdot d_{f1}^3}{3} k_1^3 + n_f \cdot A_{f1} \cdot d_{f1}^2 \cdot (1 - k_1)^2 = 23 \cdot in^4$$

The reduction coefficient related to the reduced tension stiffening is exhibited in the FRP reinforced members, computed per Equation (8-13b) of ACI 440.1R-06:

$$\beta_{d1} := \frac{1}{5} \cdot \left(\frac{\rho_{f1}}{\rho_{fb1}} \right) = 0.125$$

6.9.2 Case 2—Midspan

$$I_{cr2} := \frac{b \cdot d_{f2}^3}{3} k_2^3 + n_f \cdot A_{f2} \cdot d_{f2}^2 \cdot (1 - k_2)^2 = 23 \cdot in^4$$

$$\beta_{d2} := \frac{1}{5} \cdot \left(\frac{\rho_{f2}}{\rho_{fb2}} \right) = 0.125$$

6.9.3 Case 3—Interior support

$$I_{cr3} := \frac{b \cdot d_{f3}^3}{3} k_3^3 + n_f \cdot A_{f3} \cdot d_{f3}^2 \cdot (1 - k_3)^2 = 23 \cdot in^4$$

$$\beta_{d3} := \frac{1}{5} \cdot \left(\frac{\rho_{f3}}{\rho_{fb3}} \right) = 0.125$$

Average effective moment of inertia: The average effective moment of inertia, I_e, is computed as indicated in Section 9.5.2.4 of ACI 318-11. Branson's equation is used (Equation 9-7 of ACI 318-11).

The bending moment at midspan due to service loads is

$$M_{sPos} := M_{s2} = 3.6 \cdot kip \cdot ft$$

The bending moment at the continuous end due to service loads is

$$M_{sNeg} := M_{s3} = 5 \cdot kip \cdot ft$$

The value of I_e at midspan is

$$I_{e2} := \left(\frac{M_{crPos}}{M_{sPos}}\right)^3 \beta_{d2} \cdot I_g + \left[1 - \left(\frac{M_{crPos}}{M_{sPos}}\right)^3\right] I_{cr2} = 186 \cdot in^4$$

The value of I_e at the continuous end is

$$I_{e3} := \left(\frac{M_{crNeg}}{M_{sNeg}}\right)^3 \beta_{d3} \cdot I_g + \left[1 - \left(\frac{M_{crNeg}}{M_{sNeg}}\right)^3\right] I_{cr3} = 83 \cdot in^4$$

The average value of I_e is (as defined in Chapter 4):

$$I_e := 0.85 \cdot I_{e2} + 0.15 \cdot I_{e3} = 170 \cdot in.^4$$

Deflection at midspan: Calculate the moment at midspan due to service load on a simply supported beam, M_o:

$$M_o := \frac{w_S \cdot l_1^2}{8} = 7.2 \cdot ft \cdot kip$$

The maximum deflection under service loads is

$$\Delta_{SL} := \frac{5 \cdot M_o \cdot l_1^2}{48 E_c \cdot I_e} - \left(M_{s1} + M_{s3}\right) \cdot \frac{l_1^2}{16 E_c \cdot I_e} = 0.221 \cdot in.$$

Deflection due to dead loads only is

$$\Delta_{DL} := \Delta_{SL} \cdot \frac{w_D}{w_S} = 0.11 \cdot in.$$

Deflection due to live loads only is

$$\Delta_{LL} := \Delta_{SL} \cdot \frac{w_L}{w_S} = 0.111 \cdot in.$$

Multiplier for time-dependent deflections at 5 years (ACI 318-11) is

$$\xi := 2$$

The reduction parameter (Equation 8-14b of ACI 440.1R-06) is

$\lambda := 0.60\xi := 1.2$

Long-term deflection is

$\Delta_{LT} := \Delta_{LL} + \lambda \cdot (\Delta_{DL} + 0.20 \cdot \Delta_{LL}) = 0.27 \cdot \text{in.}$

$\Delta_{lim} = 0.425 \cdot \text{in.}$

$$\text{Check\_}\Delta_{LT} := \begin{vmatrix} \text{"OK" if } \Delta_{LT} \leq \Delta_{lim} \\ \text{"Not good" otherwise} \end{vmatrix}$$

$\text{Check\_}\Delta_{LT} = \text{"OK"}$

The midspan deflection diagram follows.

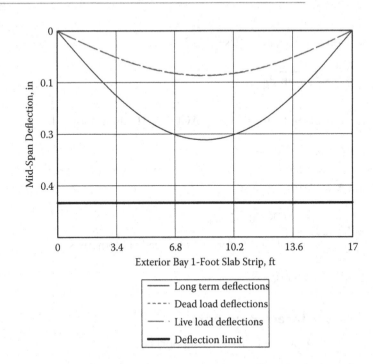

Note: deflections due to dead loads and the ones due to live loads are identical.

6.10 STEP 8—CHECK SHEAR CAPACITY

6.10.1 Case 1—Exterior support

The maximum shear at the face of the supports is

$$V_{ul} = 2.209 \cdot kip$$

The concrete shear capacity, V_c, can be calculated per Equation (9-1) of ACI 440.1R-06, where k is the ratio of depth of neutral axis to reinforcement depth, calculated per Equation (8-12):

$$V_{C1} := 5\sqrt{f_c' \cdot psib} \cdot (k_1 \cdot d_{f1}) = 3.2 \cdot kip$$

As discussed in Chapter 4, the following is verified:

$$\text{Check\_ConcreteShear1} := \begin{vmatrix} \text{"OK" if } V_{C1} \geq 0.8\sqrt{f_c' \cdot psib} \cdot d_{f1} & \left(0.8\sqrt{f_c' \cdot psib} \cdot d_{f1} = 4.8 \cdot kip\right) \\ \text{"Use } V_{cmin}\text{" otherwise} \end{vmatrix}$$

$$\text{Check\_ConcreteShear1} = \text{"Use } V_{cmin}\text{"}$$

$$V_{cmin1} := 0.8\sqrt{f_c' \cdot psib} \cdot d_{f1}$$

The shear reduction factor given by ACI 440.1R-06 is adopted:

$$\phi_v := 0.75$$

$$\phi_v \, V_{cmin1} = 3.56 \cdot kip$$

$$\text{Check\_Shear1} := \begin{vmatrix} \text{"OK" if } \phi_v \cdot V_{cmin1} \geq V_{ul} \\ \text{"Shear reinforcement is needed" otherwise} \end{vmatrix}$$

$$\text{Check\_Shear1} = \text{"OK"}$$

6.10.2 Case 2—Interior support

The maximum shear at the face of the supports is

$$\max(V_{u3}, V_{u4}) = 3.021 \cdot kip$$

The concrete shear capacity, V_c, can be calculated per Equation (9-1) of ACI 440.1R-06, where k is the ratio of depth of neutral axis to reinforcement depth, calculated per Equation (8-12):

$$V_{C3} := 5\sqrt{f_c \cdot psib} \cdot (k_3 \cdot d_{f3}) = 3.2 \cdot kip$$

As discussed in Chapter 4, the following is verified:

$$\text{Check\_ConcreteShear3} := \left|\begin{array}{l} \text{"OK" if } V_{C3} \geq 0 \cdot 8\sqrt{f_c' \cdot psib}.d_{f3} \\ \qquad \left(0 \cdot 8\sqrt{f_c' \cdot psib} \cdot d_{f3} = 4.8 \cdot kip\right) \\ \text{"Use } V_{cmin} \text{" otherwise} \end{array}\right.$$

$$\text{Check\_ConcreteShear1} = \text{"Use } V_{cmin}\text{"}$$

$$V_{cmin3} := 0.8\sqrt{f_c' \cdot psib} \cdot d_{f3}$$

The shear reduction factor given by ACI 440.1R-06 is adopted:

$$\phi_v = 0.75$$

$$\phi_v \cdot V_{cmin3} = 3.56 \cdot kip$$

$$\text{Check\_ConcreteShear3} := \left|\begin{array}{l} \text{"OK" if } \phi_v \cdot V_{cmin3} \geq \max\left(V_{u3}, V_{u4}\right) \\ \text{"Shear reinforcement is needed" otherwise} \end{array}\right.$$

$$\text{Check\_Shear3} = \text{"OK"}$$

6.11 STEP 9—DESIGN THE FRP REINFORCEMENT FOR SHRINKAGE AND TEMPERATURE

The recommended minimum FRP reinforcement ratio, $\rho_{f,smin}$, to limit cracks due to shrinkage and temperature is given by Equation (10-1) of ACI 440.1R-06. $\rho_{f,s}$ is recommended not to be smaller than 0.0014 but need not be larger than 0.0036.

Equation (10-1) of ACI 440.1R-06 is

$$\rho_{fts\_des} := \min\left(0.0018 \cdot \frac{60000 psi}{f_{fu}} \cdot \frac{29000 ksi}{E_f}, 0.0036\right) = 0.0036$$

The corresponding minimum area of GFRP reinforcement is

$$A_{fts-des} := \rho_{fts-des} \cdot b \cdot t_{slab} = 0.346 \cdot in.^2$$

The required number of no. 4 bars is

$$N_{fs\_des} := \frac{A_{fts\_des}}{A_{f\_bar}} = 1.76$$

The corresponding required spacing is

$$s_{fs\_des} := \frac{b - N_{fs\_des} \cdot \phi_{f\_bar}}{N_{fs\_des}} = 6.318 \cdot in.$$

The following bar spacing is selected:

$$s_{f-ts} := 6 \ in.$$

$$Check\_s_{f\_ts} := \begin{vmatrix} \text{``OK''} \ if \ s_{f\_ts} \leq \min(3 \cdot t_{slab}, 12 \ in) \\ \text{``Not good''} \ otherwise \end{vmatrix}$$

$$Check\_s_{f\_ts} = \text{``OK''}$$

The area of GFRP temperature and shrinkage reinforcement is

$$A_{f\_ts} := \frac{b}{s_{f\_ts}} \cdot A_{f\_bar} = 0.39 \cdot in^2$$

The FRP reinforcement ratio (Equation (8-2) of ACI 440.1R-06) is

$$\rho_{f\_ts} := \frac{A_{f\_ts}}{b \cdot t_{slab}} = 0.00409$$

No. 4 bars spaced 6 in. center to center are necessary if a single layer is used. For convenience and symmetry, top and bottom layers are used here with No. 4 bars spaced 12 in. center to center.

6.12 STEP 10—FIRE SAFETY CHECK FOR FLEXURAL STRENGTH PER NIGRO ET AL. (2014)

The fire safety check for bending moment capacity is performed following the methodology proposed by Nigro et al. (2014) and discussed in Section 4.12. The interior bay only is considered. It is assumed that the slab is unprotected when exposed to fire on the side of the fibers under tension. For the purpose of the calculations, a 60-minute fire exposure time is assumed.

Step A. Estimate the FRP bar temperature T_f (in degrees Celsius) in the event of a fire for a fire exposure time t (minutes).

$t := 60$ Fire exposure time in minutes

$c_c = 0.75 \cdot in$ Longitudinal bar clear concrete cover

Nigro et al. recommend the following expression for t = 60 min and for a concrete cover of 0.787 in.

$T_f := -4586.1 + 4221.2 \cdot t^{0.0470} = 531$ FRP bar temperature in degrees Celsius

Step B. Estimate the reduction factors of the FRP tensile strength and modulus of elasticity at the temperature T_f computed in Step A.

The following expressions are recommended to estimate the tensile strength and modulus reduction factors for GFRP bars at the temperature computed in Step A.

$$\rho_{ft} := \frac{0.05}{0.05 + 8.0 \cdot 10^{-11} \cdot T_f^{3.55}} = 0.117$$ FRP bar tensile strength reduction factor at temperature T_f (in degrees celsius)

$$\rho_{ET} := \frac{0.28}{0.28 + 6.0 \cdot 10^{-12} \cdot T_f^{4.3}} = 0.082$$ FRP bar tensile strength reduction factor at temperature T_f (in degrees celsius)

Step C. Estimate the embedment length of a bar with a temperature higher than 50 degrees Celsius at a fire exposure time t.

Nigro et al. propose the following equation to estimate the reduced embedment length of an FRP bar at 50 degrees Celsius for a concrete cover of 0.787 in. and an exposure time t (equal to 60 min in this case).

$l_{df\_50C} := (-23.43 + 15.38 \cdot t^{0.4660}) mm = 3.158 \cdot in$

Step D. Estimate the ultimate strength of the end anchorage in fire conditions at time t.

Nigro et al. propose the following equation to estimate the ultimate strength of the end anchorage in a zone with a temperature of at least 50 degrees Celsius.

$$f_{f\_50C} := \frac{l_d - l_{df\_50C}}{0.1 \cdot \phi_{f\_bar}} \, ksi = 456.8 \cdot ksi$$

Step E. Estimate the maximum bar stress.

The maximum usable FRP bar stress can be computed as the smaller of the tensile strength at a temperature T_f (computed in Step A) and the end anchorage strength at 50 degrees Celsius.

$$f_{f\_fire} := \min(\rho_{fT} \cdot f_{fu}, f_{f\_50C}) = 9 \cdot ksi$$

The reduced FRP tensile modulus of elasticity can be computed as follows using the reduction factor computed in Step B.

$$E_{f\_fire} := \rho_{ET} \cdot E_f = 468 \cdot ksi$$

The ultimate FRP tensile strain can, therefore, be computed as follows.

$$\varepsilon_{f\_fire} := \frac{f_{f\_fire}}{E_{f\_fire}} = 0.02$$

Step F. Compute the reduced bending moment capacity at a fire exposure time of 120 min.

The following stress - strain relationship is assumed for the concrete in compression:

$$\sigma_{c\_fire}(\varepsilon_c) := \left| \begin{array}{ll} f_c' \cdot \dfrac{3 \cdot \left(\dfrac{\varepsilon_c}{0.0025}\right)}{2 + \left(\dfrac{\varepsilon_c}{0.0025}\right)^3} & \text{if } \varepsilon_c \leq 0.0025 \\[6mm] f_c' \cdot \dfrac{0.02 - \varepsilon_c}{0.02 - 0.0025} & \text{if } 0.0025 \leq \varepsilon_c \wedge \varepsilon_c \leq 0.02 \end{array} \right.$$

The FRP design properties are substituted with the reduced FRP properties estimated in Step E. It is assumed that the failure is controlled by the FRP failure.

The effective concrete compressive strain at failure as a function of the neutral axis depth, x, is:

$$\varepsilon_{c2\_fire}(x,y) := \frac{\varepsilon_{f\_fire}}{d_{f2} - x} \cdot y$$

The effective tensile strain in the FRP reinforcement at failure is:

$\varepsilon_{f2\_fire} := \varepsilon_{f\_fire}$

The compressive force in the concrete as a function of the neutral axis depth, x, is:

$$C_{c2\_fire}(X) := b \cdot \int_{0in}^{\frac{(d_{f2}-x)\cdot 0.0025}{\varepsilon_{f\_fire}}} f_c' \cdot \frac{3 \cdot \left(\frac{\varepsilon_{c2\_fire}(x,y)}{0.0025}\right)}{2 + \left(\frac{\varepsilon_{c2\_fire}(x,y)}{0.0025}\right)^3} dy$$

$$+ b \cdot \int_{\frac{(d_{f2}-x)\cdot 0.0025}{\varepsilon_{f\_fire}}}^{x} f_c' \cdot \frac{0.02 - \varepsilon_{c2\_fire}(x,y)}{0.02 - 0.0025} dy$$

The tensile force in the FRP reinforcement is:

$T_{f2\_fire}(x) := A_{f2} \cdot E_{f\_fire} \cdot \varepsilon_{f2\_fire}$

The neutral axis depth, c_u, can be computed by solving the equation of equilibrium $C_c - T_f = 0$:

First guess: $x_{02\_fire} := 0.1 d_{f2}$ Given: $f_{2\_fire}(x) := C_{c2\_fire}(x) - T_{f2\_fire}(x)$
$c_{u2\_fire} := \text{root} \left(f_{2\_fire}(x_{02\_fire}), x_{02\_fire}\right)$

The neutral axis depth is:

$c_{u2\_fire} = 0.381 \cdot in$

The contribution of the concrete in compression is:

with: $\dfrac{(d_{f2} - c_{u2}) \cdot 0.0025}{\varepsilon_{f\_fire}} = 0.767 \cdot in$

$$M_{n2\_fire\_Conc} := b \cdot \int_{0in}^{0.767in} y \cdot f_c' \cdot \frac{3 \cdot \left(\frac{\varepsilon_{c2\_fire}(c_{u2},y)}{0.0025}\right)}{2 + \left(\frac{\varepsilon_{c2\_fire}(c_{u2},y)}{0.0025}\right)^3} dy$$

$$+ b \cdot \int_{0.767in}^{c_{u2\_fire}} c_{u2} \cdot f_c' \cdot \frac{0.02 - \varepsilon_{c2\_fire}(c_{u2},y)}{0.02 - 0.0025} dy$$

The contribution of the FRP reinforcement is:

$$M_{n2\_fire\_FRP} := T_{f2\_fire}\,(c_{u2\_fire}) \cdot (d_{f2} - c_{u2\_fire})$$

The total nominal bending moment is:

$$M_{n2\_fire} := M_{n2\_fire\_Conc} + M_{n2\_fire\_FRP} = 1.49 \cdot kip \cdot ft$$

The bending moment under service load at mid-span as computed in Step 3 is:

$$M_{s2} = 3.57 \cdot kip \cdot ft$$

$$\text{Check\_Bending Moment\_Fire} := \begin{cases} \text{"OK" if } M_{n2\_fire} \geq M_{s2} \\ \text{"Not good" otherwise} \end{cases}$$

Check_Bending Moment_Fire = "Not good"

The unprotected slab is not adequate to carry the service loads in the event of a fire for an exposure time of 60 minutes. A solution could be, for example, to increase the concrete cover to 1.25 inches. In this way, the total nominal bending moment would also increase to 3.9 kip-ft and exceed the service demand of 3.57 kip-ft.

REFERENCES

1. H. Jawahery Zadeh and A. Nanni. Reliability analysis of concrete beams internally reinforced with FRP bars. *ACI Structural Journal* 110 (6): 1023–1032 (2013).
2. C. E. Ospina and C. E. Bakis. Indirect flexural crack control of concrete beams and one-way slabs reinforced with FRP bars. *Proceedings of the 8th International Symposium on Fiber Reinforced Polymer Reinforcement for Concrete Structures, FRPRCS-8*, ed. T. C. Triantafillou, University of Patras, Greece, CD ROM (2007).
3. E. Nigro, G. Cefarelli, A. Bilotta, G. Manfredi, E. Cosenza. Guidelines for flexural resistance of FRP reinforced concrete slabs and beams in fire. *Composites Part B: Engineering*, 58: 103–112 (2014).

Chapter 7

Design of a T-beam

7.1 INTRODUCTION

The floor plan of a two-story medical facility building is shown in Figure 7.1. The column spacing is dictated by the size of the equipment that occupies the ground floor. The second floor system is a one-way RC slab spanning along the east–west direction. The building is located in a region of low seismicity. This example describes the procedure to design beam AB-3. Loading of each floor consists of the self-weight, a superimposed dead load of 2.5 psf, and a live load of 100 psf.

The design is discussed as a sequence of nine steps as summarized here:

Step 1   Define beam geometry and concrete properties
Step 2   Compute factored loads
Step 3   Compute ultimate and service bending moments and shear forces
Step 4   Design FRP primary reinforcement for bending moment capacity
Step 5   Check creep-rupture stress
Step 6   Check crack width
Step 7   Check maximum midspan deflection
Step 8   Design FRP reinforcement for shear capacity
Step 9   Compute FRP contribution to torsional strength

The beam design is summarized in the next section and discussed in detail in the following ones.

7.2 DESIGN SUMMARY

Considering a T-beam of dimensions 14 in. (b_w) by 28 in. (h), the following loads are considered:

Dead load	2.31 kip/ft
Live load for positive moments	1.29 kip/ft
Live load for negative moments	1.05 kip/ft

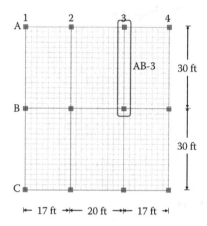

Figure 7.1 Second-floor plan.

Table 7.1 Bending moments and shear forces

| | Bending moment | | | Shear force | |
Section	Moment coefficient	Ultimate moment (kip-ft)	Service moment (kip-ft)	Shear coefficient	Ultimate shear (kip)
Exterior support	1/24	145	111	1.0	62
Midspan	1/14	271	204	0.15	10
Interior support	1/10	349	266	1.15	72

Total factored load for positive moments	4.88 kip/ft
Total factored load for negative moments	4.49 kip/ft
Service load for positive moments	3.64 kip/ft
Service load for negative moments	3.39 kip/ft

Bending moments and shear forces for beam AB-3 are computed at the support and midspan sections, as summarized in Table 7.1.

The design is limited to the first span (beam AB-3) because of symmetry. The required FRP reinforcement for both flexure and shear is given in Table 7.2, while Table 7.3 shows a summary of demands and capacities at critical sections.

The final bar layout is selected to optimize rebar production time and construction effort. Bar layout and typical details are shown in Figure 7.2(a) and 7.2(b).

Table 7.4 is provided to convert US customary units to the SI system.

Table 7.2 Beam geometry and reinforcement

Section	Height	Width	Primary reinforcement	Shear reinforcement
Exterior support			Four no. 8	No. 4 @ 3 in.
Midspan	28 in.	14 in.	Eight no. 8 in two layers	No. 4 @ 13 in. (max spacing)
Interior support			Ten No. 8	No. 4 @ 3 in.

Table 7.3 Beam design summary

	Limit state	Section	Demand/computed	Capacity/limit
Ultimate	Flexural strength	Exterior support	145 kip-ft	205 kip-ft
		Midspan	271 kip-ft	504 kip-ft
		Interior support	349 kip-ft	563 kip-ft
	Shear strength	Interior support	72 kip	80 kip
Serviceability	Creep rupture	Exterior support	6.4 ksi	12.8 ksi
		Midspan	12.0 ksi	
		Interior support	12.5 ksi	
	Crack width	Exterior support	0.016 in.	0.028 in.
		Midspan	0.013 in.	
		Interior support	0.021 in.	
	Maximum midspan deflection		0.446 in.	1.0 in.

7.3 STEP I—DEFINE BEAM GEOMETRY AND CONCRETE PROPERTIES

7.3.1 Geometry

The beam has two spans of length l_1 and l_2:

$l_1 := 30$ ft
$l_2 := 30$ ft

The widths of the two adjacent bays are

$l_{t1} := 17$ ft
$l_{t2} := 20$ ft

The slab thickness is

$t_{slab} := 8$ in.

The column width is

$b_{col} := 24$ in.

A clear top and bottom concrete cover, c_c, is

$c_c := 1.5$ in.

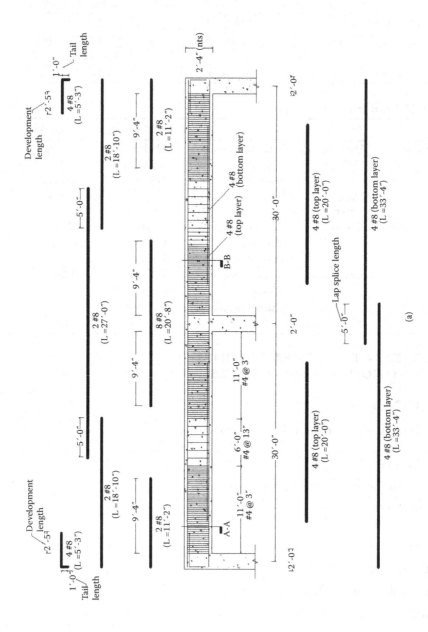

Figure 7.2 (a) Reinforcement layout.

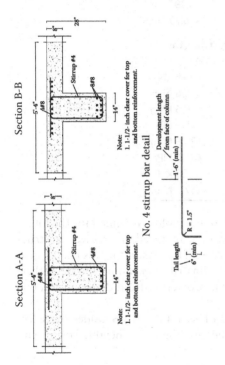

Section A-A

Section B-B

5'-4"

4#8

8"

1"

Stirrup #4

4#8

14"

Note:
1. 1-1/2- inch clear cover for top
and bottom reinforcement.

5'-4"

4#8

8"

28"

Stirrup #4

4#8

14"

Note:
1. 1-1/2- inch clear cover for top
and bottom reinforcement.

No. 4 stirrup bar detail

Development length
from face of column

Tail length

1'-6" (min)

6" (min)

R = 1.5"

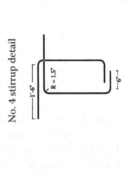

No. 4 stirrup detail

1'-6"

R = 1.5"

6"

(b)

Figure 7.2 (Continued) (b) typical details.

Table 7.4 Conversion table

US customary	SI units
Lengths, areas, section properties	
1 in.	25.4 mm
	0.025 m
1 ft	304.8 mm
	0.305 m
1 in.^{2}	645 mm^2
1 ft^2	0.093 m^2
1 in.^{3}	16,387 mm^3
1 in.^{4}	416,231 mm^4
Forces, pressures, strengths	
1 lbf	4.448 N
1 kip	4.448 kN
1 lbf-ft	1.356 N.m
1 kip-ft	1.356 kN·m
1 psi	6.895 kPa
1 psf	47.88 N/m^2
1 ksi	6.895 MPa
1 ksf	47.88 kN/m^2
1 lbf/ft^3	157.1 N/m^3

Note that the clear cover in this design example is intended for the longitudinal bars and thus is smaller than what is required in steel RC construction where it is measured from the stirrup. The reason for this choice is to stress that cover requirements in GFRP construction can be relaxed if their only purpose is to ensure reinforcement protection from deterioration due to corrosion.

Define beam dimensions: Table 8-2 of ACI 440.1R-06 guides the selection by recommending minimum values for the beam height. For an end bay, the recommended depth is

$$h_{ACI} := \text{round}\left(\frac{l_1}{12 \cdot \text{in}}\right)\text{in} = 30 \cdot \text{in.}$$

It has to be noted that these values are only a starting point for the design. If a different beam height, h, is selected, deflections need to be computed. In this example, a different height is selected:

h := 28 in.

The width of the beam is

b_w := 14 in.

The effective width of the beam flanges is computed according to Section 8.10.2 of ACI 318-11:

$$b_{eff} := \min\left(\frac{l_1}{4}, 8 \cdot t_{slab}, \frac{l_{t1}}{2} + b_w + \frac{l_{t2}}{2}\right) = 64 \cdot in.$$

7.3.2 Concrete properties

The following concrete properties are considered for the design:

$f'_c := 6000 \text{ psi}$ Compressive strength

$\varepsilon_{cu} := 0.003$ Ultimate compressive strain

$\rho_c := 145\dfrac{lbf}{ft^3}$ Density

$E_c := 33\text{psi}^{0.5}\left(\dfrac{\rho_c}{lbf \cdot ft^{-3}}\right)^{1.5} \cdot \sqrt{f'_c}$ Compressive modulus of elasticity

$E_c := 4463 \cdot ksi$ Computed as indicated in ACI 318-11

$f_r := 7.5 \cdot \text{psi}^{0.5} \cdot \sqrt{f'_c}$ Concrete tensile strength

$f_r = 581 \cdot \text{psi}$ Computed as indicated in ACI 318-11

The stress-block factor, β_1, is computed as indicated in ACI 318-11:

$$\beta_1 := \begin{vmatrix} 0.85 & \text{if } f'_c = 4000\text{psi} & = 0.75 \\ 1.05 - 0.05 \cdot \dfrac{f'_c}{1000\text{psi}} & \text{if } 4000\text{psi} < f'_c < 8000\text{psi} \\ 0.65 & \text{otherwise} \end{vmatrix}$$

7.3.3 Analytical approximations of concrete compressive stress–strain curve—Todeschini's model

Compressive strain at peak:

$$\varepsilon_{c0} := \frac{1.71 \cdot f'_c}{E_c} = 0.0023$$

Compressive stress at peak:

$$\sigma''_c := 0.5 \cdot \frac{f'_c}{ssi}$$

Stress–strain curve equation:

$$\sigma_c(\varepsilon_c) := \frac{2 \cdot \sigma''_c \cdot \left(\dfrac{\varepsilon_c}{\varepsilon_{c0}}\right)}{1 + \left(\dfrac{\varepsilon_c}{\varepsilon_{c0}}\right)^2}$$

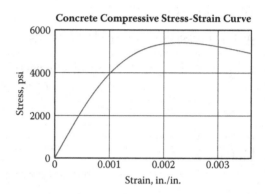

Concrete Compressive Stress-Strain Curve

7.4 STEP 2—COMPUTE FACTORED LOADS

The self-weight is computed considering a concrete density of 145 psf. Other dead loads such as floor cover (0.5 psf) and ceiling (2 psf) are considered. A live load of 100 psf was requested by the owner. The following unfactored uniform loads are considered:

$SW := t_{slab} \cdot \rho_c = 96.7 \cdot psf$	Slab self-weight
$OD := 2.5\ psf$	Other dead loads
$DL := SW + OD = 99.2 \cdot psf$	Total dead load
$LL_o := 100\ psf$	Live load

The ASCE 7-10 design-loading code allows live-load reductions based on tributary areas multiplied by a live-load element factor, $K_{LL} = 2$, to convert the tributary area to an influence area. The live-load element factor is

$$K_{LL} := 2$$

Beams 23-B and 43-B: The tributary area for positive moment is

$$A_{TPos} := (l_{t1} \cdot 0.5 + l_{t2} \cdot 0.5) \cdot l_1 = 555 \cdot ft^2$$

The tributary area for negative moment is

$$A_{TNeg} := (l_{t1} + l_{t2}) \cdot l_1 = 1110 \cdot ft^2$$

The reduced live load for positive moments is

$$LL_{Pos} := LL_o \cdot \left(0.25 + \frac{15}{\sqrt{K_{LL} \cdot A_{TPos} \cdot ft^{-2}}} \right) = 70 \cdot psf$$

The reduced live load for negative moments is

$$LL_{Neg} := LL_o \cdot \left(0.25 + \frac{15}{\sqrt{K_{LL} \cdot A_{TNeg} \cdot ft^{-2}}} \right) = 56.8 \cdot psf$$

The reduced live load cannot be less than 50% of the unreduced live load for members supporting one floor.

$$CheckLL_{Pos} := \begin{vmatrix} \text{"OK"} & \text{if } LL_{Pos} \geq 0.50 \cdot LL_o \\ \text{"Not OK"} & \text{otherwise} \end{vmatrix}$$

$$CheckLL_{Pos} = \text{"OK"}$$

$$CheckLL_{Neg} := \begin{vmatrix} \text{"OK"} & \text{if } LL_{Neg} \geq 0.50 \cdot LL_o \\ \text{"Not OK"} & \text{otherwise} \end{vmatrix}$$

$$CheckLL_{Neg} = \text{"OK"}$$

The dead load per unit length (including the beam's self-weight) is

$$w_D := DL \cdot (l_{t1} \cdot 0.5 + l_{t2} \cdot 0.5) + 1.2 \cdot (h \cdot b_w \cdot \rho_c) = 2.31 \cdot \frac{kip}{ft}$$

The live load per unit length for computing maximum positive moments is

$$w_{LPos} := LL_{Pos} \cdot (l_{t1} \cdot 0.5 + l_{t2} \cdot 0.5) = 1.3 \cdot \frac{kip}{ft}$$

The live load per unit length for computing maximum negative moments is

$$w_{LNeg} := LL_{Neg} \cdot (l_{t1} \cdot 0.5 + l_{t2} \cdot 0.5) = 1.05 \cdot \frac{kip}{ft}$$

The governing load combination for computing the total factored load, w_{TFL}, is load combination Equation (9-2) defined in ASCE 7-10. Total factored load for positive moments is

$$w_{TFLPos} := 1.2 w_D + 1.6 \cdot w_{LPos} = 4.84 \cdot \frac{kip}{ft}$$

Total factored load for negative moments is

$$w_{TFLNeg} := 1.2w_D + 1.6 \cdot w_{LNeg} = 4.45 \cdot \frac{kip}{ft}$$

Total service load for positive moments is

$$w_{SPos} := w_D + w_{LPos} = 3.6 \cdot \frac{kip}{ft}$$

Total service load for negative moments is

$$w_{SNeg} := w_D + w_{LNeg} = 3.36 \cdot \frac{kip}{ft}$$

7.5 STEP 3—COMPUTE BENDING MOMENTS AND SHEAR FORCES

Bending moments and shear forces are determined as indicated in ACI 318-11. The moment coefficients can be used as the beam satisfies the requirements specified in ACI 318-11. In fact, there are two spans; the ratio of the longer clear span to the shorter clear span is less than 1.2, the loads are uniformly distributed, the unfactored live load does not exceed three times the unfactored dead load, and the members are prismatic. Clear span values, l_n, are computed considering a column width of 24 in.:

$$l_n := l_1 - b_{col} = 28 \cdot ft$$

Bending moments

Exterior support

$$C_{mNeg1} := \frac{1}{24}$$ Moment coefficient

$$M_{uNeg1} := C_{mNeg1} \cdot W_{TFLNeg} \cdot l_n^2 = 145 \cdot ft \cdot kip$$ Ultimate bending moment

$$M_{s1} := C_{mNeg1} \cdot W_{SNeg} \cdot l_n^2 = 145 \cdot ft \cdot kip$$ Service bending moment

Midspan

$$C_{mPos} := \frac{1}{14}$$ Moment coefficient

$$M_{uPos} := C_{mPos} \cdot W_{TFLPos} \cdot l_n^2 = 271 \cdot ft \cdot kip$$ Ultimate bending moment

$$M_{s2} := C_{mPos} \cdot W_{SPos} \cdot l_n^2 = 202 \cdot ft \cdot kip$$ Service bending moment

Interior support

$$C_{mNeg2} := \frac{1}{10}$$ Moment coefficient

$M_{uNeg2} := C_{mNeg2} \cdot W_{TFLNeg} \cdot 1_n{}^2 = 349 \cdot ft \cdot kip$ Ultimate bending moment

$M_{s3} := C_{mNeg2} \cdot W_{SNeg} \cdot 1_n{}^2 = 263 \cdot ft \cdot kip$ Service bending moment

Shear forces

Exterior support

$C_{v1} := 1$ Moment coefficient

$V_{u1} := C_{v1} \cdot w_{TFLNeg} \cdot \dfrac{1_n}{2} = 62 \cdot kip$ Ultimate bending moment

Midspan

$C_{v2} := 0.15$ Moment coefficient

$V_{u2} := C_{v2} \cdot w_{TFLPos} \cdot \dfrac{1_n}{2} = 10 \cdot kip$ Ultimate bending moment

Interior support

$Cv3 := 1.15$ Moment coefficient

$V_{u3} := C_{v3} \cdot w_{TFLNeg} \cdot \dfrac{1_n}{2} = 72 \cdot kip$ Ultimate bending moment

7.6 STEP 4—DESIGN FRP PRIMARY REINFORCEMENT FOR BENDING MOMENT CAPACITY

The design is limited to the first span (beam AB-3) because of the symmetry of the structure. The following cross sections at beam AB-3 are considered:

1. Exterior support cross section
2. Midspan cross section
3. Interior support cross section

Ultimate bending moment diagram:

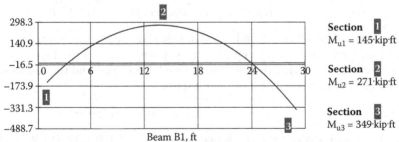

Section **1**
$M_{u1} = 145 \cdot kip \cdot ft$

Section **2**
$M_{u2} = 271 \cdot kip \cdot ft$

Section **3**
$M_{u3} = 349 \cdot kip \cdot ft$

Select the FRP reinforcement: For the purpose of this design example, it is assumed that GFRP bars of the same size are used as longitudinal reinforcement for both positive and negative moments.

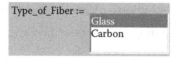 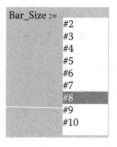

ACI 440.6-08 minimum guaranteed mechanical properties of the selected bars are

f_{fuu} = 80·ksi   Ultimate guaranteed tensile strength of the FRP
ε_{fuu} = 0.014   Ultimate guaranteed rupture strain of the FRP
E_f = 5700·ksi   Guaranteed tensile modulus of elasticity of the FRP
$n_f := \dfrac{E_f}{E_c} = 1.277$ Ratio of modulus of elasticity of bars to modulus of elasticity of concrete

The geometrical properties of the selected bars are

ϕ_{f-bar} = 1·in.   Bar diameter
A_{f-bar} = 0.785·in.^{2} Bar area

FRP reduction factors: Table 7-1 of ACI 440.1R-06 is used to define the environmental reduction factor, C_E. The type of exposure has to be selected:

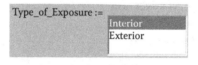

C_E = 0.8   Environmental reduction factor for GFRP

Table 8-3 of ACI 440.1R-06 is used to define the reduction factor to take into account the FRP creep-rupture stress. Creep stress in the FRP has to be evaluated considering the total unfactored dead loads and the sustained portion of the live load (20% of the total live load):

$k_{creep-R}$ = 0.2   Creep-rupture stress limitation factor

Crack width is checked using Equation (8-9) of ACI 440.1R-06. A crack width limit, w_{lim}, of 0.028 in. is used for interior exposure, while 0.020 in. is used for exterior exposure:

w_{lim} = 0.028·in.   Crack width limit

FRP ultimate design properties: The ultimate design properties are calculated per Section 7.2 of ACI 440.1R-06:

$$f_{fu} := C_E \cdot f_{fuu} = 64 \cdot ksi \qquad \text{Design tensile strength}$$
$$\varepsilon_{fu} = C_E \cdot \varepsilon_{fuu} = 0.0112 \qquad \text{Design rupture strain}$$

FRP creep-rupture limit stress: The FRP creep-rupture limit stress is calculated per Section 8.4 of ACI 440.1R-06:

$$f_{f-creep} := k_{creep\_R} \cdot f_{fu} = 12.8 \cdot ksi$$

7.6.1 Case I—Exterior support

Reinforcement required to resist bending moments: As discussed in Chapter 4, the following design condition shall be satisfied: $\phi M_n > M_u$. When the failure is due to concrete crushing: $\phi M_n = \phi A_f f_{fu}\left(d_f - \dfrac{\beta_1}{2}c_u\right)$. Considering the lower bound condition ($\phi M_n = M_u$) and solving for A_f, the following can be written:

$$A_{f-req-bend} = M_u \Big/ \left(\phi A_f f_{fu}\left(d_f - \frac{\beta_1}{2}c_u\right)\right) = \text{,}$$

$$\Rightarrow \rho_{f-req-bend} = A_{f-req-bend}/bd_f = M_u \Big/ \left(\phi A_f f_{fu}\left(d_f - \frac{\beta_1}{2}c_u\right)bd_f\right)$$

Assuming a ϕ-factor of 0.65 and a neutral axis depth equal to 15% of the effective reinforcement depth, the longitudinal reinforcement ratio required for bending, $\rho_{f-req-bend}$, can be estimated. The effective reinforcement depth is

$$d_{f1} := h - c_c - \frac{\phi_{f\_bar}}{2} = 26 \cdot in.$$

The longitudinal reinforcement ratio required for bending is

$$\rho_{f\_req\_bend1} := \frac{M_{u1}}{0.65 \cdot f_{fu} \cdot \left(d_{f1} - \dfrac{\beta_1}{2} \cdot 0.15 d_{f1}\right)} \cdot \frac{1}{b_w \cdot d_{f1}} = 0.00472$$

The minimum reinforcement requirement has to be verified. Equation (8–8) of ACI 440.1R-06 is used. If the failure is not governed by FRP rupture, this requirement is automatically achieved:

$$A_{f\_min1} := \min\left[\frac{4.9 \cdot \sqrt{f'_c \cdot psi}}{f_{fu}} b_w \cdot d_{f1}, \frac{300 psi}{f_{fu}}(b_w \cdot d_{f1})\right] = 1.706 \cdot in^2$$

$$\rho_{f\_min1} := \frac{A_{f\_min1}}{b_w \cdot d_{f1}} = 0.004687$$

The design reinforcement ratio required for bending is taken as

$$\rho_{f-bend1} := \max(\rho_{f-req-bend1}, \rho_{f-min1}) = 0.004716$$

Reinforcement ratio required for creep-rupture stress check: For the case of GFRP reinforcement, the limitation to the creep-rupture stress could be governing for the design. The bending moment to consider for the creep-rupture stress check is obtained for the combination DL + 0.20 LL, as shown:

$$M_{1\_creep0} := M_{s1} \cdot \frac{w_D + 0.20 w_{LNeg}}{w_{SNeg}} = 82.3 \cdot ft \cdot kip$$

The limit stress is

$$f_{f\_creep} = 12.8 \cdot ksi$$

The ratio of depth of neutral axis to reinforcement depth, k_f, can be written as a function of the reinforcement ratio, $\rho_{f-creep}$:

$$k_{f\_creep1}(\rho_{f\_creep}) := \sqrt{2 \cdot \rho_{f\_creep} \cdot n_f + (\rho_{f\_creep} \cdot n_f)^2} - \rho_{f\_creep} \cdot n_f$$

The tensile stress in the FRP can also be expressed as a function of the reinforcement ratio, $\rho_{f-creep}$:

$$f_{f1\_creep0}(\rho_{f\_creep}) := \frac{M_{1\_creep0}}{\rho_{f\_creep} \cdot b_w \cdot d_{f1}^2 \cdot \left(1 - \frac{k_{f\_creep1}(\rho_{f\_creep})}{3}\right)}$$

Solving for $\rho_{f-creep}$:
First guess:

$$\rho_{f-creep0} := 0.002$$

Given:

$$f\rho_1(\rho_{f-creep}) := f_{f1-creep0}(\rho_{f-creep}) - (f_{f-creep})$$

$$\rho_{f-creep1} := root(f\rho_1(\rho_{f-creep0}), \rho_{f-creep0})$$

The reinforcement ratio required for creep rupture is

$$\rho_{f-creep1} = 0.0085$$

FRP longitudinal reinforcement design: The design reinforcement ratio can be selected as the maximum between ρ_{f-bend} and $\rho_{f-creep}$:

$$\rho_{f-des1} := \max(\rho_{f-bend1}, \rho_{f-creep1}) = 0.0085$$

This reinforcement ratio corresponds to an area of

$$A_{f-des1} := \rho_{f-des1} \cdot b_w \cdot d_{f1} = 3.109 \cdot in.^2$$

The required number of bars is

$$N_{f\_des1} := \frac{A_{f\_des1}}{A_{f\_bar}} = 3.958$$

As discussed in Chapter 4, the failure mode depends on the amount of FRP reinforcement. If ρ_f is larger than the balanced reinforcement ratio, ρ_{fb}, then concrete crushing is the failure mode. If ρ_f is smaller than the balanced reinforcement ratio, ρ_{fb}, then FRP rupture is the failure mode.
Equation (8–3) of ACI 440.1R-06:

$$\rho_{fb1} := 0.85\beta_1 \cdot \frac{f'_c}{f_{fu}} \cdot \frac{E_f \cdot \varepsilon_{cu}}{E_f \cdot \varepsilon_{cu} \cdot + f_{fu}} = 0.0126$$

The selected reinforcement ratio is smaller than the ratio corresponding to the balanced conditions, as shown:

$$\frac{\rho_{fb1}}{\rho_{f\_des1}} = 1.476$$

The following number of FRP bars is selected:

$$N_{bar1} := 4$$

$$A_{f1} := N_{bar1} \cdot A_{f-bar} = 3.142 \cdot in^2$$

The corresponding FRP reinforcement ratio is (Equation 8-2 of ACI 440.1R-06):

$$\rho_{f1} := \frac{A_{f1}}{b_w \cdot d_{f1}} = 0.00863$$

FRP bar spacing: The clear bar spacing is

$$s_{f10\_clear} := \frac{b_{eff} - N_{bar1} \cdot \phi_{f\_bar} - c_c - 1in}{N_{bar1} - 1} = 19.167 \cdot in.$$

The clear bar spacing is taken equal to

$$s_{f1-clear} := 3 \ in.$$

The minimum required bar spacing is

$$s_{f-min1} := max(1 \cdot in, \phi_{f-bar}) = 1 \cdot in.$$

$$\text{Check\_BarSpacing1} := \begin{vmatrix} \text{"OK"} & \text{if} \ s_{f1\_clear} \geq s_{f\_min1} \\ \text{"Too many bars"} & \text{otherwise} \end{vmatrix}$$

$$\text{Check\_BarSpacing1} = \text{"OK"}$$

The center-to-center bar spacing is

$$s_{f1} := s_{f1-clear} + \phi_{f-bar} = 4 \cdot \text{in.}$$

Design flexural strength: Based on cross-section compatibility, the effective concrete compressive strain distribution at failure can be computed as a function of the neutral axis depth, x:

$$\varepsilon_{c1}(x,y) := \begin{vmatrix} \dfrac{\varepsilon_{cu}}{x} y & \text{if} \ \rho_{f1} \geq \rho_{fb1} \\[4mm] \dfrac{\varepsilon_{fu}}{d_{f1} - x} \cdot y & \text{if} \ \rho_{f1} < \rho_{fb1} \end{vmatrix}$$

Based on cross-section compatibility, the effective tensile strain in the FRP reinforcement can be computed as a function of the neutral axis depth, x:

$$\varepsilon_{f1}(x) := \begin{vmatrix} \varepsilon_{fu} \ \text{if} \ \rho_{f1} < \rho_{fb1} \\[4mm] \min\left[\dfrac{\varepsilon_{cu}}{x} \cdot (d_{f1} - x), \varepsilon_{fu}\right] & \text{if} \ \rho_{f1} \geq \rho_{fb1} \end{vmatrix}$$

The compressive force in the concrete as a function of the neutral axis depth, x, is

$$C_{c1}(x) := b_w \cdot \int_{0in}^{x} \frac{2 \cdot \rho''_c \cdot \left(\dfrac{\varepsilon_{c1}(x,y)}{\varepsilon_{c0}}\right)}{1 + \left(\dfrac{\varepsilon_{c1}(x,y)}{\varepsilon_{c0}}\right)^2} \text{psi dy}$$

The tensile force in the FRP reinforcement as a function of the neutral axis depth, x, is

$$T_{f1}(x) := A_{f1} \cdot E_f \cdot \varepsilon_{f1}(x)$$

The neutral axis depth, c_u, can be computed by solving the equation of equilibrium $C_c - T_f = 0$:

First guess:

$$x_{01} := 0.1 d_{f1}$$

Given:

$$f_1(x) := C_{c1}(x) - T_{f1}(x)$$

$$c_{u1} := \text{root}\,(f_1(x_{01}), x_{01})$$

The neutral axis depth is

$$c_{u1} = 4.038 \cdot in.$$

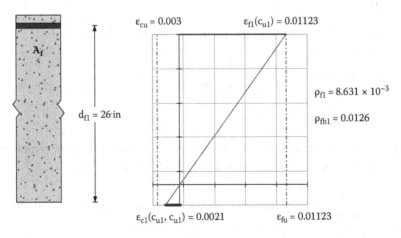

The nominal bending moment capacity can be computed as follows:

$$M_{n1} := b_w \cdot \int_0^{c_{u1}} y \cdot \frac{\left(2 \cdot \sigma''_c \cdot \dfrac{\varepsilon_{c1}(c_{u1}, y)}{\varepsilon_{c0}} \right)}{\left[1 + \left(\dfrac{\varepsilon_{c1}(c_{u1}, y)}{\varepsilon_{c0}} \right)^2 \right]} psi \; dy + T_{f1}(c_{u1}) \cdot (d_{f1} - c_{u1}) = 410 \cdot ft \cdot kip$$

The strain distribution over the cross section is shown next.

$\varepsilon_{cu} = 0.003$ $\qquad$ $\varepsilon_{f1}(c_{u1}) = 0.01123$

A_f

$d_{f1} = 26 \cdot in$

$\rho_{f1} = 8.631 \times 10^{-3}$

$\rho_{fb1} = 0.0126$

$\varepsilon_{c1}(c_{u1}, c_{u1}) = 0.0021$ $\qquad$ $\varepsilon_{fu} = 0.01123$

The concrete crushing failure mode is less brittle than the one due to FRP rupture. The ϕ-factor is calculated according to Jawahery and Nanni [1]:

$$\phi_{b1} := \begin{vmatrix} 0.65 & \text{if} & 1.15 - \dfrac{\varepsilon_{f1}(c_{u1})}{2\varepsilon_{fu}} \le 0.65 \\[2mm] 0.75 & \text{if} & 1.15 - \dfrac{\varepsilon_{f1}(c_{u1})}{2\varepsilon_{fu}} \ge 0.75 \\[2mm] 1.15 - \dfrac{\varepsilon_{f1}(c_{u1})}{2\varepsilon_{fu}} & \text{otherwise} \end{vmatrix}$$

$$\phi_{b1} := 0.65$$

The design flexural strength equation is computed per Equation (8-1) of ACI 440.1R-06:

$$\phi_{b1} \cdot M_{n1} = 267 \cdot \text{kip} \cdot \text{ft}$$

$$\text{Check\_Flexure1} := \begin{vmatrix} \text{"OK"} & \text{if} & \phi_{b1} \cdot M_{n1} \ge M_{u1} & (\phi_{b1} \cdot M_{n1} = 267 \cdot \text{kip} \cdot \text{f}) \\ \text{"Not good"} & \text{otherwise} & & (M_{b1} := 145.44 \cdot \text{kip} \cdot \text{ft}) \end{vmatrix}$$

$$\text{Check\_Flexure1} = \text{"OK"}$$

Flexural strength computed per ACI 440.1R-06: The tensile stress in the FRP is computed per Equation (8-4c) when $\rho_f > \rho_{fb}$, or is f_{fu} if $\rho_f < \rho_{fb}$.

$$f_{f1} := \begin{vmatrix} \sqrt{\dfrac{(E_f \cdot \varepsilon_{cu})^2}{4} + \dfrac{0.85\beta_1 \cdot f'_c}{\rho_{f1}} E_f \cdot \varepsilon_{cu}} - 0.5 E_f \cdot \varepsilon_{cu} & \text{if } \rho_{f1} \ge \rho_{fb1} = 64 \cdot \text{ksi} \ (f_{fu} = 64 \cdot \text{ksi}) \\ f_{fu} & \text{otherwise} \end{vmatrix}$$

f_f cannot exceed f_{fu}; therefore, the following has to be checked:

$$\text{CheckMaxStress1} := \begin{vmatrix} \text{"OK"} & \text{if } f_{f1} \le f_{fu} \\ \text{"Reduce bar spacing or increase bar size"} & \text{otherwise} \end{vmatrix}$$

$$\text{CheckMaxStress1} = \text{"OK"}$$

The stress-block depth is computed per Equation (8-4b) or Equation (8-6c) depending on whether $\rho_f > \rho_{fb}$ or $\rho_f < \rho_{fb}$, respectively.

$$a_{f1} := \begin{vmatrix} \dfrac{A_{f1} \cdot f_{f1}}{0.85 \cdot f'_c \cdot b_w} & \text{if } \rho_{f1} \ge \rho_{fb1} & \text{Equation (8-4b) of ACI 440.1R-06} \\[3mm] \left[\beta_1 \cdot \left(\dfrac{\varepsilon_{cu}}{\varepsilon_{cu} + \varepsilon_{fu}}\right) d_{f1}\right] & \text{otherwise} & \text{Equation (8-6c) of ACI 440.1R-06} \end{vmatrix}$$

$a_{f1} = 4.112 \cdot \text{in.}$

Neutral axis depth is

$$c_{f1} := \frac{a_{f1}}{\beta_1} = 5.482 \cdot \text{in.}$$

The nominal moment capacity is

$$M_{nACI\_1} := A_{f1} \cdot f_{f1} \cdot \left(d_{f1} - \frac{a_{f1}}{2} \right) = 401 \cdot \text{ft} \cdot \text{kip}$$

The concrete crushing failure mode is less brittle than the one due to FRP rupture. The ϕ-factor is computed according to Equation (8-7) of ACI 440.1R-06:

$$\phi_{bACI\_1} := \left| \begin{array}{l} 0.55 \text{ if } \rho_{f1} \leq \rho_{fb1} \qquad = 0.55 \\[2mm] 0.30 + 0.25 \cdot \dfrac{\rho_{f1}}{\rho_{fb1}} \text{ if } \rho_{fb1} < \rho_{f1} < 1.4 \cdot \rho_{fb1} \\[2mm] 0.65 \text{ otherwise} \end{array} \right.$$

The design flexural strength equation is computed per Equation (8-1) of ACI 440.1R-06:

$$\phi_{bACI-1} \cdot M_{nACI} = 221 \cdot \text{kip} \cdot \text{ft}$$

$$\text{Check\_FlexureACI\_1} := \left| \begin{array}{l} \text{"OK" if } \phi_{bACI\_1} \cdot M_{nACI\_1} \geq M_{u1} \quad (M_{u1} = 145 \cdot \text{kip} \cdot \text{ft}) \\[2mm] \text{"Not good" otherwise} \end{array} \right.$$

$$\text{Check\_FlexureACI\_1} = \text{"OK"}$$

Embedment length: Because this is a case of negative reinforcement, it has to be checked if adequate moment capacity can be achieved at the end of the embedment length. The available length for embedment is

$$l_{emb1} := 12 \text{ in.}$$

The developable tensile stress is calculated per Equation (11-3) of ACI 440.1R-06. Minimum between cover to bar center and half of the center-to-center bar spacing is

$$C_{b1} := \min\left(c_c + \frac{\phi_{f\_bar}}{2}, \frac{s_{f1}}{2} \right) = 2 \cdot \text{in}$$

Bar location modification factor for top reinforcement is

$$\alpha_{Neg1} := 1.5$$

Required stress in the FRP is

$$f_{fr1} := E_f \cdot \varepsilon_{f1}(c_{u1}) = 64 \cdot ksi$$

The developable tensile stress (ACI 440.1R-06 Equation 11-3) is

$$f_{fd11} := \begin{vmatrix} f_{fu} & \text{if } \dfrac{\sqrt{f_c \cdot psi}}{\alpha_{Neg1}} \cdot \left(13.6 \cdot \dfrac{l_{emb1}}{\phi_{f\_bar}} + \dfrac{C_{b1}}{\phi_{f\_bar}} \cdot \dfrac{l_{emb1}}{\phi_{f\_bar}} + 340 \right) \geq f_{fu} = 27.2 \cdot ksi \\[2em] \left[\dfrac{\sqrt{f'_c \cdot psi}}{\alpha_{Neg1}} \cdot \left(13.6 \cdot \dfrac{l_{emb1}}{\phi_{f\_bar}} + \dfrac{C_{b1}}{\phi_{f\_bar}} \cdot \dfrac{l_{emb1}}{\phi_{f\_bar}} + 340 \right) \right] & \text{otherwise} \end{vmatrix}$$

$$CheckFailure1 := \begin{vmatrix} \text{"Bar ultimate strength"} & \text{if } f_{fd11} \geq f_{fr1} \\ \text{"Bar strength"} & \text{otherwise} \end{vmatrix}$$

$$CheckFailure1 = \text{"Bond strength"}$$

The cross section of interest is a bond-critical section. The nominal moment capacity, therefore, has to be computed as per ACI 440.1R-06 Equation (8-5) or ACI 440.1R-06 Equation (8-6b) when the failure mode is concrete crushing or bond, respectively.

$$M_{n1b1} := \begin{vmatrix} \left[A_{f1} \cdot f_{fd11} \cdot \left(d_{f1} - \dfrac{1}{2} \dfrac{A_{f1} \cdot f_{f1}}{0.85 \cdot f'_c \cdot b_w} \right) \right] \text{if CheckFailure} = \text{"Bond strength"} \\ M_{n1} \quad \text{if checkFailure} = \text{"Bar ultimate strength"} \end{vmatrix}$$

$$M_{n1b1} = 175 \cdot ft \cdot kip$$

The ultimate moment is

$$M_{u1} = 145 \cdot ft \cdot kip$$

The strength-reduction factor when failure is controlled by bond is

$$\phi_{b1\_bond} := \begin{vmatrix} 0.55 \text{ if CheckFailure1} = \text{"Bond strength"} & = 0.55 \\ \phi_{b1} \quad \text{if CheckFailure1} = \text{"Bar ultimate strength"} \end{vmatrix}$$

The design flexural strength is therefore:

$\phi_{b1\_bond} \cdot M_{n1b1} = 96 \cdot kip \cdot ft$

$$Check\_FlexureNeg1b := \begin{vmatrix} \text{"OK"} & \text{if } \phi_{b1\_bond} \cdot M_{n1b1} \geq M_{u1} \\ \text{"Not good"} & \text{otherwise} \end{vmatrix}$$

$Check\_FlexureNeg1b = \text{"Not good"}$

Because the embedment length is not adequate, the addition of bent bars is required. The geometrical properties of the selected bent bars are

Bent_Bar_Size :=
#2
#3
#4
#5
#6
#7
#8
#9
#10

$\phi_{f\_bent} = 1 \cdot in.$ Bar diameter
$A_{f\_bent} = 0.785 \cdot in.^2$ Bar area
$r_b := 3\phi_{f\_bent} = 3 \cdot in.$ Radius of the bend

The minimum guaranteed ultimate tensile strength of the selected bar is

$f_{fuu\_bent} = 80 \cdot ksi$

The design strength of the selected bar is

$f_{fuu\_bent} := C_E \cdot f_{fuu\_bent} = 64 \cdot ksi$

The design tensile strength of the bend of FRP bar is

$$f_{f\_bent} := \left(0.05 \cdot \frac{r_b}{\phi_{f\_bent}} + 0.3 \right) f_{fu\_bent} = 28.8 \cdot ksi \quad \text{ACI 440.1R-06 Eq} \cdot (7\text{-}3)$$

Determine which is smaller between the strength of the bend and the bond strength of the end of the longitudinal bar.

$$Check\_BentBar := \begin{vmatrix} \text{"strength of bend controls"} & \text{if } f_{f\_bent} \geq f_{fd11} \quad (f_{fd11} = 27.2 \cdot ksi) \\ \text{"strength of bend controls"} & \text{otherwise} \end{vmatrix}$$

Check_BentBar = "Strength of bend controls"

The development length of the bent bar is computed per ACI 440.1R-06 Equation (11-5):

$$l_{bhf\_min} := \begin{vmatrix} 2000\sqrt{psi} \cdot \dfrac{\phi_{f\_bent}}{\sqrt{f_c'}} & \text{if } f_{fu} \leq 75000 \text{ psi} & = 25.8 \cdot in \\[2ex] \dfrac{f_{fu}}{37.5\sqrt{psi}} \cdot \dfrac{\phi_{f\_bent}}{\sqrt{f_c'}} & \text{if } 75000 \text{ psi} < f_{fu} < 150000 \text{ psi} \\[2ex] 2000\sqrt{psi} \cdot \dfrac{\phi_{f\_bent}}{\sqrt{f_c'}} & \text{if } f_{fu} \geq 150000 \text{ psi} \end{vmatrix}$$

The following development length, l_{bhf}, is considered.

$$l_{bhf} := \left(round\left(\frac{l_{bhf\_min}}{in} \right) + 1 \right) in = 27 \cdot in.$$

The tail of the bent bar is to be at least 12 bar diameters:

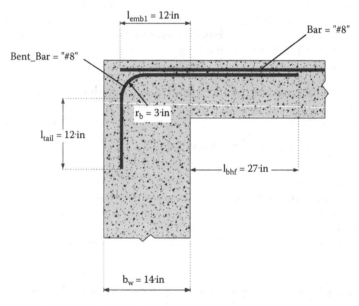

$l_{emb1} = 12 \cdot in$

Bar = "#8"

Bent_Bar = "#8"

$r_b = 3 \cdot in$

$l_{tail} = 12 \cdot in$

$l_{bhf} = 27 \cdot in$

$b_w = 14 \cdot in$

$$l_{tail} := 12 \cdot \phi_{f\_bent} = 12 \cdot in.$$

Recompute the new area of reinforcement and its corresponding ratio for straight bars and bends:

$$A_{f1new} := A_{f1} + A_{f1}$$

$$\rho_{f1new} := \frac{A_{f1new}}{b_w \cdot d_{f1}} = 0.017261$$

The stress-block depth is recomputed based on the presence of straight bars and an equal number of bends:

$$a_{f1new} := \begin{vmatrix} \dfrac{A_{f1new} \cdot f_{f\_bent}}{0.85 \cdot f_c' \cdot b_w} & \text{if } \rho_{f1new} \geq \rho_{fb1} & \text{Equation (8-4b) of ACI 440.1R-06} \\[2em] \left[\beta_1 \cdot \left(\dfrac{\varepsilon_{cu}}{\varepsilon_{cu} + \varepsilon_{fu}} \right) d_{f1} \right] & \text{otherwise} & \text{Equation (8-4c) of ACI 440.1R-06} \end{vmatrix}$$

$$a_{f1new} = 2.534 \cdot \text{in.}$$

Neutral axis depth is

$$c_{f1new} := \frac{a_{f1new}}{\beta_1} = 3.379 \cdot \text{in.}$$

The nominal moment capacity is

$$M_{nnew} := A_{f1new} \cdot f_{f\_bent} \cdot \left(d_{f1} - \frac{a_{f1new}}{2} \right) = 373 \cdot \text{ft} \cdot \text{kip}$$

The ϕ-factor is 0.55 when bond controls:

$$\phi_{bond} := 0.55$$

The design flexural strength equation is computed per Equation (8-1) of ACI 440.1R-06:

$$\phi_{bond} \cdot M_{nnew} = 205 \text{ kip} \cdot \text{fit}$$

$$\text{Check\_Flexure\_bond\_1} := \begin{vmatrix} \text{"OK" if } \phi_{bond} \cdot M_{nnew} \geq M_{u1} & (M_{u1} = 145 \cdot \text{kip} \cdot \text{ft}) \\ \text{"Not good" otherwise} \end{vmatrix}$$

$$\text{Check\_Flexure\_bond\_1} = \text{"OK"}$$

7.6.2 Case 2—Midspan

Following the same approach discussed for Case 1, the minimum reinforcement ratio required for bending can be selected. Because this section is at midspan, a T-section is considered. The effective reinforcement depth is

$$d_{f2} := h - c_c - \frac{\phi_{f\_bar}}{2} = 26 \cdot in.$$

The gross-sectional area is

$$A_{c2} := t_{slab} \cdot b_{eff} + b_w \cdot (d_{f2} - t_{slab}) = 764 \cdot in.^2$$

The longitudinal reinforcement ratio required for bending is

$$\rho_{f\_-req\_bend2} := \frac{M_{u2}}{0.65 \cdot f_{fu} \cdot \left(d_{f2} - \frac{\beta_1}{2} \cdot 0.15 d_{f2} \right)} \cdot \frac{1}{A_{c2}} = 0.00419$$

The minimum reinforcement ratio is computed per Equation (8-8) of ACI 440.1R-06:

$$A_{f\_min2} := min\left[\frac{4.9 \cdot \sqrt{f_c' \cdot psi}}{f_{fu}} b_w \cdot d_{f2}, \frac{300psi}{f_{fu}} (b_w \cdot d_{f2}) \right] = 1.706 \cdot in^2$$

$$\rho_{f\_min2} := \frac{A_{f\_min2}}{A_{c2}} = 0.002233$$

The design reinforcement ratio required for bending is taken as

$$\rho_{f\_bend2} := max(\rho_{freq\_bend2}, \rho_{f\_kmin2}) = 0.004189$$

Reinforcement ratio required for creep-rupture stress check: For the case of GFRP reinforcement, the limitation to the creep-rupture stress could be governing for the design. The bending moment to consider for the creep-rupture stress check is obtained for the combination DL + 0.20 LL, as shown:

$$M_{2\_creep0} := M_{s2} \cdot \frac{w_D + 0.20 w_{LPos}}{w_{SPos}} = 143.8 \cdot ft \cdot kip$$

The limit stress is

$$f_{f\_creep} = 12.8 \cdot ksi$$

The ratio of depth of neutral axis to reinforcement depth, k_f, can be written as a function of the reinforcement ratio, ρ_{f_creep}:

$$k_{f\_creep2}\left(\rho_{f\_creep}\right) := \sqrt{2 \cdot \rho_{f\_creep} \cdot n_f + \left(\rho_{f\_creep} \cdot n_f\right)^2} - \rho_{f\_creep} \cdot n_f$$

The tensile stress in the FRP can also be expressed as a function of the reinforcement ratio, ρ_{f_creep}:

$$f_{f2\_creep0}\left(\rho_{f\_creep}\right) := \frac{M_{2\_creep0}}{\rho_{f\_creep} \cdot A_{c2} \cdot d_{f2} \cdot \left(1 - \dfrac{k_{f\_creep2}\left(\rho_{f\_creep}\right)}{3}\right)}$$

Solving for ρ_{f_creep}:
First guess:

$$\rho_{f\_creep20} := 0.002$$

Given:

$$f_{p2}(\rho_{f\_creep}) := f_{f2\_creep0}(\rho_{f\_creep}) - (f_{f\_creep})$$

$$\rho_{f\_creep2} := \text{root}(f_{p2}(\rho_{f\_creep20}), \rho_{f\_creep20})$$

The reinforcement ratio required for creep-rupture is

$$\rho_{f\_creep2} := 0.0071$$

FRP longitudinal reinforcement design: The design reinforcement ratio can be selected as the maximum between ρ_{f_bend} and ρ_{f_creep}:

$$\rho_{f\_creep2} := \max(\rho_{f\_bend2}, \rho_{f\_creep2}) = 0.0071$$

This reinforcement ratio corresponds to an area of

$$A_{f\_des1} := \rho_{f\_des2} \cdot A_{c2} = 5.411 \cdot \text{in.}^2$$

The required number of bars is

$$N_{f\_des2} := \frac{A_{f\_des2}}{A_{f\_bar}} = 6.889$$

As discussed in Chapter 4, the failure mode depends on the amount of FRP reinforcement. If ρ_f is larger than the balanced reinforcement ratio, ρ_{fb}, then concrete crushing is the failure mode. If ρ_f is smaller than the balanced reinforcement ratio, ρ_{fb}, then FRP rupture is the failure mode.

Equation (8-3) of ACI 440.1R-06 is

$$\rho_{fb2} := 0.85\beta_1 \cdot \frac{f'_c}{f_{fu}} \cdot \frac{E_f \cdot \varepsilon_{cu}}{E_f \cdot \varepsilon_{cu} + f_{fu}} = 0.0126$$

The selected reinforcement ratio is smaller than the ratio corresponding to the balanced conditions, as shown:

$$\frac{\rho_{fb2}}{\rho_{f\_des2}} = 1.779$$

Because the reinforcement is to be placed on two levels, an even number of FRP bars is selected:

$$N_{bar2} := 8$$

The total area of FRP reinforcement is

$$A_{f2} := N_{bar2} \cdot A_{f\_bar} = 6.283 \cdot in.^2$$

The corresponding FRP reinforcement ratio is (Equation 8-2 of ACI 440.1R-06)

$$\rho_{f2} := \frac{A_{f2}}{A_{c2}} = 0.00822$$

FRP bar spacing: The clear bar spacing is

$$s_{f20\_clear} := \frac{b_w - N_{bar2} \cdot \phi_{f\_bar} - c_c - 1in}{N_{bar2} - 1} = 0.5 \cdot in$$

The minimum required bar spacing is

$$S_{f\_min2} := \max(1 \cdot in, \phi_{f\_bar}) = 1 \cdot in.$$

$$\text{Check\_BarSpacing2} := \begin{vmatrix} \text{``OK'' if } s_{f20\_clear} > s_{f\_min2} \\ \text{``Too many bars''} \qquad \text{otherwise} \end{vmatrix}$$

Check_BarSpacing = "Too many bars"

TwoManyBars := ✓ ☐Check Box

The flexural FRP reinforcement is placed in two layers:

First layer of FRP reinforcement

$N_{bar2I} := 4$ Number of bars

$A_{f2I} := N_{bar2I} \cdot A_{f\_bar} = 3.142 \cdot in.^2$ Area of FRP reinforcement

$d_{f2I} := h - c_c - \dfrac{\phi_{f\_bar}}{2} = 26 \cdot in$ Effective depth

Second layer of FRP reinforcement

$N_{bar2II} := 4$ Number of bars

$A_{f2II} := N_{bar2II} \cdot A_{f\_bar} = 3.142 \cdot in.^2$ Area of FRP reinforcement

$d_{f2II} := h - c_c - \phi_{f\_bar} - s_{f\_min2} - \dfrac{\phi_{f\_bar}}{2} = 24 \cdot in$ Effective depth

$$\text{Check\_FRPreinforcement} := \begin{vmatrix} \text{``OK'' if } N_{bar2I} + N_{bar2II} \geq N_{bar2} \\ \text{``Increase the number of bars''} \qquad \text{otherwise} \end{vmatrix}$$

$\text{Check\_FRPreinforcement} = \text{``OK''}$

The clear spacing for the first layer is

$$s_{f21\_clear} := \dfrac{b_w - 2c_c - N_{bar2I} \cdot \phi_{f\_bar} - 1in}{N_{bar2I} - 1} = 2 \cdot in$$

and for the second layer is

$$s_{f22\_clear} := \dfrac{b_w - 2c_c - N_{bar2II} \cdot \phi_{f\_bar} - 1in}{N_{bar2II} - 1} = 2 \cdot in$$

$$\text{Check\_BarSpacing22} := \begin{vmatrix} \text{``OK'' if } \min(s_{f21\_clear}, s_{f22\_clear}) > s_{f\_min2} \\ \text{``Too many bars''} \qquad \text{otherwise} \end{vmatrix}$$

$\text{Check\_BarSpacing22} = \text{``OK''}$

The center-to-center bar spacing for the first layer is

$$s_{f21} := s_{f21\_clear} + \phi_{f\_bar} = 3 \cdot in.$$

and for the second layer is

$$s_{f22} := s_{f22\_clear} + \phi_{f\_bar} = 3 \cdot in.$$

Design flexural strength: The effective concrete compressive strain at failure as a function of the neutral axis depth, x, is

$$\varepsilon_{c2}(x,y) := \left| \begin{array}{l} \dfrac{\varepsilon_{cu}}{x} \cdot y \ \text{if} \ \rho_{f2} \geq \rho_{fb2} \\[3mm] \dfrac{\varepsilon_{fu}}{d_{f2} - x} \cdot y \ \text{if} \ \rho_{f2} < \rho_{fb2} \end{array} \right.$$

The effective tensile strain in the first layer of FRP reinforcement as a function of the neutral axis depth, x, is

$$\varepsilon_{f2I}(x) := \left| \begin{array}{l} \varepsilon_{fu} \ \text{if} \ \rho_{f2} < \rho_{fb2} \\[3mm] \min\left[\dfrac{\varepsilon_{cu}}{x} \cdot (d_{f2I} - x), \varepsilon_{fu} \right] \ \text{if} \ \rho_{f2} \geq \rho_{fb2} \end{array} \right.$$

The compressive force in the concrete as a function of the neutral axis depth, x, is

$$\varepsilon_{f2II}(x) := \left| \begin{array}{l} \varepsilon_{fu} - \dfrac{\varepsilon_{c2}(x,x)}{x} \cdot \left(\phi_{f\_bar} + s_{f\_min2} \right) \ \text{if} \ \rho_{f2} < \rho_{fb2} \wedge N_{bar2II} > 0 \\[3mm] \min\left[\dfrac{\varepsilon_{cu}}{x} \cdot (d_{f2II} - x), \varepsilon_{fu} - \dfrac{\varepsilon_{cu}}{x} \cdot \left(\phi_{f\_bar} + s_{f\_min2} \right) \right] \ \text{if} \ \rho_{f2} \geq \rho_{fb2} \wedge N_{bar2II} > 0 \\[3mm] 0 \ \text{otherwise} \end{array} \right.$$

The compressive force in the concrete as a function of the neutral axis depth, x, is

$$C_{c2}(x) := b_{eff} \cdot \int_{0in}^{x} \frac{2 \cdot \sigma''_{c} \cdot \left(\dfrac{\varepsilon_{c2}(x,y)}{\varepsilon_{c0}} \right)}{1 + \left(\dfrac{\varepsilon_{c2}(x,y)}{\varepsilon_{c0}} \right)^{2}} \, psi \, dy$$

The tensile force in the first layer of FRP reinforcement as a function of the neutral axis depth, x, is

$$T_{f2I}(x) := A_{f2I} \cdot E_{f} \cdot \varepsilon_{f2I}(x)$$

The tensile force in the second layer of FRP reinforcement as a function of the neutral axis depth, x, is

$$T_{f2II}(x) := A_{f2II} \cdot E_f \cdot \varepsilon_{f2II}(x)$$

The neutral axis depth, c_u, can be computed by solving the equation of equilibrium $C_c - T_f = 0$:

First guess:

$$x_{02} := 0.1 d_{f2}$$

Given:

$$f_2(x) := C_{c2}(x) - (T_{f2I}(x) + T_{f2II}(x))$$

$$c_{u2} := \text{root}(f_2(x_{02}), x_{02})$$

The neutral axis depth is

$$c_{u2} = 2.458 \cdot \text{in.}$$

Check_Netral Axis Depth :=

$$\left| \begin{array}{ll} \text{"The neutral axis falls within the flange depth"} & \text{if } c_{u2} \leq t_{slab} \\ \text{"A T-section analysis has to be conducted "} & \text{otherwise} \end{array} \right.$$

Check_NetralAxisDepth = "The neutral axis falls within the flange depth"

The nominal bending moment capacity can be computed as follows:

$$M_{n2} := b_{eff} \cdot \int_0^{c_{u2}} y \cdot \left[\dfrac{2 \cdot \sigma''_c \cdot \dfrac{\varepsilon_{c2}(c_{u2},y)}{\varepsilon_{c0}}}{1 + \left(\dfrac{\varepsilon_{c2}(c_{u2},y)}{\varepsilon_{c0}} \right)^2} \right] \text{psi} \, dy + T_{f2I}(c_{u2}) \cdot (d_{f2I} - c_{u2})$$

$$+ T_{f2II}(c_{u2}) \cdot (d_{f2II} - c_{u2})$$

$$M_{n2} = 776 \cdot \text{kip} \cdot \text{ft}$$

The strain distribution over the cross section is shown next.

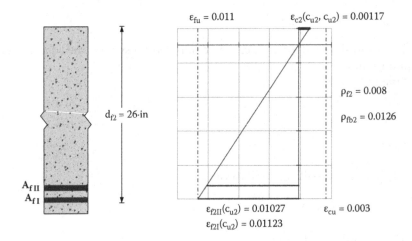

The concrete crushing failure mode is less brittle than the one due to FRP rupture. The ϕ-factor is calculated according to Jawahery and Nanni [1]:

$$\phi_{b2I} := \begin{vmatrix} 0.65 \text{ if } 1.15 - \dfrac{\varepsilon_{f2I}(c_{u2})}{2\varepsilon_{fu}} \le 0.65 \\[2ex] 0.75 \text{ if } 1.15 - \dfrac{\varepsilon_{f2I}(c_{u2})}{2\varepsilon_{fu}} \ge 0.75 \\[2ex] 1.15 - \dfrac{\varepsilon_{f2I}(c_{u2})}{2\varepsilon_{fu}} \end{vmatrix}$$

$$\phi_{b2I} = 0.65$$

$$\phi_{b2II} := \begin{vmatrix} 0.65 \text{ if } 1.15 - \dfrac{\varepsilon_{f2II}(c_{u2})}{2\varepsilon_{fu}} \le 0.65 \\[2ex] 0.75 \text{ if } 1.15 - \dfrac{\varepsilon_{f2II}(c_{u2})}{2\varepsilon_{fu}} \ge 0.75 \\[2ex] 1.15 - \dfrac{\varepsilon_{f2II}(c_{u2})}{2\varepsilon_{fu}} \end{vmatrix}$$

$$\phi_{b2II} = 0.692$$

$$\phi_{b2} := \min(\phi_{b2I}, \phi_{b2II}) = 0.65$$

The design flexural strength equation is computed per Equation (8-1) of ACI 440.1R-06:

$$\phi_{b2} \cdot M_{n2} = 504 \cdot \text{kip} \cdot \text{ft}$$

$$\text{Check\_Flexure2} := \begin{vmatrix} \text{"OK" if } \phi_{b2} \cdot M_{n2} \geq M_{u2} & (M_{u2} = 271 \cdot \text{kip} \cdot \text{ft}) \\ \text{"Not good"} & \text{otherwise} \end{vmatrix}$$

Check_Flexure2 = "OK"

Flexural strength computed per ACI 440.1R-06: The tensile stress in the FRP is computed per Equation (8-4c) when $\rho_f > \rho_{fb}$, or is f_{fu} if $\rho_f < \rho_{fb}$:

$$f_{f2} := \begin{vmatrix} \sqrt{\dfrac{(E_f \cdot \varepsilon_{cu})^2}{4} + \dfrac{0.85\beta_1 \cdot f'_c}{\rho_{f2}}} E_f \cdot \varepsilon_{cu} - 0.5 E_f \cdot \varepsilon_{cu} \text{ if } \rho_{f2} \geq \rho_{fb2} &= 64 \cdot \text{ksi} \\ f_{fu} & \text{otherwise} \end{vmatrix}$$

f_f cannot exceed f_{fu}; therefore, the following has to be checked:

$$\text{CheckMaxStress2} := \begin{vmatrix} \text{"OK" if } f_{f2} \leq f_{fu} & (f_{fu} = 64 \cdot \text{ks}) \\ \text{"Reduce bar spacing or increase bar size"} & \text{otherwise} \end{vmatrix}$$

CheckMaxStress1 = "OK"

The stress-block depth is computed per Equation (8-4b) or Equation (8-6c) depending on whether $\rho_f > \rho_{fb}$ or $\rho_f < \rho_{fb}$, respectively.

$$a_{f2} := \begin{vmatrix} \dfrac{A_{f2} f_{f2}}{0.85 \cdot f' \cdot b_{eff}} \text{ if } \rho_{f2} \geq \rho_{fb2} & \text{Equation(8-4b) of ACI 440.IR-06} \\ \left[\beta_1 \cdot \left(\dfrac{\varepsilon_{cu}}{\varepsilon_u + \varepsilon_{fu}} \right) \dfrac{d_{f2I} + d_{f2II}}{2} \right] \text{otherwise} & \text{Equation(8-6c) of ACI 440.IR-06} \end{vmatrix}$$

$a_{f2} = 3.953 \cdot \text{in}$

Neutral axis depth is

$$c_{f2} := \frac{a_{f1}}{\beta_1} = 5.482 \cdot \text{in.}$$

The nominal moment capacity is

$$M_{nACI\_2} := A_{f2} \cdot f_{f2} \cdot \left(d_{f2} - \frac{a_{f2}}{2} \right) = 805 \cdot \text{ft} \cdot \text{kip}$$

The concrete crushing failure mode is less brittle than the one due to FRP rupture. The ϕ-factor is computed according to Equation (8-7) of ACI 440.1R-06:

$$\phi_{bACI\_2} := \begin{vmatrix} 0.55 & \text{if } \rho_{fb2} && = 0.55 \\[2mm] 0.30 + 0.25 \cdot \dfrac{\rho_{f2}}{\rho_{fb2}} & \text{if } \rho_{fb2} < \rho_{f2} < 1.4 \cdot \rho_{fb2} \\[2mm] 0.65 & \text{otherwise} \end{vmatrix}$$

The design flexural strength equation is computed per Equation (8-1) of ACI 440.1R-06:

$$\phi_{bACI-2} \cdot M_{nACI-2} = 443 \cdot \text{kip} \cdot \text{ft}$$

$$\text{Check\_FlexureACI\_2} := \begin{vmatrix} \text{``OK''} & \text{if } \phi_{bACI\_2} \cdot M_{nACI\_2} \geq M_{u2} \;\; (M_{u2} = 271 \cdot \text{kip} \cdot \text{ft}) \\[2mm] \text{``Not good''} & \text{otherwise} \end{vmatrix}$$

Check_FlexureACI_2 = "OK"

Development of positive moment reinforcement: The development length, l_d, for straight bars can be calculated using Equation (11-3) of ACI 440.1R-06:

Minimum between cover to bar center and half of the center-to-center bar spacing is

$$C_{b2} := \min\left(c_c + \frac{\phi_{f\_bar}}{2}, \frac{s_{f21}}{2}, \frac{s_{f22}}{2} \right) = 1.5 \cdot \text{in.}$$

Bar location modification factor for bottom reinforcement is

$$\alpha_{Pos} := 1$$

The minimum development length is computed according to ACI 440.1R-06 Equation (11-6):

$$l_{d\_min} := \frac{\alpha_{Pos} \cdot \dfrac{f_{fu}}{\sqrt{f_c'} \cdot \text{psi}} - 340}{13.6 + \dfrac{C_{b2}}{\phi_{f\_bar}}} \phi_{f\_bar} = 32.201 \cdot \text{in.}$$

The following development length is used and is available to develop the required moment capacity:

$$l_d := 33 \text{ in.}$$

7.6.3 Case 3—Interior support

Reinforcement required to resist bending moments: The same approach discussed for Case 1 is followed. The effective reinforcement depth is

$$d_{f3} := h - c_c - \frac{\phi_{f\_bar}}{2} = 26 \cdot in.$$

The longitudinal reinforcement ratio required for bending is

$$\rho_{f\_req\_bend3} := \frac{M_{u3}}{0.65 \cdot f_{fu} \cdot \left(d_{f3} - \frac{\beta_1}{2} \cdot 0.15 d_{f3}\right)} \cdot \frac{1}{b_w \cdot d_{f3}} = 0.01132 \quad (M_{u3} = 349 \cdot kip \cdot ft)$$

The minimum reinforcement requirement has to be verified. Equation (8-8) of ACI 440.1R-06 is used. If the failure is not governed by FRP rupture, this requirement is automatically achieved:

$$A_{f\_min3} := min\left[\frac{4.9 \cdot \sqrt{f_c' \cdot psi}}{f_{fu}} b_w \cdot d_{f3}, \frac{300 psi}{f_{fu}}(b_w \cdot d_{f3})\right] = 1.706 \cdot in^2$$

$$\rho_{f\_min3} := \frac{A_{f\_min3}}{b_w \cdot d_{f3}} = 0.004687$$

The design reinforcement ratio required for bending is taken as

$$\rho_{f\_bend3} := max(\rho_{f\_req\_bend3}, \rho_{f\_min3}) = 0.0113$$

Reinforcement ratio required for creep-rupture stress check: The bending moment to consider for the creep-rupture stress check is

$$M_{3\_creep0} := M_{s3} \cdot \frac{w_D + 0.20 w_{LNeg}}{w_{SNeg}} = 197.5 \cdot ft \cdot kip$$

The limit stress is

$$f_{f-creep} = 12.8 \cdot ksi$$

The ratio of depth of neutral axis to reinforcement depth, k_f, can be written as a function of the reinforcement ratio, $\rho_{f-creep}$:

$$k_{f\_creep3}(\rho_{f\_creep}) := \sqrt{2 \cdot \rho_{f\_creep} \cdot n_f + (\rho_{f\_creep} \cdot n_f)^2} - \rho_{f\_creep} \cdot n_f$$

The tensile stress in the FRP can also be expressed as a function of the reinforcement ratio, $\rho_{f-creep}$:

$$f_{f3\_creep0}\left(\rho_{f\_creep}\right) := \frac{M_{3\_creep0}}{\rho_{f\_creep} \cdot b_w \cdot d_{f3}^2 \cdot \left(1 - \frac{k_{f\_creep3}\left(\rho_{f\_creep}\right)}{3}\right)}$$

Solving for $\rho_{f-creep}$:

$$\rho_{f-creep0} = 0.002$$

Given:

$$fp_3(\rho_{f-creep}) := f_{f3-creep0}(\rho_{f-creep}) - (f_{f-creep})$$

$$\rho_{f-creep3} := root(fp_3(\rho_{f-creep0}), \rho_{f-creep0})$$

The reinforcement ratio required for creep rupture is

$$\rho_{f-creep3} := 0.021$$

FRP longitudinal reinforcement design: The design reinforcement ratio can be selected as the maximum between ρ_{f-bend} and $\rho_{f-creep}$:

$$\rho_{f-des3} := max(\rho_{f-bend3}, \rho_{f-creep3}) = 0.021$$

This reinforcement ratio corresponds to an area of

$$A_{f-des3} := \rho_{f-des3} \cdot b_w \cdot d_{f3} = 7.646 \cdot in.^2$$

The required number is

$$N_{f\_des3} := \frac{A_{f\_des3}}{A_{f\_bar}} = 9.735$$

As discussed in Chapter 4, the failure mode depends on the amount of FRP reinforcement. If ρ_f is larger than the balanced reinforcement ratio, ρ_{fb}, then concrete crushing is the failure mode. If ρ_f is smaller than the balanced reinforcement ratio, ρ_{fb}, then FRP rupture is the failure mode.

Equation (8-3) of ACI 440.1R-06 is

$$\rho_{fb3} := 0.85\beta_1 \cdot \frac{f_c'}{f_{fu}} \cdot \frac{E_f \cdot \varepsilon_{cu}}{E_f \cdot \varepsilon_{cu} + f_{fu}} = 0.0126$$

The selected reinforcement ratio is larger than the ratio corresponding to the balanced conditions, as shown:

$$\frac{\rho_{fb3}}{\rho_{f\_des3}} = 0.6$$

The following number of FRP bars is selected:

$$N_{bar3} := 10$$

$$A_{f3} := N_{bar3} \cdot A_{f-bar} = 7.854 \cdot in.^2$$

The corresponding FRP reinforcement ratio is (Equation 8-2 of ACI 440.1R):

$$\rho_{f3} := \frac{A_{f3}}{b_w \cdot d_{f3}} = 0.02158$$

FRP bar spacing: The clear bar spacing is

$$s_{f30\_clear} := \frac{b_{eff} - N_{bar3} \cdot \phi_{f\_bar} - c_c - 1in}{N_{bar3} - 1} = 5.722 \cdot in.$$

The clear bar spacing is taken equal to

$$S_{f3-clear} := 5 \text{ in.}$$

The minimum required bar spacing is

$$S_{f-min3} := mac(1 \cdot in, \phi_{f-bar}) = 1 \cdot in.$$

$$\text{Check\_BarSpacing3} := \begin{vmatrix} \text{"OK" if } s_{f3\_clear} \geq s_{f\_min3} \\ \text{"Too many bars" otherwise} \end{vmatrix}$$

$$\text{Check\_BarSpacing3} = \text{"OK"}$$

The center-to-center bar spacing is

$$S_{f3} := S_{f3-clear} + \phi_{f-bar} = 6 \text{ in.}$$

Design flexural strength: The effective concrete compressive strain at failure as a function of the neutral axis depth, x, is

$$\varepsilon_{c3}(x,y) := \begin{vmatrix} \dfrac{\varepsilon_{cu}}{x} y \text{ if } \rho_{f3} \geq \rho_{fb3} \\ \dfrac{\varepsilon_{cu}}{d_{f3} - x} \cdot x \text{ if } \rho_{f3} < \rho_{fb3} \end{vmatrix}$$

The effective tensile strain in the FRP reinforcement as a function of the neutral axis depth, x, is

$$\varepsilon_{f3}(x) := \begin{vmatrix} \varepsilon_{fu} \text{ if } \rho_{f3} < \rho_{fb3} \\ \min\left[\dfrac{\varepsilon_{cu}}{x} \cdot (d_{f3} - x), \varepsilon_{fu}\right] & \text{if } \rho_{f3} \geq \rho_{fb3} \end{vmatrix}$$

The compressive force in the concrete as a function of the neutral axis depth, x, is

$$C_{c3}(x) := b_w \cdot \int_{0in}^{x} \frac{2 \cdot \sigma''_c \cdot \left(\dfrac{\varepsilon_{c3}(x,y)}{\varepsilon_{c0}}\right)}{1 + \left(\dfrac{\varepsilon_{c3}(x,y)}{\varepsilon_{c0}}\right)^2} \text{psi} \, dy$$

The tensile force in the FRP reinforcement as a function of the neutral axis depth, x, is

$$T_{f3}(x) := A_{f3} \cdot E_f \cdot \varepsilon_{f3}(x)$$

The neutral axis depth, c_u, can be computed by solving the equation of equilibrium $C_c - T_f = 0$:

First guess:

$$x_{03} := 0.1 d_{f3}$$

Given:

$$f_3(x) := C_{c3}(x) - T_{f3}(x)$$

$$c_{u3} := \text{root}(f_3(x_{03}), x_{03})$$

The neutral axis depth is

$$c_{u3} = 6.7 \text{ in.}$$

The nominal bending moment capacity can be computed as follows:

$$M_{n3} := b_w \cdot \int_0^{c_{u3}} y \cdot \frac{2 \cdot \sigma''_c \cdot \left(\dfrac{\varepsilon_{c3}(c_{u3},y)}{\varepsilon_{c0}}\right)}{\left[1 + \left(\dfrac{\varepsilon_{c3}(c_{u3},y)}{\varepsilon_{c0}}\right)^2\right]} \text{psi} \, dy + T_{f3}(c_{u3}) \cdot (d_{f3} - c_{u3}) = 750 \cdot \text{ft} \cdot \text{kip}$$

The strain distribution over the cross section is shown next.

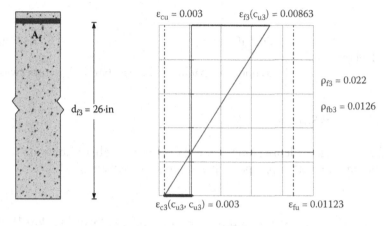

The φ-factor is calculated according to Jawahery and Nanni [1].

$$\phi_{b3} := \begin{vmatrix} 0.65 & \text{if } 1.15 - \dfrac{\varepsilon_{f3}(c_{u3})}{2\varepsilon_{fu}} \le 0.65 \\[2mm] 0.75 & \text{if } 1.15 - \dfrac{\varepsilon_{f3}(c_{u3})}{2\varepsilon_{fu}} \ge 0.75 \quad = 0.75 \\[2mm] 1.15 - \dfrac{\varepsilon_{f3}(c_{u3})}{2\varepsilon_{fu}} & \text{otherwise} \end{vmatrix}$$

The design flexural strength equation is computed per Equation (8-1) of ACI 440.1R-06:

$$\phi_{b3} \cdot M_{n3} = 563 \cdot kip \cdot ft$$

$$\text{Check\_Flexure3} := \begin{vmatrix} \text{``OK''} & \text{if } \phi_{b3} \cdot M_{n3} \ge M_{u3} \quad (M_{u3} = 349 \cdot kip \cdot ft) \\[2mm] \text{``Not good''} & \text{otherwise} \end{vmatrix}$$

Check_Flexure3 = "OK"

Flexural strength computed per ACI 440.1R-06: The tensile stress in the FRP is computed per Equation (8-4c) when $\rho_f > \rho_{fb}$ or is f_{fu} if $\rho_f < \rho_{fb}$.

$$f_{f3} := \begin{vmatrix} \sqrt{\dfrac{(E_f \cdot \varepsilon_{cu})^2}{4} + \dfrac{0.85\beta_1 \cdot f'_c}{\rho_{f3}} E_f \cdot \varepsilon_{cu}} - 0.5 E_f \cdot \varepsilon_{cu} & \text{if } \rho_{f3} \ge \rho_{fb3} = 47.2 \cdot ksi \\[2mm] f_{fu} & \text{otherwise} \end{vmatrix}$$

f_f cannot exceed f_{fu}; therefore, the following has to be checked:

$$\text{CheckMaxStress3} := \begin{array}{|ll} \text{"OK" if } f_{f3} \leq f_{fu} & \left(f_{fu} = 64 \cdot \text{ksi}\right) \\ \\ \text{"Reduce bar spacing or increase bar size"} & \text{otherwise} \end{array}$$

CheckMaxStress3 = "OK"

The stress-block depth computed per Equation (8-4b) or Equation (8-6c) depending on whether $\rho_f > \rho_{fb}$ or $\rho_f < \rho_{fb}$, respectively, is

$$a_{f3} := \begin{array}{|ll} \dfrac{A_{f3} \cdot f_{f3}}{0.85 \cdot f'_c \cdot b_w} \text{ if } \rho_{f3} \geq \rho_{fb3} & \text{Equation (8-4b) of ACI 440.1R-06} \\ \\ \left[\beta_1 \cdot \left(\dfrac{\varepsilon_{cu}}{\varepsilon_{cu} + \varepsilon_{fu}} \right) d_{f3} \right] \text{otherwise} & \text{Equation (8-6c) of ACI 440.1R-06} \end{array}$$

$a_{f3} = 5.188 \cdot \text{in.}$

Neutral axis depth is

$$c_{f3} := \frac{a_{f3}}{\beta_1} = 6.918 \cdot \text{in.}$$

The nominal moment capacity is

$$M_{nACI\_3} := A_{f3} \cdot f_{f3} \cdot \left(d_{f3} - \frac{a_{f3}}{2} \right) = 723 \cdot \text{ft} \cdot \text{kip}$$

The concrete crushing failure mode is less brittle than the one due to FRP rupture. The ϕ-factor is computed according to Equation (8-7) of ACI 440.1R-06:

$$\phi_{bACI\_3} := \begin{array}{|l} 0.55 \text{ if } \rho_{f3} \leq \rho_{fb3} \\ \\ 0.30 + 0.25 \cdot \dfrac{\rho_{f3}}{\rho_{fb3}} \text{ if } \rho_{fb3} < \rho_{f3} < 1.4 \cdot \rho_{fb3} \quad = 0.65 \\ \\ 0.65 \text{ otherwise} \end{array}$$

The design flexural strength equation is computed per Equation (8-1) of ACI 440.1R-06:

$$\phi_{bACI-3} \cdot M_{nACI-3} = 470 \cdot \text{kip} \cdot \text{ft}$$

$$\text{Check\_FlexureACI\_3} := \begin{vmatrix} \text{"OK"} & \text{if } \phi_{bACI\_3} \cdot M_{nACI\_3} \geq M_{u3} & \left(M_{u3} = 349 \cdot \text{kip} \cdot \text{ft} \right) \\ \text{"Not good"} & \text{otherwise} \end{vmatrix}$$

$$\text{Check\_FlexureACI\_3} = \text{"OK"}$$

▶

Tension lap splice: The recommended development length of FRP tension lap splices is $1.3 l_d$ (Section 11.4 of ACI 440.1R-06). The minimum recommended tension lap splice development length is

$$1.3 l_{d-\min} = 41.861 \cdot \text{in.}$$

where $l_{d\text{-}min} = 32.2$ in. as computed for the case of midspan.

The following minimum tension lap splice development length is considered and adopted:

$$l_{tls} := 42 \text{ in.}$$

Embedment length: Because this is a case of negative reinforcement, it has to be checked if adequate moment capacity can be achieved at the end of the embedment length. The maximum available length for embedment is equal to half the length of the adjacent span.

In this case, a quarter of the adjacent span is considered.

$$l_{emb3} := 0.25 \cdot l_2 = 90 \cdot \text{in.}$$

The developable tensile stress is calculated per Equation (11-3) of ACI 440.1R-06. Minimum between cover to bar center and half of the center-to-center bar spacing is

$$C_{b3} := \min\left(c_c + \frac{\phi_{f\_bar}}{2}, \frac{s_{f3}}{2} \right) = 2 \cdot \text{in.}$$

Bar location modification factor for top reinforcement is

$$\alpha_{Neg3} := 1.5$$

Required stress in the FRP:

$$f_{fr3} := E_f \cdot \varepsilon_{f3}(c_{u3}) = 49.191 \cdot \text{ksi}$$

The developable tensile stress is (ACI 440.1R-06 Equation 11-3)

$$f_{fd13} := \begin{vmatrix} f_{fu} \text{ if } \dfrac{\sqrt{f'_c \cdot psi}}{\alpha_{Neg3}} \cdot \left(13 \cdot 6 \cdot \dfrac{l_{emb3}}{\phi_{f\_bar}} + \dfrac{c_{b3}}{\phi_{f\_bar}} \cdot \dfrac{l_{emb3}}{\phi_{f\_bar}} + 340 \right) \geq f_{fu} \\[2em] \left[\dfrac{\sqrt{f'_c \cdot psi}}{\alpha_{Neg3}} \cdot \left(13 \cdot 6 \cdot \dfrac{l_{emb3}}{\phi_{f\_bar}} + \dfrac{c_{b3}}{\phi_{f\_bar}} \cdot \dfrac{l_{emb3}}{\phi_{f\_bar}} + 340 \right) \right] \text{ otherwise} \end{vmatrix}$$

$f_{fd13} = 64 \cdot ksi$

$$\text{CheckFailure3} := \begin{vmatrix} \text{"Bar ultimate strength"} & \text{if } f_{fd13} \geq f_{fr3} \\[1em] \text{"Bond strength"} & \text{otherwise} \end{vmatrix}$$

CheckFailure3 = "Bar ultimate strength"

The cross section of interest is not a bond-critical section and adequate moment capacity can be achieved.

Ultimate bending moment diagram—Exterior bay: Exterior bay 1-foot slab strip, ft:

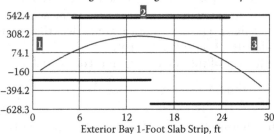

- **Ultimate Bending Moment Diagram - Exterior Bay**

Exterior Bay 1-Foot Slab Strip, ft

Section ▮1	Section ▮2	Section ▮3
$M_{u1} = 145.4 \cdot kip \cdot ft$	$M_{u2} = 271 \cdot 2 \cdot kip \cdot ft$	$M_{u3} = 349.1 \cdot kip \cdot ft$
$\phi_{bond} \cdot M_{nnew} = 205 \cdot kip \cdot ft$	$\phi_{b2} \cdot M_{n2} = 504 \cdot kip \cdot ft$	$\phi_{b3} \cdot M_{n3} = 563 \cdot kip \cdot ft$

7.7 STEP 5—CHECK CREEP-RUPTURE STRESS

7.7.1 Case 1—Exterior support

Creep-rupture stress in the FRP has to be evaluated considering the total unfactored dead loads and the sustained portion of the live load (20% of the total live load).

Bending moment due to dead load plus 20% of live load is

$$M_{1\_creep} := M_{s1} \cdot \frac{w_D + 0.20 w_{LNeg}}{w_{SNeg}} = 82.3 \cdot ft \cdot kip$$

Ratio of modulus of elasticity of bars to modulus of elasticity of concrete is

$$n_f = 1.277$$

Ratio of depth of neutral axis to reinforcement depth, calculated per Equation (8-12):

$$k_{1new} := \sqrt{2 \rho_{f1new} \cdot n_f + \left(\rho_{f1new} \cdot n_f\right)^2} - \rho_{f1new} \cdot n_f = 0.189$$

The tensile stress in the FRP is

$$f_{f1\_creep} := \frac{M_{1\_creep}}{A_{f1new} \cdot d_{f1} \cdot \left(1 - \dfrac{k_{1new}}{3}\right)} = 6.4 \cdot ksi$$

$$\text{Check\_Creep1} := \begin{vmatrix} \text{``OK''} & \text{if } f_{f1\_creep} \le f_{f\_creep} \\ \text{``Not good''} & \text{otherwise} \end{vmatrix} \qquad \left(f_{f\_creep} = 12.8 \cdot ksi\right)$$

$$\text{Check\_Creep1} = \text{``OK''}$$

The strain distribution is shown next.

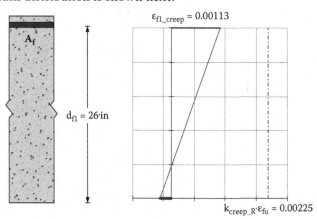

$$\varepsilon_{f1\_creep} = 0.00113$$

$$d_{f1} = 26 \cdot in$$

$$k_{creep\_R} \cdot \varepsilon_{fu} = 0.00225$$

Navier's equation is applicable because the maximum concrete stress is smaller than $0.45 f'c$:

$$E_c \cdot \varepsilon_{c\_creepl} = 1178 \cdot psi \quad < 0.45 f'_c = 2700 \cdot psi$$

7.7.2 Case 2—Midspan

Bending moment due to dead load plus 20% of live load is

$$M_{2\_creep} := M_{s2} \cdot \frac{w_D + 0.20 w_{LPos}}{w_{SPos}} = 143.8 \cdot ft \cdot kip$$

Ratio of modulus of elasticity of bars to modulus of elasticity of concrete:

$$n_f = 1.277$$

Ratio of depth of neutral axis to reinforcement depth, calculated per Equation (8-12):

$$k_2 := \sqrt{2\rho_{f2} \cdot n_f + \left(\rho_{f2} \cdot n_f\right)^2} - \rho_{f2} \cdot n_f = 0.135$$

The tensile stress in the FRP is computed, conservatively, assuming one single layer of reinforcement placed at the depth of the second layer, d_{f2II}:

$$f_{f2\_creep} := \frac{M_{2\_creep}}{\left(A_{f2I} + A_{f2II}\right) \cdot d_{f2II} \cdot \left(1 - \dfrac{k_2}{3}\right)} = 12.0 \cdot ksi$$

$$Check\_Creep2 := \begin{vmatrix} \text{"OK" if } f_{f2\_creep} \le f_{f\_creep} & \left(f_{f\_creep} = 12.8 \cdot ksi\right) \\ \text{"Not good" otherwise} \end{vmatrix}$$

$$Check\_Creep2 = \text{"OK"}$$

$k_{creep\_R} \cdot \varepsilon_{fu} = 0.00225$ $\varepsilon_{check\_creep2} = 0.000327$

$d_{f2II} = 24 \cdot in$

$\varepsilon_{f2\_creep} = 0.0021$

Navier's equation is applicable because the maximum concrete stress is smaller than $0.45 f'_c$:

$$E_c \cdot \varepsilon_{c\_creep2} = 1462 \cdot psi \quad < 0.45 f'_c = 2700 \cdot psi$$

7.7.3 Case 3—Interior support

Bending moment due to dead load plus 20% of live load is

$$M_{3\_creep} := M_{s3} \cdot \frac{w_D + 0.20 w_{LNeg}}{w_{SNeg}} = 197.5 \cdot ft \cdot kip$$

Ratio of modulus of elasticity of bars to modulus of elasticity of concrete is

$$n_f = 1.277$$

Ratio of depth of neutral axis to reinforcement depth, calculated per Equation (8-12), is

$$k_3 := \sqrt{2\rho_{f3} \cdot n_f + \left(\rho_{f3} \cdot n_f\right)^2} - \rho_{f3} \cdot n_f = 0.209$$

The tensile stress in the FRP is

$$f_{f3\_creep} := \frac{M_{3\_creep}}{A_{f3} \cdot d_{f3} \cdot \left(1 - \dfrac{k_3}{3}\right)} = 12.5 \cdot ksi$$

$$Check\_Creep3 := \left| \begin{array}{l} \text{“OK” if } f_{f3\_creep} \le f_{f\_creep} \\ \text{“Not good” otherwise} \end{array} \right. \qquad \left(f_{f\_creep} = 12.8 \cdot ksi\right)$$

$$Check\_Creep3 = \text{“OK”}$$

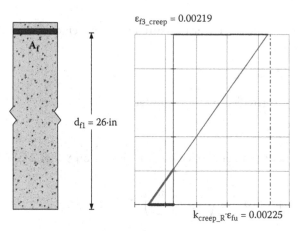

$$\varepsilon_{f3\_creep} = 0.00219$$

$$A_f$$

$$d_{f1} = 26 \cdot in$$

$$k_{creep\_R} \cdot \varepsilon_{fu} = 0.00225$$

Navier's equation is applicable because the maximum concrete stress is smaller than $0.45f'_c$:

$$E_c \cdot \varepsilon_{c\_creep3} = 2577 \cdot psi \quad < 0.45 f'_c = 2700 \cdot psi$$

7.8 STEP 6—CHECK CRACK WIDTH

Crack width is checked using Equation (8-9) of ACI 440.1R-06. A crack width limit, w_{lim}, of 0.028 in. is used for interior exposure.

7.8.1 Case 1—Exterior support

Ratio of modulus of elasticity of bars to modulus of elasticity of concrete is

$$n_f = 1.277$$

Ratio of depth of neutral axis to reinforcement depth is

$$k_{1new} = 0.189$$

Tensile stress in FRP under service loads is

$$f_{fs1new} := \frac{M_{s1}}{A_{f1new} \cdot d_{f1} \cdot \left(1 - \frac{k_{1new}}{3}\right)} = 8.604 \cdot ksi$$

Ratio of distance from neutral axis to extreme tension fiber to distance from neutral axis to center of tensile reinforcement is

$$\beta_{11} := \frac{h - k_{1new} \cdot d_{f1}}{d_{f1} \cdot (1 - k_{1new})} = 1.095$$

Thickness of concrete cover measured from extreme tension fiber to center of bar is

$$d_{c1} := h - d_{f1} = 2 \cdot in.$$

Bond factor (provided by the manufacturer):

$$k_b := 0.9$$

The crack width under service loads is (Equation 8-9 of ACI 440.1R-06):

$$w_1 := 2 \frac{f_{fs1new}}{E_f} \beta_{11} \cdot k_b \cdot \sqrt{d_{c1}^2 + \left(\frac{s_{f1}}{2}\right)^2} = 0.008 \cdot in.$$

The crack width limit for the selected exposure is

$$w_{lim} = 0.028 \cdot in.$$

$$\text{Check\_Crack1} := \begin{vmatrix} \text{"OK" if } w_1 \le w_{lim} \\ \text{"Not good" otherwise} \end{vmatrix}$$

Check_Crack1 = "OK"

7.8.2 Case 2—Midspan

Ratio of modulus of elasticity of bars to modulus of elasticity of concrete is

$n_f = 1.277$

Ratio of depth of neutral axis to reinforcement depth is

$k_2 = 0.135$

Tensile stress in FRP under service loads assuming one layer of reinforcement is

$$f_{fs2} := \frac{M_{s2}}{\left(A_{f2I} + A_{f2II}\right) \cdot d_{f2I} \cdot \left(1 - \dfrac{k_2}{3}\right)} = 15.5 \cdot ksi$$

Ratio of distance from neutral axis to extreme tension fiber to distance from neutral axis to center of tensile reinforcement is

$$\beta_{12} := \frac{h - k_2 \cdot d_{f2}}{d_{f2} \cdot \left(1 - k_2\right)} = 1.089$$

Thickness of concrete cover measured from extreme tension fiber to center of bar is

$d_{c2} := h - d_{f2} = 2 \cdot in.$

Bond factor (provided by the manufacturer):

$k_b = 0.9$

The crack width under service loads is (Equation 8-9 of ACI 440.1R-06):

$$w_2 := 2 \frac{f_{fs2}}{E_f} \beta_{12} \cdot k_b \cdot \sqrt{d_{c2}^2 + \left(\frac{s_{f22}}{2}\right)^2} = 0.013 \cdot in.$$

The crack width limit for the selected exposure is

$w_{lim} = 0.028 \cdot in.$

$$\text{Check\_Crack2} := \begin{vmatrix} \text{``OK'' if } w_2 \le w_{lim} \\ \text{``Not good''} \quad \text{otherwise} \end{vmatrix}$$

$$\text{Check\_Crack2} = \text{``OK''}$$

7.8.3 Case 3—Interior support

Ratio of modulus of elasticity of bars to modulus of elasticity of concrete is

$$n_f = 1.277$$

Ratio of depth of neutral axis to reinforcement depth is

$$k_3 = 0.209$$

Tensile stress in FRP under service loads is

$$f_{fs3} := \frac{M_{s3}}{A_{f3} \cdot d_{f3} \cdot \left(1 - \dfrac{k_3}{3}\right)} = 16.637 \cdot ksi$$

Ratio of distance from neutral axis to extreme tension fiber to distance from neutral axis to center of tensile reinforcement

$$\beta_{13} := \frac{h - k_3 \cdot d_{f3}}{d_{f3} \cdot (1 - k_3)} = 1.097$$

Thickness of concrete cover measured from extreme tension fiber to center of bar is

$$d_{c3} := h - d_{f3} = 2 \cdot in.$$

Bond factor (provided by the manufacturer) is

$$k_b = 0.9$$

The crack width under service loads (Equation 8-9 of ACI 440.1R-06) is

$$w_3 := 2 \frac{f_{fs3}}{E_f} \beta_{13} \cdot k_b \cdot \sqrt{d_{c3}^2 + \left(\frac{s_{f3}}{2}\right)^2} = 0.021 \cdot in.$$

The crack width limit for the selected exposure is

$$w_{lim} = 0.028 \cdot in.$$

$$\text{Check\_Crack3} := \begin{vmatrix} \text{"OK" if } w_3 \leq w_{lim} \\ \text{"Not good" otherwise} \end{vmatrix}$$

$$\text{Check\_Crack3} = \text{"OK"}$$

7.9 STEP 7—CHECK MAXIMUM MIDSPAN DEFLECTION

The service bending moment diagram is shown next.

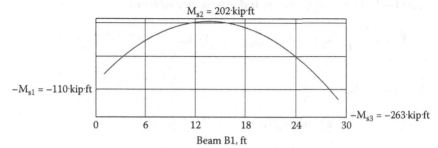

Beam B1, ft

Select the maximum allowable deflection:

$$\Delta_{lim} := \frac{l_1}{360} = 1 \cdot in.$$

Preliminary calculations: The depth of the neutral axis of the gross section is

$$d_g := \frac{b_{eff} \cdot \dfrac{t_{slab}^2}{2} + b_w \cdot \dfrac{h^2 - t_{slab}^2}{2}}{b_{eff} \cdot t_{slab} + b_w \cdot (h - t_{slab})} = 8.9 \cdot in$$

The gross moment of inertia is

$$I_g := \frac{b_{eff} \cdot t_{slab}^3}{12} + b_{eff} \cdot t_{slab} \cdot \left(d_g - \frac{t_{slab}}{2}\right)^2 + \frac{b_w \cdot (h - t_{slab})^3}{12}$$

$$+ b_w \cdot (h - t_{slab}) \cdot \left[d_g - (h - t_{slab})\right]^2$$

$$I_g = 58799 \cdot in.^4$$

The negative cracking moment is

$$M_{crNeg} := \frac{f_r \cdot I_g}{d_g} = 318 \cdot kip \cdot ft$$

The positive cracking moment is

$$M_{crPos} := \frac{f_r \cdot I_g}{h - d_g} = 149 \cdot kip \cdot ft$$

Cracked moment of inertia: The cracked moment of inertia, I_{cr}, is computed per Equation (8-11) of ACI 440.1R-06 at the following locations.

Case 1—Exterior support (includes presence of bends)

$$I_{cr1} := \frac{b_w \cdot d_{f1}^3}{3} k_{1new}^3 + n_f \cdot A_{f1new} \cdot d_{f1}^2 \cdot (1 - k_{1new})^2 = 4122 \cdot in^4$$

Case 2—Midspan

$$I_{cr2} := \frac{b_w \cdot d_{f2}^3}{3} k_2^3 + n_f \cdot A_{f2} \cdot d_{f2}^2 \cdot (1 - k_2)^2 = 4261 \cdot in^4$$

Case 3—Interior support

$$I_{cr3} := \frac{b_w \cdot d_{f3}^3}{3} k_3^3 + n_f \cdot A_{f3} \cdot d_{f3}^2 \cdot (1 - k_3)^2 = 4991 \cdot in^4$$

Reduction coefficients (modified Branson's method of ACI 440.1R-06):

Case 1—Exterior support

The reduction coefficient related to the reduced tension stiffening exhibited in the FRP reinforced members, computed per Equation (8-13b) of ACI 440.1R is

$$\beta_{d1} := \frac{1}{5} \cdot \left(\frac{\rho_{f1}}{\rho_{fb1}} \right) = 0.137$$

Case 2—Midspan

$$\beta_{d2} := \frac{1}{5} \cdot \left(\frac{\rho_{f2}}{\rho_{fb2}} \right) = 0.131$$

Case 3—Interior support

$$\beta_{d3} := \frac{1}{5} \cdot \left(\frac{\rho_{f3}}{\rho_{fb3}} \right) = 0.342$$

Average effective moment of inertia (modified Branson's method of ACI 440.1R-06): The bending moment at midspan due to service loads is

$$M_{sPos} = M_{s2} = 202 \cdot kip \cdot ft$$

The bending moment at the continuous end due to service loads is

$$M_{sNeg} := M_{s3} = 263 \cdot kip \cdot ft$$

The value of I_{e_Br} at midspan is

$$I_{e\_Br2} := \left(\frac{M_{crPos}}{M_{sPos}} \right)^3 \beta_{d2} \cdot I_g + \left[1 - \left(\frac{M_{crPos}}{M_{sPos}} \right)^3 \right] I_{cr2} = 5647 \cdot in^4$$

The value of I_{e_Br} at the continuous end is

$$I_{e\_Br3} := \left(\frac{M_{crNeg}}{M_{sNeg}} \right)^3 \beta_{d3} \cdot I_g + \left[1 - \left(\frac{M_{crNeg}}{M_{sNeg}} \right)^3 \right] I_{cr3} = 31657 \cdot in^4$$

The average effective moment of inertia, I_{e_Br}, is computed as indicated in Section 9.5.2.4 of ACI 318. The average value of I_{e_Br} is

$$I_{e\_Br} := 0.85 \cdot I_{e\_Br2} + 0.15 \cdot I_{e\_Br3} = 9549 \cdot in.^4$$

Deflection at midspan (modified Branson's method of ACI 440.1R-06): Calculate the moment at midspan due to service load on a simply supported beam, M_o:

$$M_o := \frac{w_{sPos} \cdot 1_1^2}{8} = 405.412 \cdot ft \cdot kip$$

The maximum deflection under service loads is

$$\Delta_{SL\_Br} := \frac{5M_o \cdot 1_1^2}{48E_c \cdot I_{e\_Br}} - \left(M_{s1} + M_{s3} \right) \cdot \frac{1_1^2}{16E_c \cdot I_{e\_Br}} = 0.69 \cdot in.$$

Deflection due to dead loads only is

$$\Delta_{DL\_Br} := \Delta_{SL\_Br} \cdot \frac{w_D}{w_{sPos}} = 0.442 \cdot in.$$

Deflection due to live loads only is

$$\Delta_{LL\_Br} := \Delta_{SL\_Br} \cdot \frac{W_{LPos}}{W_{SPos}} = 0.248 \cdot in.$$

The multiplier for time-dependent deflection at 5 years (ACI 318-11) is

$$\xi := 2$$

The reduction parameter, Equation (8-14b) of ACI 440.1R-06, is

$$\lambda := 0.60\xi = 1.2$$

$$\Delta_{LT-Br} := \Delta_{LL-Br} + \lambda \cdot (\Delta_{DL-Br} + 0.20 \cdot \Delta_{LL\_Br}) = 0.838 \text{ in long-term deflection}$$

$$\text{Check\_}\Delta_{LT\_Br} := \left| \begin{array}{l} \text{"OK" if } \Delta_{LT\_Br} \leq \Delta_{lim} \\ \text{"Not good" otherwise} \end{array} \right.$$

$$\text{Check\_}\Delta_{LT\_Br} = \text{"OK"}$$

Average effective moment of inertia [2]: The value of I_e at midspan is computed as follows according to the procedure defined by Bischoff [2] and discussed in Chapter 4. The ratio between the cracking moment and the applied moment is smaller than 1.0:

$$\frac{M_{crPos}}{M_{sPos}} = 0.74$$

$$\gamma_{Bischoff2} := 1.72 - 0.72 \cdot \left(\frac{M_{crPos}}{M_{sPos}} \right) = 1.187$$

$$I_{e2} := \frac{I_{cr2}}{1 - \gamma_{Bischoff2} \cdot \left(\dfrac{M_{crPos}}{M_{sPos}} \right)^2 \cdot \left(1 - \dfrac{I_{cr2}}{I_g} \right)} = 10749 \cdot in^4$$

The value of I_e at the continuous end is taken equal to the gross moment of inertia because the ratio between the cracking moment and the applied moment is larger than 1.0:

$$\frac{M_{crNeg}}{M_{sNeg}} = 1.208$$

$$I_{e3} := I_g = 58799 \cdot in.^4$$

The average value of I_e is

$$I_e := 0.85 \cdot I_{e2} + 0.15 \cdot I_{e3} = 17956 \cdot in.^4$$

Deflection at midspan [2]: Calculate the moment at midspan due to service load on a simply supported beam, M_o:

$$M_o = 405.4 \cdot \text{ft} \cdot \text{kip}$$

The maximum deflection under service loads is $\xi = 2$ and $\lambda = 1.2$. Maximum deflection under service loads is

$$\Delta_{SL} := \frac{5 \cdot M_o \cdot 1_1^2}{48 E_c \cdot I_e} - \left(M_{s1} + M_{s3}\right) \cdot \frac{1_1^2}{16 E_c \cdot I_e} = 0.367 \cdot \text{in.}$$

Deflection due to dead loads only is

$$\Delta_{DL} := \Delta_{SL} \cdot \frac{w_D}{w_{SPos}} = 0.235 \cdot \text{in.}$$

Deflection due to live loads only is

$$\Delta_{LL} := \Delta_{SL} \cdot \frac{w_{LPos}}{w_{SPos}} = 0.132 \cdot \text{in.}$$

Long-term deflection is

$$\Delta_{LT} := \Delta_{LL} + \lambda \cdot (\Delta_{DL} + 0.20 \cdot \Delta_{LL}) = 0.446 \cdot \text{in.}$$

$$\text{Check\_}\Delta_{LT} := \begin{vmatrix} \text{"OK" if} \leq \Delta_{LT} \leq \Delta_{\lim} \\ \text{"Not good" otherwise} \end{vmatrix}$$

$$\text{Check\_}\Delta_{LT} = \text{"OK"}$$

Midspan deflection diagram [2] and ACI 440.1R–06:

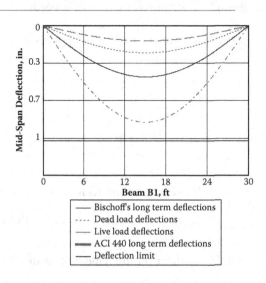

7.10 STEP 8—DESIGN FRP REINFORCEMENT FOR SHEAR CAPACITY

Preliminary calculations: The maximum shear values at the face of the external and internal supports and midspan are

$$V_{u1} = 62 \cdot \text{kip} \qquad V_{u2} = 10.2 \cdot \text{kip} \qquad V_{u3} = 72 \cdot \text{kip}$$

The concrete shear capacity, V_c, can be calculated per Equation (9-1) of ACI 440.1R-06, where k is the ratio of depth of neutral axis to reinforcement depth, calculated per Equation (8-12). As discussed in Chapter 4, V_c minimum value of $0.8\sqrt{f'_c} \cdot b \cdot d_f$ is not applicable here.

The shear strength-reduction factor is

$$\phi_v := 0.75$$

The concrete shear capacity at the exterior support is

$$k_{1new} = 0.189$$

$$d_{f1} = 26 \cdot \text{in.}$$

$$V_{c1} := 5\sqrt{f'_c \cdot \text{psib}_w} \cdot \left(k_{1new} \cdot d_{f1}\right) = 26.7 \cdot \text{kip}$$

$$\text{Check\_ShearReinf1} := \begin{vmatrix} \text{"OK. No shear reinforcement is needed."} & \text{if } \phi_v \cdot \dfrac{V_{c1}}{2} \geq V_{u1} \\ \text{"Shear reinforcement is needed"} & \text{otherwise} \end{vmatrix}$$

$$\text{Check\_ShearReinf1} = \text{"Shear reinforcement is needed"}$$

The concrete shear capacity at the interior support is

$$k_3 = 0.209$$

$$d_{f3} = 26 \cdot \text{in.}$$

$$V_{c3} := 5\sqrt{f'_c \text{psib}_w} \cdot \left(k_3 \cdot d_{f3}\right) = 29.4 \cdot \text{kip}$$

$$\text{Check\_ShearReinf3} := \begin{vmatrix} \text{"OK. No shear reinforcement is needed."} & \text{if } \phi_v \cdot \dfrac{V_{c3}}{2} \geq V_{u3} \\ \text{"Shear reinforcement is needed"} & \text{otherwise} \end{vmatrix}$$

$$\text{Check\_ShearReinf3} = \text{"Shear reinforcement is needed"}$$

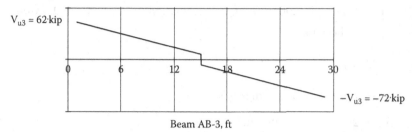

Shear diagram

$V_{u3} = 62 \cdot kip$

$-V_{u3} = -72 \cdot kip$

Beam AB-3, ft

Select the GFRP reinforcement

Stirrup_Size :=

#2
#3
#4
#5
#6
#7
#8
#9
#10

The ACI 440.6 minimum manufacturer's guaranteed mechanical properties of the selected bars are

$f_{fuuv} = 100 \cdot ksi$ Ultimate guaranteed tensile strength of the FRP
$\varepsilon_{fuuv} = 0.018$ Ultimate guaranteed rupture strain of the FRP
$E_f = 5700 \cdot ksi$ Guaranteed tensile modulus of elasticity of the FRP

The geometrical properties of the selected bars are

$\phi_{f-stirrup} = 0.5 \cdot in.$ Bar diameter
$A_{f-stirrup} = 0.196 \cdot in.^2$ Bar area
$r_{bsv} := 3\phi_{f-stirrup} = 1.5 \cdot in.$ Radius of the bend

The tail of the stirrup is to be at least 12 bar diameters:

$1_{vtail} := 12 \cdot \phi_{f-stirrup} = 6 \cdot in.$

The design tensile strength of the bend of FRP bar is

$$f_{fbsv} := \left(0.05 \cdot \frac{r_{bsv}}{\phi_{f\_stirrup}} + 0.3 \right) f_{fu} = 28.8 \cdot ksi$$ Equation (7-3) of ACI 440.1R-06

The stress level in the shear reinforcement is limited by Equation (9-3) of ACI 440.1R-06:

$$f_{fv} := \begin{vmatrix} (0.004 \cdot E_f) & \text{if } 0.004 \cdot E_f \leq f_{fbsv} \\ f_{fbsv} & \text{otherwise} \end{vmatrix} \qquad \text{Equation (9-3) of ACI 440.1R-06:}$$

$f_{fv} = 22.8 \cdot \text{ksi}$

The area of the two-leg stirrup is

$A_{fv} := 2 \cdot A_{f\_stirrup} = 0.393 \cdot \text{in.}^2$

FRP shear reinforcement at interior support (same reinforcement is used for the exterior support): The maximum stirrup spacing, S_{sv}, is (Equation 9-4 of ACI 440.1R-06):

$$S_{sv3\_max} := \frac{A_{fv} \cdot \phi_v \cdot f_{fv} \cdot d_{f3}}{V_{u3} - \phi_v \cdot V_{c3}} = 3.52 \cdot \text{in.}$$

The selected spacing is

$S_{sv3} := 3 \text{ in.}$

The shear resistance provided by FRP stirrups, V_f, is

$$V_{f30} := \frac{A_{fv} \cdot f_{fv} \cdot d_{f3}}{S_{sv3}} = 78 \cdot \text{kip} \qquad \text{Equation (9-2) of ACI 440.1R-06}$$

As discussed in Chapter 4, V_f cannot exceed $3V_c$:

$3V_{c3} = 88 \cdot \text{kip}$

$V_{f3} := \min(V_{f30}, 3V_{c3}) = 78 \cdot \text{kip}$

The nominal shear capacity, V_n, is

$V_{n3} := V_{c3} + V_{f3} = 107 \cdot \text{kip}$

$\phi_v \cdot V_{n3} = 80 \cdot \text{kip}$

$$\text{Check\_Shear3} := \begin{vmatrix} \text{"OK"} & \text{if } \phi_v V_{n3} \geq V_{u3} \\ \text{"Not good"} & \text{otherwise} \end{vmatrix} \qquad (V_{u3} = 71.7 \text{kip})$$

Check_Shear3 = "OK"

The maximum stirrup spacing is 24 in. or d/2:

$$s_{svmax} := min\left(24in, \frac{d_{f3}}{2}\right) = 13 \cdot in.$$

$$Check\_Spacing := \begin{vmatrix} \text{"OK" if } s_{sv3\_max} \geq s_{sv3} \wedge s_{svmax} \geq s_{sv3} \\ \text{"Not good!" otherwise} \end{vmatrix}$$

$$Check\_Spacing = \text{"OK"}$$

FRP shear reinforcement at midspan: When the maximum spacing is used, the shear resistance provided by FRP stirrups, V_f, is

$$s_{svmax} = 13 \text{ in.}$$

$$V_{fmin} := \frac{A_{fv} \cdot f_{fv} \cdot d_{f2}}{s_{svmax}} = 18 \cdot kip \qquad \text{Equation (9-2) of ACI 440.1R-06}$$

$$V_{c2} := 5\sqrt{f'_c \cdot psi} b_w \cdot (k_2 \cdot d_{f2}) = 19 \cdot kip$$

The nominal shear capacity, V_n, is

$$V_{n2} := V_{c2} + V_{fmin} = 37 \cdot kip$$

$$\phi_v \cdot V_{n2} = 28 \cdot kip$$

$$Check\_Shear2 := \begin{vmatrix} \text{"OK" if } \phi_v \cdot V_{n2} \geq V_{u2} \qquad (V_{u2} = 10.2 \cdot kip) \\ \text{"Not good" otherwise} \end{vmatrix}$$

$$Check\_Shear2 = \text{"OK"}$$

The FRP stirrups are placed at the maximum spacing in the center part of the span over a length that extends for

$$l_{min-shear} := 0.25 \cdot l_1 = 7.5 \cdot ft$$

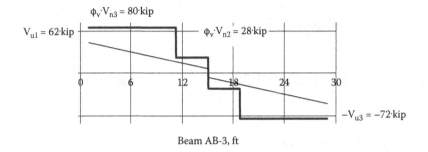

Beam AB-3, ft

7.II STEP 9—COMPUTE FRP CONTRIBUTION TO TORSIONAL STRENGTH

In this section, an attempt is made to discuss how to compute the contribution to the torsional strength provided by the FRP lateral reinforcement and close FRP stirrups.

Preliminary calculations: The area enclosed by centerline of the outermost closed transverse torsional reinforcement is

$$A_{oh} := \left(b_w - 2c_c - 2\frac{\phi_{f\_stirrup}}{2}\right) \cdot \left(h - 2c_c - 2\frac{\phi_{f\_stirrup}}{2}\right) = 257 \cdot in^2$$

Perimeter of centerline of outermost closed transverse torsional reinforcement is

$$p_h := 2 \cdot \left(b_w - 2c_c - 2\frac{\phi_{f\_stirrup}}{2}\right) + 2 \cdot \left(h - 2c_c - 2\frac{\phi_{f\_stirrup}}{2}\right) = 70 \cdot in.$$

Area enclosed by outside perimeter of concrete cross section is

$$A_{cp} := b_{eff} \cdot t_{slab} + b_w \cdot (h - t_{slab}) = 792 \cdot in.^2$$

Outside perimeter of concrete cross section is

$$P_{cp} := 2b_{eff} + 2 \cdot t_{slab} + 2 \cdot (h - t_{slab}) = 184 \cdot in.$$

Gross sectional area is

$$A_g := b_{eff} \cdot t_{slab} + b_w \cdot (h - t_{slab}) = 792 \cdot in.^2$$

- Compatibility torsion—ACI 318-11, Section 11.6.2:

$$T_{cr} := \phi_v \cdot 4 \cdot \sqrt{f_c' \cdot psi} \cdot \left(\frac{A_{cp}^2}{P_{cp}} \right) = 66 \cdot kip \cdot ft$$

- Threshold torsion—ACI 318-11, Section 11.6.1:

$$T_{tr} := \phi_v \cdot \sqrt{f_c' \cdot psi} \cdot \left(\frac{A_{cp}^2}{P_{cp}} \right) = 17 \cdot kip \cdot ft$$

- FRP reinforcement contribution to torsional strength—following ACI 318-11, Equation (11-21):
  The angle of the torsional cracks is taken equal to 45°:

$$\theta := 45° = 0.785$$

The cross-sectional area resisting to torsion can be computed as

$$A_o := 0.85 \cdot A_{oh} = 219 \cdot in.^2$$

The area of the FRP reinforcement resisting to torsion is

$$A_t := A_{f-stirrup} = 0.196 \cdot in.^2$$

The spacing of the FRP torsional reinforcement is

$$s_{st} := s_{sv3} = 3 \cdot in.$$

Provided that closed FRP stirrups are used, the FRP contribution to resist torsion could be computed as follows:

$$T_n := \frac{2 A_o \cdot A_t \cdot f_{fv}}{s_{st}} \cot(\theta) = 54 \cdot kip \cdot ft$$

- Limitation on cross section: ACI 318-11, Section 11.6.3.1:
  The following ACI 318-11 limitation to the cross-section geometry should be satisfied:

$$\sqrt{\left(\max\left(\frac{V_{u1}}{b_{w1} \cdot d_1} \right) \right)^2 + \left(\frac{T_u \cdot p_h}{1.7 \cdot A_{oh}^2} \right)^2} \leq \phi \cdot \left(\max\left(\frac{V_{c1}}{b_{w1} \cdot d_1} \right) + 8 \cdot \sqrt{f_c' \cdot psi} \right)$$

REFERENCES

1. H. Jawahery Zadeh and A. Nanni. Reliability analysis of concrete beams internally reinforced with FRP bars. *ACI Structural Journal* 110 (6): 1023–1032 (2013).
2. P. H. Bischoff. Reevaluation of deflection prediction for concrete beams reinforced with steel and fiber reinforced polymer bars. *Journal of Structural Engineering* 131 (5): 752–767 (2005).

Chapter 8

Design of a two-way slab

8.I INTRODUCTION

The floor plan of a two-story medical facility building is shown in Figure 8.1. The column spacing is dictated by the size of the equipment that occupies the ground floor. The second-floor system is a two-way reinforced concrete (RC) slab with each panel supported by perimeter beams. The building is located in a region of low seismicity. Loading of each floor consists of the self-weight, a superimposed dead load of 2.5 psf, and a live load of 100 psf.

This example describes the procedure to design the two-way slab of the second floor. The design is presented as a sequence of eight steps as summarized here:

Step 1   Define slab geometry and concrete properties
Step 2   Compute factored loads
Step 3   Compute ultimate and service bending moments and shear forces
Step 4   Design FRP reinforcement for bending moment capacity (with shear check)
Step 5   Check creep-rupture stress
Step 6   Check crack width
Step 7   Check maximum midspan deflection
Step 8   Check for punching shear (no perimeter beams)

Because of the symmetry, the design refers to the panel included between gridlines A, B, 1, and 2 (Figure 8.1). The results of the design are summarized next to facilitate the understanding of the eight sequential steps devoted to calculations.

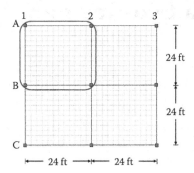

Figure 8.1 Floor plan.

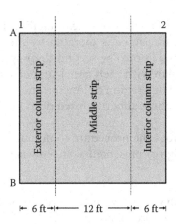

Figure 8.2 Column and middle strips relative to the design panel.

8.2 DESIGN SUMMARY

Based on a slab thickness of 8 in., the following loads are considered:

Slab self-weight	96.7 psf	
Superimposed dead load	2.5 psf	
Live load	100 psf	
Total factored load	279 psf	(0.279 kip/ft)
Service load	199 psf	(0.199 kip/ft)

Bending moments and shear forces per unit strip are computed for the column and middle strips (see Figure 8.2). A commercial finite element modeling software was used to compute bending moment and shear forces for the various strips as summarized in Table 8.1. Because of the symmetry, only the forces in one direction are shown.

The required fiber-reinforced polymer (FRP) reinforcement for a slab thickness of 8 in. is summarized in Table 8.2.

Table 8.3 shows a summary of demands and capacities at critical sections. It should be noted that minimum shrinkage and temperature reinforcement provide sufficient flexural strength everywhere.

The final bar layout is selected to optimize bar production time and construction effort.

Bar layout and typical details are shown in Figures 8.3(a) and 8.3(b).

Table 8.4 is provided to convert US customary units to the SI system.

Table 8.1 Bending moments and shear force

| Section | Bending moment | | Shear force |
	Ultimate moment (kip-ft/ft)	Service moment (kip-ft/ft)	Ultimate shear (kip/ft)
Ext. col strip			
Exterior support	2.79	2.00	1.19
Midspan	1.60	1.14	0.25
Interior support	5.09	3.64	1.83
Middle strip			
Exterior support	1.91	3.12	1.91
Midspan	3.79	2.71	0.25
Interior support	7.51	5.37	2.35
Int. col strip			
Exterior support	2.47	1.77	0.81
Midspan	1.45	1.04	0.23
Interior support	4.62	4.62	1.40

Table 8.2 Slab geometry and reinforcement

Slab thickness	Reinforcement
8 in.	No. 4 @ 5 in. (top and bottom and both directions)

Table 8.3 Slab design summary

Limit state	Section	Demand/computed	Capacity/limit
Ultimate			
Flexural strength		7.51 kip-ft/ft	12.5 kip-ft/ft
Shear strength	Interior support Middle strip	2.35 kip/ft	3.31 kip/ft
Serviceability			
Creep rupture		13.1 ksi	16 ksi
Crack width		0.025 in.	0.028 in.
Maximum panel center deflection		0.232 in.	0.600 in.

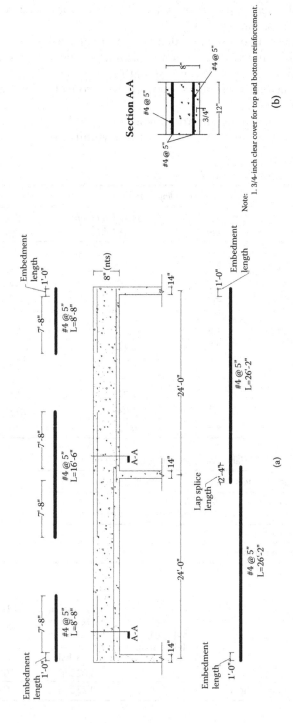

Section A-A

#4 @ 5"

#4 @ 5"

3/4"

8"

12"

#4 @ 5"

#4 @ 5"

Note:

1. 3/4-inch clear cover for top and bottom reinforcement.

(b)

Embedment length 1'-0"

7'-8"

#4 @ 5"
L=8'-8"

7'-8"

#4 @ 5"
L=16'-6"

7'-8"

1'-0"

14"

A-A

14"

24'-0"

8" (nts)

Lap splice length

2'-4"

#4 @ 5"
L=26'-2"

Embedment length

14"

A-A

14"

24'-0"

#4 @ 5"
L=26'-2"

Embedment length 1'-0"

7'-8"

#4 @ 5"
L=8'-8"

Embedment length 1'-0"

Embedment length 1'-0"

(a)

Figure 8.3 (a) Reinforcement layout; (b) typical details.

Table 8.4 Conversion table

US customary	SI units
Lengths, areas, section properties	
1 in.	25.4 mm
	0.025 m
1 ft	304.8 mm
	0.305 m
1 in.^{2}	645 mm^2
1 ft^2	0.093 m^2
1 in.^{3}	16,387 mm^3
1 in.^{4}	416,231 mm^4
Forces, pressures, strengths	
1 lbf	4.448 N
1 kip	4.448 kN
1 lbf-ft	1.356 N·m
1 kip-ft	1.356 kN·m
1 psi	6.895 kPa
1 psf	47.88 N/m^2
1 ksi	6.895 MPa
1 ksf	47.88 kN/m^2
1 lbf/ft^3	157.1 N/m^3

8.3 STEP 1—DEFINE SLAB GEOMETRY AND CONCRETE PROPERTIES

8.3.1 Geometry

The two-way slab spans over two bays along both directions:

$l_1 := 24$ ft
$l_2 := 24$ ft

A clear top and bottom concrete cover, c_c, is

$c_c := 0.75$ in.

The width of the perimeter beams is

$b_{beam} := 14$ in.

8.3.2 Concrete properties

The following concrete properties are considered for the design:

$f'_c := 5000 \text{ psi}$ Compressive strength

$\varepsilon_{cu} := 0.003$ Ultimate compressive strain

$\rho_c := 145 \dfrac{1bf}{ft^3}$ Density

$E_c := 33 \text{ psi}^{0.5} \left(\dfrac{\rho_c}{1bf \cdot ft^{-3}} \right)^{1.5} \cdot \sqrt{f'_c}$ Compressive modulus of elasticity

$E_c = 4074 \cdot \text{ksi}$ Computed as indicated in ACI 318-11

$f_{ct} := 7.5 \text{ psi}^{0.5} \cdot \sqrt{f'_c}$ Concrete tensile strength

$f_{ct} = 530 \cdot \text{psi}$ Computed as indicated in ACI 318-11

The stress-block factor, β_1, is computed as indicated in ACI 318-11:

$$\beta_1 := \begin{vmatrix} 0.85 \text{ if } f'_c = 4000 \text{ psi} \\ 1.05 - 0.05 \cdot \dfrac{f'_c}{1000 \text{ psi}} \quad \text{if } 4000 \text{ psi} < f'_c < 8000 \text{ psi} \quad = 0.8 \\ 0.65 \text{ otherwise} \end{vmatrix}$$

8.3.3 Analytical approximations of concrete compressive stress–strain curve—Todeschini's model

Compressive strain at peak:

$$\varepsilon_{c0} := \frac{1.71 \cdot f'_c}{E_c} = 0.0021$$

Compressive stress at peak:

$$\sigma''_c := 0.9 \cdot \frac{f''_c}{\text{psi}}$$

Stress–strain curve equation:

$$\sigma_c(\varepsilon_c) := \frac{2 \cdot \sigma''_c \cdot \left(\dfrac{\varepsilon_c}{\varepsilon_{c0}}\right)}{1 + \left(\dfrac{\varepsilon_c}{\varepsilon_{c0}}\right)^2}$$

Concrete Compressive Stress-Strain Curve

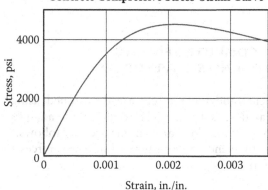

Strain, in./in.

8.4 STEP 2—COMPUTE THE FACTORED LOADS

As a first guess, the following slab thickness, t_{slab}, was selected:

$t_{slab} := 8$ in.

The self-weight of the slab is computed considering a concrete density of 145 psf. Other dead loads such as floor cover (0.5 psf) and ceiling (2 psf) are considered. A live load of 100 psf was requested by the owner. The following unfactored uniform loads are considered:

$SW := t_{slab} \cdot \rho_c = 96.7 \cdot psf$ Slab self-weight
$OD := 2.5$ psf   Other dead loads
$DL := SW + OD = 99.2 \cdot psf$ Total dead load
$LL_o := 100$ psf   Live load

The governing load combination for computing the total factored load, TFL, is load combination Equation (9-2) defined in Section 9.2.1 of ACI 318-11:

$TFL := 1.2 \cdot DL + 1.6 \cdot LL = 279 \cdot psf$

The total service load, SL, is

$SL := DL + LL = 199 \cdot psf$

For the unit-width slab strip, the design loads per unit width are w_u and w_s, for ultimate and service limit state, respectively:

$w_u := TFL \cdot 1 \ ft = 279 \cdot plf$

$w_s := SL \cdot 1 \ ft = 199 \cdot plf$

8.5 STEP 3—COMPUTE BENDING MOMENTS AND SHEAR FORCES

Bending moments and shear forces are determined using a commercial finite element analysis software. The results of the analysis relative to the panel for ultimate conditions are summarized as follows. Because of the symmetry, only the bending moments and the shear forces along one direction are considered:

Exterior column strip	Middle strip	Interior column strip
$V_{u11} := 7.14 \ kip$	$V_{u21} := 22.9 \ kip$	$V_{u31} := 9.67 \ kip$
$M_{uNeg11} := 16.7 \ kip \cdot ft$	$M_{uNeg21} := 52.4 \ kip \cdot ft$	$M_{uNeg31} := 29.6 \ kip \cdot ft$
$V_{u12} := 1.50 \ kip$	$V_{u22} := 3.05 \ kip$	$V_{u32} := 2.73 \ kip$
$M_{uPos12} := 9.60 \ kip \cdot ft$	$M_{uPos22} := 45.4 \ kip \cdot ft$	$M_{uPos32} := 17.4 \ kip \cdot ft$
$V_{u13} := 11.0 \ kip$	$V_{u23} := 28.2 \ kip$	$V_{u33} := 16.9 \ kip$
$M_{uNeg13} := 30.5 \ kip \cdot ft$	$M_{uNeg23} := 90.2 \ kip \cdot ft$	$M_{uNeg33} := 55.4 \ kip \cdot ft$

$$v_{u11} := 1.19 \frac{kip}{ft} \qquad v_{u21} := 1.19 \frac{kip}{ft} \qquad v_{u31} := 0.81 \frac{kip}{ft}$$

$$m_{uNeg11} := 2.79 \cdot \frac{kip \cdot ft}{ft} \qquad m_{uNeg21} := 4.36 \cdot \frac{kip \cdot ft}{ft} \qquad m_{uNeg31} := 2.47 \cdot \frac{kip \cdot ft}{ft}$$

$$v_{u12} := 0.25 \frac{kip}{ft} \qquad v_{u22} := 0.25 \frac{kip}{ft} \qquad v_{u32} := 0.23 \frac{kip}{ft}$$

$$m_{uPos12} := 1.60 \frac{kip \cdot ft}{ft} \qquad m_{uPos22} := 3.79 \frac{kip \cdot ft}{ft} \qquad m_{uPos32} := 1.45 \frac{kip \cdot ft}{ft}$$

$$v_{u13} := 1.83 \frac{kip}{ft} \qquad v_{u23} := 2.35 \frac{kip}{ft} \qquad v_{u33} := 1.40 \frac{kip}{ft}$$

$$m_{uNeg13} := 5.09 \cdot \frac{kip \cdot ft}{ft} \qquad m_{uNeg23} := 7.51 \cdot \frac{kip \cdot ft}{ft} \qquad m_{uNeg33} := 4.62 \cdot \frac{kip \cdot ft}{ft}$$

The values of the service bending moments per linear foot are summarized next:

Exterior column strip   Middle strip   Interior column strip

$$m_{sNeg11} := 2.00 \cdot \frac{kip \cdot ft}{ft} \qquad m_{sNeg21} := 3.12 \frac{kip \cdot ft}{ft} \qquad m_{sNeg31} := 1.77 \frac{kip \cdot ft}{ft}$$

$$m_{sPos12} := 1.14 \frac{kip \cdot ft}{ft} \qquad m_{sPos22} := 2.71 \frac{kip \cdot ft}{ft} \qquad m_{sPos32} := 1.04 \frac{kip \cdot ft}{ft}$$

$$m_{sNeg13} := 3.64 \frac{kip \cdot ft}{ft} \qquad m_{sNeg23} := 5.37 \frac{kip \cdot ft}{ft} \qquad m_{sNeg33} := 4.62 \frac{kip \cdot ft}{ft}$$

8.6 STEP 4—DESIGN FRP REINFORCEMENT FOR BENDING MOMENT CAPACITY

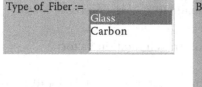
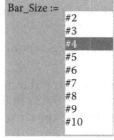

Select the FRP reinforcement: For the purpose of this design example, it is assumed that glass FRP (GFRP) bars of the same size are used everywhere in the slab.

The ACI 440.6 minimum manufacturer's guaranteed mechanical properties of the selected bars are

$f_{fuu} = 100 \cdot ksi$ Ultimate guaranteed tensile strength of the FRP

$\varepsilon_{fuu} = 0.018$ Ultimate guaranteed rupture strain of the FRP

$E_f = 5700 \cdot ksi$ Guaranteed tensile modulus of elasticity of the FRP

The geometrical properties of the selected bars are the following:

$\phi_{f\_bar} = 0.5 \cdot$ in. Bar diameter
$A_{f\_bar} = 0.196 \cdot$ in.^{2} Bar area

FRP reduction factors: Table 7-1 of ACI 440.1R-06 is used to define the environmental reduction factor, C_E. The type of exposure has to be selected:

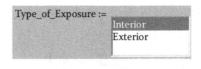

$C_E = 0.8$ Environmental reduction factor for GFRP

Table 8-3 in ACI 440.1R-06 is used to define the reduction factor to take into account the FRP creep-rupture stress. Creep-rupture stress in the FRP has to be evaluated considering the total unfactored dead loads and the sustained portion of the live load (20% of the total live load):

$k_{creep} = 0.2$ Creep-rupture stress limitation factor

Crack width is checked using Equation (8-9) of ACI 440.1R-06. A crack width limit, w_{lim}, of 0.028 in. is used for interior exposure, while 0.020 in. is used for exterior exposure:

$w_{lim} = 0.028$ in. Crack width limit

FRP ultimate design properties: The ultimate design properties are calculated per Section 7.2 of ACI 440.1R-06:

$f_{fu} := C_E \cdot f_{fuu} = 80 \cdot$ ksi Design tensile strength
$\varepsilon_{fu} = C_E \cdot \varepsilon_{fuu} = 0.014$ Design rupture strain

FRP creep-rupture limit stress: The FRP creep-rupture limit stress is calculated per Section 8.4 of ACI 440.1R-06:

$f_{f\_creep} := k_{creep} \cdot f_{fu} = 16 \cdot$ ksi

8.6.1 Thickness control

One-way shear strength is generally a controlling factor in determining the thickness of a two-way slab with perimeter beams inasmuch as the reinforcing level is normally close to minimum. The effective depths of the FRP reinforcement along the two directions, d_{f1} and d_{f2}, are

$$d_{f1} := t_{slab} - c_c - \frac{\phi_{f\_bar}}{2} = 7 \cdot in.$$

$$d_{f2} := t_{slab} - c_c - \frac{3 \cdot \phi_{f\_bar}}{2} = 6.5 \cdot in.$$

The maximum value of shear force per unit width, v_{uMax}, is the following:

$$v_{uMax} := v_{u23} = 2.35 \cdot \frac{kip}{ft}$$

The shear strength of the slab per unit width, ϕV_c, can be computed as follows. It is assumed that, for the cracked transformed section, the ratio of depth of neutral axis to reinforcement depth, k, is 0.16:

$$k := 0.16$$

The concrete one-way shear capacity per unit width, V_c, can be calculated per Equation (9-1) of ACI 440.1R-06 with k = 0.16. This means that its minimum value is $0.8\sqrt{f_c'}bd$ per Chapter 4:

$$V_{c0} := \left(\frac{5}{2}k\right) 2 \cdot \sqrt{f_c' \cdot psi} \cdot (1ft) \cdot \min(d_{f1}, d_{f2}) = 4.412 \cdot kip$$

The shear reduction factor given by ACI 440.1R-06 is adopted:

$$\phi_v := 0.75$$

$$\phi_v \cdot V_{c0} = 3.31 \cdot kip$$

$$CheckThickness := \begin{vmatrix} \text{``OK''} & \text{if } \phi_v \cdot V_{c0} \geq v_{uMax} \cdot 1ft \\ \text{``Inadequate shear strength''} & \text{otherwise} \end{vmatrix}$$

$$CheckThickness = \text{``OK''}$$

8.6.2 Temperature and shrinkage FRP reinforcement

A simple method of designing the reinforcement for two-way slabs is to reinforce the slab with a uniform mesh equal to what is required for shrinkage and temperature $(\rho_{f,ts})$, and then add extra bars at sections that might be deficient. The minimum flexural reinforcement can be calculated as

$$\rho_{f\_tsMin} := \left(0.0018 \cdot \frac{60\text{ksi}}{f_{fu}} \cdot \frac{29000\text{ksi}}{E_f}, 0.0036 \right) = 0.0036$$

The minimum FRP area per unit width is

$$A_{f\_tsMin} := \rho_{f\_tsMin} \cdot t_{slab} = 0.346 \cdot \frac{\text{in}^2}{\text{ft}}$$

The selected spacing of the FRP bars is

$$s_{f\_ts} := 5 \text{ in.}$$

The spacing also satisfies the crack width limitation; see Step 6.

The selected area of FRP temperature and shrinkage reinforcement per unit width is

$$A_f := \frac{A_{f\_bar}}{s_{f\_ts}} = 0.471 \cdot \frac{\text{in}^2}{\text{ft}}$$

No. 4 GFRP bars spaced 5 in. center to center in both directions are considered.

Minimum FRP reinforcement: The minimum reinforcement requirement has to be verified. Equation (8-8) of ACI 440.1R-06 is used. If the failure is not governed by FRP rupture, this requirement is automatically achieved:

$$A_{f\_min1} := \min\left(\frac{4.9 \cdot \sqrt{f_c'\text{psi}}}{f_{fu}} \min\left(d_{f1}, d_2\right), \frac{300\text{psi}}{f_{fu}} \min\left(d_{f1}, d_{f2}\right) \right) = 0.292 \cdot \frac{\text{in}^2}{\text{ft}}$$

$$\text{Check\_MinReinf1} := \begin{vmatrix} \text{``OK'' if } A_f \geq A_{f\_min1} \\ \\ \text{``Not satisfied'' otherwise} \end{vmatrix} \quad \left(A_f = 0.471 \cdot \frac{\text{in}^2}{\text{ft}} \right)$$

$$\text{Check\_MinReinf1} = \text{``OK''}$$

8.6.3 Bending moment capacity

The maximum value of bending moment per unit width, m_{uMax}, (middle strip, interior support) is the following:

$$m_{uMax} := m_{uNeg23} = 7.51 \cdot \frac{kip \cdot ft}{ft}$$

The FRP reinforcement ratio, ρ_f, is (Equation 8-2 of ACI 440.1R-06):

$$\rho_f := \frac{A_f}{min(d_{f1}, d_{f2})} = 0.006042$$

The failure mode depends on the amount of FRP reinforcement. If ρ_f is larger than the balanced reinforcement ratio, ρ_{fb}, then concrete crushing is the failure mode. If ρ_f is smaller than the balanced reinforcement ratio, ρ_{fb}, then FRP rupture is the failure mode.

ρ_{fb} is computed per Equation (8-3) of ACI 440.1R-06:

$$\rho_{fb} := 0.85 \beta_1 \cdot \frac{f_c'}{f_{fu}} \cdot \frac{E_f \cdot \varepsilon_{cu}}{E_f \cdot \varepsilon_{cu} + f_{fu}} = 0.00748$$

The effective concrete compressive strain at failure as a function of the neutral axis depth, x, is

$$\varepsilon_c(x,y) := \begin{vmatrix} \dfrac{\varepsilon_{cu}}{x} \cdot y & \text{if } \rho_f \geq \rho_{fb} \\ \\ \dfrac{\varepsilon_{fu}}{d_{f2} - x} \cdot y & \text{if } \rho_f < \rho_{fb} \end{vmatrix}$$

The effective tensile strain in the FRP reinforcement as a function of the neutral axis depth, x, is

$$\varepsilon_f(x) := \begin{vmatrix} \varepsilon_{fu} & \text{if } \rho_f < \rho_{fb} \\ \\ min\left[\dfrac{\varepsilon_{cu}}{x} \cdot (d_{f2} - x), \varepsilon_{fu} \right] & \text{if } \rho_f \geq \rho_{fb} \end{vmatrix}$$

The compressive force in the concrete as a function of the neutral axis depth, x, is

$$C_c(x) := \int_{0\,in}^{x} \frac{2\sigma''_c \cdot \left(\dfrac{\varepsilon_c(x,y)}{\varepsilon_{c0}}\right)}{1 + \left(\dfrac{\varepsilon_c(x,y)}{\varepsilon_{c0}}\right)^2} \, psi \, dy$$

The tensile force in the first layer of FRP reinforcement as a function of the neutral axis depth, x, is

$$T_f(x) := A_f E_f \varepsilon_f(x)$$

The neutral axis depth, c_u, can be computed by solving the equation of equilibrium $C_c - T_f = 0$:
First guess:

$$x_0 := 0.1 \, \min(d_{f1}, d_{f2})$$

Given:

$$f_0(x) := C_c(x) - T_f(x)$$

$$C_u := root(f_0(x_0), x_0)$$

The neutral axis depth is

$$cu = 0.953 \cdot in.$$

The nominal bending moment capacity per unit width can be computed as follows:

$$m_n := \int_0^{c_u} y \cdot \frac{\left(2 \cdot \sigma''_c \cdot \dfrac{\varepsilon_c(c_u,y)}{\varepsilon_{c0}}\right)}{\left[1 + \left(\dfrac{\varepsilon_c(c_u,y)}{\varepsilon_{c0}}\right)^2\right]} \, psi \, dy + T_f(c_u) \cdot \left(\min(d_{f1}, d_{f2}) - c_u\right) = 19 \cdot \frac{ft \cdot kip}{ft}$$

The strain distribution over the cross section is shown next.

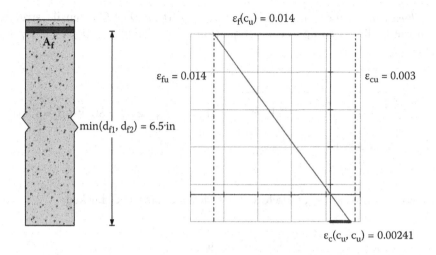

8.6.4 Flexural strength with newly proposed φ-factors

The φ-factor is calculated according to Jawahery and Nanni [1]:

$$\phi_b := \begin{vmatrix} 0.65 & \text{if} & 1.15 - \dfrac{\varepsilon_f(c_u)}{2\varepsilon_{fu}} \leq 0.65 = 0.65 \\[2ex] 0.75 & \text{if} & 1.15 - \dfrac{\varepsilon_f(c_u)}{2\varepsilon_{fu}} \geq 0.75 \\[2ex] 1.15 - \dfrac{\varepsilon_f(c_u)}{2\varepsilon_{fu}} & \text{otherwise} \end{vmatrix}$$

The design flexural strength is computed per Equation (8-1) of ACI 440.1R-06:

$$\phi_b \cdot m_n = 12.5 \frac{\text{kip} \cdot \text{ft}}{\text{ft}}$$

$$\text{Check\_Flexure} := \begin{vmatrix} \text{"OK"} & \text{if } \phi_b \cdot m_n \geq m_{uMax} \\[2ex] \text{"Not good"} & \text{otherwise} \end{vmatrix} \qquad \left(m_{uMax} = 7.5 \cdot \frac{\text{kip} \cdot \text{ft}}{\text{ft}} \right)$$

$$\text{Check\_Flexure} = \text{"OK"}$$

Flexural strength computed per ACI 440.1R-06: The tensile stress in the GFRP is computed per Equation (8-4c) when $\rho_f > \rho_{fb}$, or is f_{fu} if $\rho_f < \rho_{fb}$:

$$f_f := \left| \begin{array}{l} \sqrt{\dfrac{(E_f \cdot \varepsilon_{cu})^2}{4} + \dfrac{0.85\beta_1 \cdot f_c'}{\rho_f} E_f \cdot \varepsilon_{cu}} - 0.5E_f \cdot \varepsilon_{cu} \ \text{ if } \rho_f \geq \rho_{fb} = 80 \cdot \text{ksi } (f_{fu} = 80\text{ksi}) \\ \\ f_{fu} \text{ otherwise} \end{array} \right.$$

f_f cannot exceed f_{fu}; therefore, the following has to be checked:

$$\text{CheckMaxStress} := \left| \begin{array}{l} \text{"OK" if } f_f \leq f_{fu} \\ \\ \text{"Reduce bar spacing or increase bar size" otherwise} \end{array} \right.$$

$$\text{CheckMaxStress} = \text{"OK"}$$

The stress-block depth is computed per Equation (8-4b) or Equation (8-6c) depending on whether $\rho_f > \rho_{fb}$ or $\rho_f < \rho_{fb}$, respectively.

$$a_f := \left| \begin{array}{ll} \dfrac{A_f \cdot f_f}{0.85 \, f_c'} \text{ if } \rho_f \geq \rho_{fb} & \text{Equation (8-4b) of ACI 440.1R-06} \\ \\ \left[\beta_1 \cdot \left(\dfrac{\varepsilon_{cu}}{\varepsilon_{cu} + \varepsilon_{fu}} \right) \min(d_{f1}, d_{f2}) \right] & \text{other Equation (8-6c) of ACI 440.1R-06} \end{array} \right.$$

$$a_f = 0.916 \cdot \text{in.}$$

$$c_f := \frac{a_f}{\beta_1} = 1.145 \cdot \text{in} \qquad\qquad \text{Neutral axis depth}$$

The nominal moment capacity is

$$m_{nACI} := A_f \cdot f_f \left(\min(d_{f1}, d_{f2}) - \frac{a_f}{2} \right) = 19 \cdot \frac{\text{ft} \cdot \text{kip}}{\text{ft}}$$

The concrete crushing failure mode is less brittle than the one due to GFRP rupture. The φ-factor is computed according to Equation (8-7) of ACI 440.1R-06:

$$\phi_{bACI} \begin{vmatrix} 0.55 & \text{if } \rho_f \leq \rho_{fb} & =0.55 \\ 0.30+0.25 \cdot \dfrac{\rho_f}{\rho_{fb}} & \text{if } \rho_{fb} < \rho_f < 1.4 \cdot \rho_{fb} \\ 0.65 & \text{otherwise} \end{vmatrix}$$

The design flexural strength equation is computed per Equation (8-1) of ACI 440.1R-06:

$$\phi_{bACI} \cdot m_{nACI} = 10 \cdot \frac{(kip \cdot ft)}{ft}$$

$$\text{Check\_FlexureACI} := \begin{vmatrix} \text{"OK" if } \phi_{bACI} \cdot m_{nACI} \geq m_{uMax} \\ (m_{uMax} = 7.5 \cdot \dfrac{kip \cdot ft}{ft}) \\ \text{"Not good" otherwise} \end{vmatrix}$$

$$\text{Check\_FlexureACI} = \text{"OK"}$$

8.6.5 Embedment length at exterior support

Because this is a case of negative reinforcement, it has to be checked if adequate moment capacity can be achieved at the end of the embedment length. The available length for embedment is

$$l_{emb} := 12 \text{ in.}$$

The developable tensile stress is calculated per Equation (11-3) of ACI 440.1R-06:

Minimum between cover to bar center and half of the center-to-center bar spacing is

$$C_b := \min\left(c_c + \frac{\phi_{f\_bar}}{2}, \frac{s_{f\_ts}}{2} \right) = 1 \cdot \text{in.}$$

Bar location modification factor for top reinforcement with less than 12 in. of concrete below it is

$$\alpha_{Neg} := 1.0$$

Required stress in the FRP is

$$f_{fr} := E_f \varepsilon_f(cu) \; 80 \cdot ksi$$

The developable tensile stress is (ACI 440.1R-06 Equation 11-3):

$$f_{fd} := \begin{vmatrix} f_{fu} \; if \; \dfrac{\sqrt{f'_c \cdot psi}}{\alpha_{Neg}} \cdot \left(13.6 \cdot \dfrac{1_{emb}}{\phi_{t\_bar}} + \dfrac{C_b}{\phi_{t\_bar}} \cdot \dfrac{1_{emb}}{\phi_{t\_bar}} + 340 \right) \ge f_{fu} = 33.677 \cdot ksi \\[2em] \left[\sqrt{\dfrac{f'_c \cdot psi}{\alpha_{Neg}}} \cdot \left(13.6 \cdot \dfrac{1_{emb}}{\phi_{t\_bar}} + \dfrac{C_b}{\phi_{t\_bar}} \cdot \dfrac{1_{emb}}{\phi_{t\_bar}} + 340 \right) \right] \; otherwise \end{vmatrix}$$

$$CheckFailure := \begin{vmatrix} \text{“Bar ultimate strength” if } f_{fd} \ge f_{fr} \\ \text{“Bond strength otherwise”} \end{vmatrix}$$

$$CheckFailure = \text{“Bond strength”}$$

The cross section of interest is a bond-critical section. The nominal moment capacity, therefore, has to be computed per ACI 440.1R-06 Equation (8-5) or ACI 440.1R-06 Equation (8-6b) when the failure mode is vconcrete crushing or bond, respectively:

$$m_{nb} := \begin{vmatrix} \left[A_f \cdot f_{fd} \cdot \left(\min(d_{f1}, d_{f2}) - \dfrac{1}{2} \cdot \dfrac{A_f \cdot f_f}{0.85 \cdot f'_c} \right) \right] \\ \quad if \; Check \; Failure = \text{“Bold strength”} \\ m_n \; if \; CheckFailure = \text{“Bar ultimate strength”} \end{vmatrix}$$

$$m_{nb} = 8.11 \cdot \frac{ft \cdot kip}{ft}$$

The maximum ultimate moment at the exterior support is

$$m_{uNegMax} := 4.36 \cdot \frac{ft \cdot kip}{ft}$$

The strength-reduction factor when failure is controlled by bond is

$$\phi_{b\_bond} := \begin{vmatrix} 0.55 \text{ if CheckFailure} = \text{``Bond strength''} & =0.55 \\ \phi_b \text{ if CheckFailure} = \text{``Bar ultimate strength''} \end{vmatrix}$$

The design flexural strength is, therefore,

$$\phi_{b\_bond} \cdot m_{nb} = 4.46 \cdot \frac{ft \cdot kip}{ft}$$

$$\text{Check\_FlexureNeg} := \begin{vmatrix} \text{``OK''} & \text{if} \phi_{b\_bond} \cdot m_{nb} \geq m_{uNegMax} \\ \text{``Not good'' otherwise} \end{vmatrix}$$

Check_FlexureNeg = "OK"

The embedment length is adequate. If the embedment length had not been adequate, a bent bar could have been used as indicated in the following.

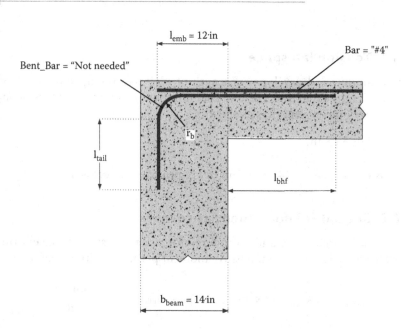

8.6.6 Development length for positive moment reinforcement

The development length, l_d, for straight bars can be calculated using Equation (11-3) of ACI 440.1R-06:

Minimum between cover to bar center and half of the center-to-center bar spacing is

$$C_{b2} := \min\left(c_c + \frac{\phi_{f\_bar}}{2}, \frac{s_{f\_ts}}{2}\right) = 1 \cdot in.$$

Bar location modification factor for bottom reinforcement is

$$\alpha_{Pos} := 1$$

The minimum development length is computed according to ACI 440.1R-06 Equation 11-6:

$$l_{d\_min} := \frac{\alpha_{Pos} \cdot \dfrac{f_{fu}}{\sqrt{f'_c \cdot psi}} - 340}{13.6 + \dfrac{C_{b2}}{\phi_{f\_bar}}} \phi_{f\_bar} = 25.364 \cdot in$$

The following development length is considered and provided to develop the required moment capacity:

$$l_d := 26 \ in.$$

8.6.7 Tension lap splice

The recommended development length of FRP tension lap splices is $1.3l_d$ (Section 11.4 of ACI 440.1R-06). The minimum recommended tension lap splice development length is as shown:

$$1.3l_d = 33.8 \cdot in.$$

Tension lap splices 34 in. in length are provided if necessary.

8.6.8 Special reinforcement at corners

At exterior corners, top and bottom reinforcement must resist the existing moment plus the maximum positive moment per unit width of the slab:

$$m_{uTopMax} := m_{uNeg11} + \max\left(m_{uPos12}, m_{uPos22}, m_{uPos32}\right) = 6.58 \frac{kip \cdot ft}{ft}$$

No. 4 GFRP bars spaced 5 in. center to center in both directions are considered.

$$\text{Check} := \begin{array}{|l} \text{"No additional reinforcement at corner is required"} \\ \qquad\qquad\qquad\qquad\qquad \text{if } m_{uTopmax} \le \phi_b \cdot m_n \\ \text{"Additional reinforcement at corner is required" otherwise} \end{array}$$

Check = "No additional reinforcement at corner is required"

8.6.9 Check for shear capacity

The maximum shear at the face of the supports is

$$V_u := v_{uMax} \cdot (1 \text{ ft}) = 2.35 \cdot \text{kip}$$

The concrete shear capacity, V_c, can be calculated per Equation (9-1) of ACI 440.1R-06, where k is the ratio of depth of neutral axis to reinforcement depth, calculated per Equation (8-12). Also, the limitation discussed in Chapter 4 should be verified $(V_c > 0.8\sqrt{f'_{ci}} b \cdot d)$:

$$V_c := \max \begin{bmatrix} 5\sqrt{f'_c \cdot \text{psi}} \cdot (1\text{ft}) \cdot \left(k \cdot \min\left(d_{f1}, d_{f2}\right)\right), \\ 0.8\sqrt{f'_c \cdot \text{psi}} \left(1\text{ft}\right) \cdot \left(\min\left(d_{f1}, d_{f2}\right)\right) \end{bmatrix} = 4.4 \cdot \text{kip}$$

The shear reduction factor given by ACI 440.1R-06 is adopted:

$$\phi_v = 0.75$$

$$\phi_v \cdot V_c = 3.31 \cdot \text{kip}$$

$$\text{Check\_Shear1:} \begin{array}{|l} \text{"OK" if } \phi_v \cdot V_c \ge V_u \\ \text{"Shear reinforcement is needed" otherwise} \end{array}$$

Check_Shear 1 = "OK"

8.7 STEP 5—CHECK CREEP-RUPTURE STRESS

Creep-rupture stress in the FRP has to be evaluated considering the total unfactored dead loads and the sustained portion of the live load (20% of the total live load). The maximum value of bending moment per unit width, m_{uMax}, is the following:

$$m_{sMax} := m_{sNeg23} = 5.37 \cdot \frac{\text{kip} \cdot \text{ft}}{\text{ft}}$$

Bending moment due to dead load plus 20% of live load is

$$m_{creep} := m_{sMax} \cdot \frac{DL+0.20LL}{SL} = 3.2 \cdot \frac{kip \cdot ft}{ft}$$

Ratio of modulus of elasticity of bars to modulus of elasticity of concrete is

$$n_f := \frac{E_f}{E_c} = 1.399$$

Ratio of depth of neutral axis to reinforcement depth, calculated per Equation (8-12), is

$$k_f := \sqrt{2\rho_f \cdot n_f + \left(\rho_f \cdot n_f\right)^2} - \rho_f \cdot n_f = 0.122$$

The tensile stress in the FRP is computed as

$$f_{fcreep} := \frac{m_{creep}}{A_f \cdot \min(d_{f1}, d_{f2}) \cdot \left(1 - \frac{k_f}{3}\right)} = 13.1 \cdot ksi$$

$$\text{Check\_Creep} := \begin{vmatrix} \text{"OK" if } f_{fcreep} \leq f_{f\_creep} \\ \\ \text{"Not good" otherwise} \end{vmatrix} \qquad (f_{f\_creep} = 16 \cdot ksi)$$

Check_Creep = "OK"

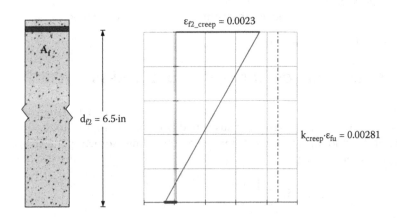

8.8 STEP 6—CHECK CRACK WIDTH

Crack width is checked using Equation (8-9) of ACI 440.1R-06. A crack width limit, w_{lim}, of 0.028 in. is used for interior exposure, while 0.020 in. is used for exterior exposure. The middle strip interior support is considered.

Ratio of modulus of elasticity of bars to modulus of elasticity of concrete is

$$n_f = 1.399$$

Ratio of depth of neutral axis to reinforcement depth, calculated per Equation (8-12), is

$$k_f = 0.122$$

Tensile stress in GFRP under service loads is

$$f_{fs} := \frac{m_{sMax}}{A_f \cdot \min(d_{f1}, d_{f2}) \cdot \left(1 - \dfrac{k_f}{3}\right)} = 21.928 \cdot \text{ksi}$$

Ratio of distance from neutral axis to extreme tension fiber to distance from neutral axis to center of tensile reinforcement is

$$\beta_{11} := \frac{t_{slab} - k_f \cdot \min(d_{f1}, d_{f2})}{\min(d_{f1}, d_{f2}) \cdot (1 - k_f)} = 1.263$$

Thickness of concrete cover measured from extreme tension fiber to center of bar is

$$d_c := t_{slab} - \min(d_{f1}, d_{f2}) = 1.5 \cdot \text{in.}$$

Bond factor (provided by the manufacturer) is

$$K_b := 0.9$$

The crack width under service loads is

$$w := 2\frac{f_{fs}}{E_f}\beta_{11} \cdot k_b \cdot \sqrt{d_c^2 + \left(\frac{s_{f\_ts}}{2}\right)^2} = 0.025 \cdot \text{in Equation (8-9) of ACI 440.1R-06}$$

The crack width limit for the selected exposure is

$$w_{lim} = 0.028 \cdot \text{in.}$$

$$\text{Check\_Crack1} := \begin{vmatrix} \text{"OK" if } w \le w_{lim} \\ \text{"Not good" otherwise} \end{vmatrix}$$

Check_Crack1 = "OK"

8.9 STEP 7—CHECK DEFLECTIONS

Maximum long-term midspan deflection is checked. The maximum allowable deflection is l/480.

The gross moment of inertia is

$$I_g := \frac{(1ft) \cdot t_{slab}^3}{12} = 512 \cdot in^4$$

The cracking moment is

$$M_{cr} := \frac{2f_{ct} \cdot I_g}{t_{slab}} = 5.657 \cdot kip \cdot ft$$

The cracking moment per unit width is

$$m_{cr} := \frac{2f_{ct} \cdot I_g}{t_{slab} \cdot (12in)} = 5.657 \cdot \frac{kip \cdot ft}{ft}$$

Because the service bending moments are smaller than the cracking moment, the section remains uncracked and the gross moment of inertia may be used to calculate the deflections.

From the finite element analysis the following results can be obtained:

$\Delta_{DL} := 0.10$ in.   Deflection due to dead loads only

$\Delta_{LL} := 0.09$ in.   Deflection due to live loads only

$\xi := 2$ Multiplier for time-dependent deflection at 5 years (ACI 318-11)

$\lambda := 0.60\xi = 1.2$ Reduction parameter, Equation (8-14b) of ACI 440.1R-06

$\Delta_{LT} := \Delta_{LL} + \lambda \cdot (\Delta_{DL} + 0.20 \cdot \Delta_{LL}) = 0.232 \cdot in.$ Long-term deflection

$\Delta_{limit} := \frac{l_1}{480} = 0.6 \cdot in.$ Deflection limit

$$\text{Check\_}\Delta_{\text{LT}} := \begin{vmatrix} \text{"OK" if } \Delta_{\text{LT}} \leq \Delta_{\text{limit}} \\ \text{"Not good" otherwise} \end{vmatrix}$$

$$\text{Check\_}\Delta_{\text{LT}} = \text{"OK"}$$

8.10 STEP 8—CHECK FOR PUNCHING SHEAR (NO PERIMETER BEAMS)

Assuming that no slab-perimeter beams are present, the slab at columns A1, B1, and B2 are checked for punching shear. It is assumed that columns are square with sides equal to $b_{\text{col}} := 20$ in.

8.10.1 Check at column A1

The area of influence is

$$A_{I1} := (0.45 \cdot 1_1) \cdot (0.45 \cdot 1_2) - b_{\text{col}}^2 = 113.862 \cdot \text{ft}^2$$

The factored shear force is

$$V_{u1} := A_{I1} \cdot \text{TFL} = 31.8 \cdot \text{kip}$$

The perimeter of critical section is

$$b_{01} := 2 \left(b_{\text{col}} + \frac{\min(d_{f1}, d_{f2})}{2} \right) = 46.5 \cdot \text{in}.$$

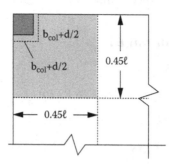

The ratio of modulus of elasticity of bars to modulus of elasticity of concrete is

$$n_f = 1.399$$

The ratio of depth of neutral axis to reinforcement depth, calculated per Equation (8-12) of ACI 440.1R-06, is

$$k_f = 0.122$$

The concrete two-way capacity to resist punching shear is from Equation (9-8a) of ACI 440.1R-06:

$$V_{c1} := \left(\frac{5}{2} k_f \right) 4 \cdot \sqrt{f_c' \cdot psi} \cdot (b_{01}) \cdot \min(d_{f1}, d_{f2}) = 26.04 \cdot kip$$

The concrete shear capacity per unit width, V_c, has a minimum value is $1.6 \sqrt{f_c'} b_o d$ per Chapter 4; therefore:

$$V_{c1min} := 1.6 \cdot \sqrt{f_c' \cdot psi} \cdot (b_{01}) \cdot \min(d_{f1}, d_{f2}) = 34.2 \cdot kip$$

$$\phi V_{n1} := \phi_v V_{c1min} = 25.647 \cdot kip$$

$$\text{Check\_PunchingShear1} := \begin{vmatrix} \text{"OK" if } V_{u1} \le \phi V_{n1} \\ \text{"Not good" otherwise} \end{vmatrix}$$

$$\text{Check\_PunchingShear1} = \text{"Not good"}$$

For example, including a drop panel at the corner columns could be a solution to prevent punching shear failure.

8.10.2 Check at column B1

The area of influence is

$$A_{12} := (0.45 \cdot l_1) \cdot (0.5 \cdot l_2) - b_{col}^2 = 126.822 \cdot ft^2$$

The factored shear force is

$$V_{u2} := A_{12} \cdot TFL = 35.4 \cdot kip$$

The perimeter of critical section is

$$b_{02} := 2 \left(b_{col} + \frac{\min(d_{f1}, d_{f2})}{2} \right) + \left(b_{col} + \min(d_{f1}, d_{f2}) \right) = 73 \cdot in.$$

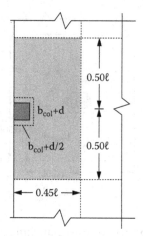

The concrete shear capacity to resist punching shear is

$$V_{c2} := \left(\frac{5}{2}k_f\right)4 \cdot \sqrt{f_c' \cdot psi} \cdot (b_{02}) \cdot \min(d_{f1}, d_{f2}) = 40.9 \cdot kip$$

$$V_{c2min} := 1.6 \cdot \sqrt{f_c' \cdot psi} \cdot (b_{02}) \cdot \min(d_{f1}, d_{f2}) = 53.68 \cdot kip$$

$$\phi V_{n2} := \phi_v \cdot V_{c2min} = 40.3 \cdot kip$$

$$\text{Check\_PunchingShear2} := \begin{vmatrix} \text{``OK'' if } V_{u2} \leq \phi V_{n2} \\ \text{``Not good'' otherwise} \end{vmatrix}$$

$$\text{Check\_PunchingShear2} = \text{``OK''}$$

8.10.3 Check at column B2

The area of influence is

$$A_{I3} := (0.5 \cdot l_1) \cdot (0.5 \cdot l_2) - b_{col}^2 = 141.222 \cdot ft^2$$

The factored shear force is

$$V_{u3} := A_{I3} \cdot TFL = 39.4 \cdot kip$$

The perimeter of critical section is

$$b_{03} := 4(b_{col} + \min(d_{f1}, d_{f2})) = 106 \cdot in.$$

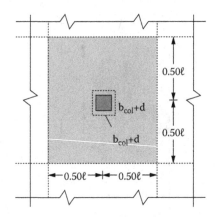

The concrete shear capacity to resist punching shear is

$$V_{c3} := \left(\frac{5}{2}k_f\right)4 \cdot \sqrt{f'_c \cdot psi} \cdot (b_{03}) \cdot \min(d_{f1}, d_{f2}) = 59.4 \cdot kip$$

$$V_{c3min} := 1.6 \cdot \sqrt{f'_c \cdot psi} \cdot (b_{03}) \cdot \min(d_{f1}, d_{f2}) = 77.95 \cdot kip$$

$$\phi V_{n3} := \phi_v \cdot V_{c3min} = 58.5 \cdot kip$$

$$\text{Check\_PunchingShear3} := \begin{vmatrix} \text{"OK" if } V_{u3} \leq \phi V_{n3} \\ \text{"Not good" otherwise} \end{vmatrix}$$

Check_PunchingShear3 = "Ok"

REFERENCE

1. H. Jawahery Zadeh and A. Nanni. Reliability analysis of concrete beams internally reinforced with FRP bars. *ACI Structural Journal* 110(6): 1023–1032 (2013).

Chapter 9

Design of a column

9.1 INTRODUCTION

The floor plan of a two-story medical facility building is shown in Figure 9.1. The building hosts two MRI units at the second level that are located in bays AB-34 and BC-34. The floor system is a one-way fiber-reinforced polymer (FRP) reinforced concrete (RC) slab spanning along the east–west plan direction. The location of the building excluded the presence of any snow load. The building is located in a region of low seismicity. This design example describes the procedure to design column B3 (lower level). The frame considered in the analysis is displayed in Figure 9.2.

Loads on each floor consist of the self-weight, a superimposed dead load of 2.5 psf, and a live load of 100 psf (on both levels). The MRI equipment load of 815 psf is included in the self-weight and is considered as uniformly distributed over an area of 10 by 10 ft² (Figure 9.1). It is assumed that the lateral load effects on the building are caused by a wind force of 100 kip applied conservatively at the roof level, and that the dead and live loads are the only sustained loads. A roof uplift pressure of 35 psf is also considered. Wind loads are computed according to ASCE 7-10.[1]

This example describes the procedure to design the first-story portion of column B3 (Figure 9.3). The design is presented as a sequence of five steps as summarized here:

Step 1   Define column geometry and concrete properties
Step 2   Compute factored loads
Step 3   Design FRP longitudinal reinforcement
Step 4   Design FRP shear reinforcement
Step 5   Check creep-rupture stress

---

[1] ASCE 7-10, 2010, "Minimum Design Loads for Buildings and Other Structures," American Society of Civil Engineers, Washington, DC.

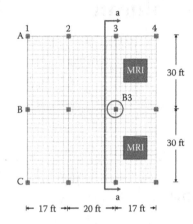

Figure 9.1 Floor plan.

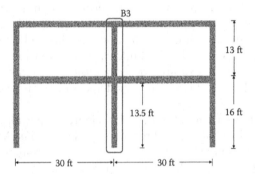

Figure 9.2 Frame section.

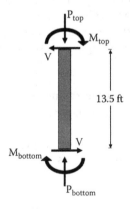

Figure 9.3 Column B3 (lower level).

The results of the column design are summarized next to facilitate the understanding of the five sequential steps devoted to calculations.

9.2 DESIGN SUMMARY

The slab thickness is 8 in., the dimensions of the beams supported by column B3 are 14 by 28 in., and the column dimensions are 20 by 20 in. The following loads are considered:

Slab self-weight	96.7 psf
Superimposed dead load	2.5 psf
MRI dead load	815 psf
Live load	100 psf
Roof uplift	35 psf
Wind lateral force at the roof	100 kip

To analyze the structure and calculate the lateral deflections, the stiffness of each member type is modified as

$$I_{beam} = [0.075 + 0.275(E_f/E_s)]I_g = 0.139I_g$$

$$I_{slab} = [0.10 + 0.15(E_f/E_s)]I_g = 0.135I_g$$

$$I_{column} = [0.40 + 0.30(E_f/E_s)]I_g = 0.469I_g$$

The structural analysis is conducted for the frame displayed in Figure 9.2 using the modified member stiffnesses. Table 9.1 presents the results of this analysis for column B3 when individual unfactored loads are applied.

It is assumed that the column is subject to monoaxial bending around the B axis line. Negative values of the axial load signify tension. The difference in the moment sign at the top and bottom of the column indicates double curvature (Figure 9.3). The forces and moments of Table 9.1 can be combined according to ACI 318-11 to obtain the ultimate factored loads that are used for design. The ultimate factored loads are listed in Table 9.2.

Table 9.1 Analysis results for unfactored single load cases (column B3)

Load case	Axial load: P (kip)	Bending moment: M (ft-kip)		Shear: V (kip)
		Top	Bottom	
Dead (D)	186.9	52.1	−36.1	6.6
Live (L)	121.9	0.0	0.0	0.0
Wind (W⁺)	−34.7	−32.0	98.2	9.6
Wind (W⁻)	−34.7	32.0	98.2	9.7

Table 9.2 Analysis results for ultimate load combinations (column B3)

| Load combination | Axial load: P_u (kip) | Bending moment: M_u (ft-kip) | | Shear: V_u (kip) |
		Top	Bottom	
1.4D	261.7	73.0	−50.6	9.2
1.2D + 1.6L	419.3	62.5	−43.3	7.8
1.2D + 1.6W⁺ + 1.0L	290.8	11.4	113.7	7.6
1.2D + 1.6W⁻ + 1.0L	290.8	113.7	−200.4	23.3
0.9D + 1.6W⁺	112.8	−4.3	124.6	9.5
0.9D + 1.6W⁻	112.8	98.1	−189.6	21.3

Table 9.3 Calculation of stability index, Q

Load combination	ΣP_u (kip)	V_u (kip)	Δ_0 (in.)	Q	Frame type
1.4D	1587	0.0	0.0	0.000	No sway
1.2D + 1.6L	2397	0.0	0.0	0.000	No sway
1.2D + 1.6W⁺ + 1.0L	1718	160.0	0.65	0.039	No sway
1.2D + 1.6W⁻ + 1.0L	1718	−160.0	−0.65	0.039	No sway
0.9D + 1.6W⁺	639	160.0	0.65	0.015	No sway
0.9D + 1.6W⁻	639	−160.0	−0.65	0.015	No sway

Frame type (sway or non-sway): According to ACI 318-11, a frame can be considered non-sway if the column end moments due to second-order effects do not exceed 5% of the first-order end moments:

$$Q = (\Sigma P_u \Delta_0)/(V_u l_c) \leq 0.05 \qquad (9.1)$$

where:

Q = Stability index
ΣP_u = Total vertical load in the story
V_u = Total story shear
Δ_0 = First-order relative displacement between top and bottom of the story
l_c = Length of the column center to center of joints, (16) (12)-(30/2) = 177 in.

Table 9.3 shows that under no load combination is the frame sway type ($Q \leq 0.05$).

Slenderness effects: For compression members in a non-sway frame, effects of slenderness in a column may be neglected when

$$kl_u/r \leq (kl_u/r)_{max} = 29-12(M_1/M_2) \leq 35 \qquad (9.2)$$

where:

k = 1.0: effective length factor for non-sway frames
l_u = 162 in.: clear height of the column

r = (0.3)(20 in.) = 6.0 in.: radius of gyration of the column
$kl_u/r = 27.0$

From Table 9.2 it can be concluded that the maximum value of the $(kl_u/r)_{max}$ is associated with the case of single curvature; therefore:

$(kl_u/r)_{max} = 29{-}12(11.4/113.7) = 27.8$

$kl_u/r = 27 < (kl_u/r)_{max}$

Hence slenderness effects may be disregarded.

P–M diagram: The interaction diagram was built following the procedure discussed in Chapter 5. The assumed parameters and the geometry of a generic rectangular section are displayed in Figure 9.4.

For the analysis, it is assumed that the cross-section reinforcement is applied at *i* different levels (varying from 0 to $N_f - 1$) of depth $d_f(i)$, and the spacing of the side reinforcement, s_f, is constant.

The interaction diagram is constructed by locating the critical points listed next.

Condition 1: pure compression (c = +∞). The strain and stress distributions corresponding to point 1 are shown in Figure 9.5. Because the entire cross section is in compression, the contribution of the FRP reinforcement is neglected and reflected with an equivalent area of concrete.

The combined nominal axial and moment capacities, P_n and M_{nx}, are

$$P_n = 0.85f_c'\, bh$$

$$M_{nx} = 0$$

(9.3)

Condition 2: neutral axis intersecting the bottom fiber (c = h). The strain and stress distributions for this condition are shown in Figure 9.6. The concrete stress distribution, $\sigma_c(x)$, is approximated by the stress block. The

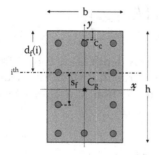

Figure 9.4 Generic rectangular cross section.

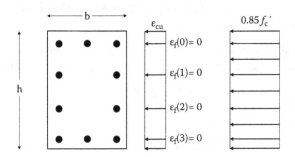

Figure 9.5 Pure compression.

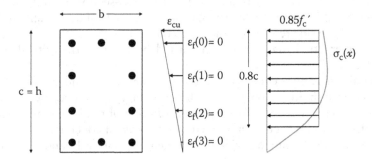

Figure 9.6 Neutral axis at the lowest level of fibers.

contribution of the FRP reinforcement is still null because the section is fully compressed.

The combined nominal axial and moment capacities, P_n and M_{nx}, for $\beta_1 = 0.8$ are

$$P_n = 0.85 f_c' \, b(0.8c)$$

$$(9.4)$$

$$M_{nx} = 0.85 f_c' \, b(0.8c) \cdot \left(\frac{h}{2} - 0.4c \right)$$

Condition 3: neutral axis depth h ≥ c ≥ c_{bal}. The cross-section failure is controlled by concrete crushing when the neutral axis depth is situated below the balanced location (c ≥ c_{bal}), which can be computed as follows:

$$c_{bal} = \frac{\varepsilon_{cu}}{\varepsilon_{cu} + \varepsilon_{fd}} d_f (N_f - 1)$$

$$(9.5)$$

For any arbitrary neutral axis depth, c, falling in the range $h \geq c \geq c_{bal}$, strain and stress distributions are shown in Figure 9.7. The balanced condition is represented by the case of ($\varepsilon_{f3} = \varepsilon_{fd}$).

The combined nominal axial and moment capacities, P_n and M_{nx}, are

$$P_n = 0.85 f_c' b(0.8c) + \sum_{i=0}^{N_f-1} A_f(i) E_f \varepsilon_f(i)$$

$$M_{nx} = 0.85 f_c' b(0.8c) \cdot \left(\frac{h}{2} - 0.4c \right) + \sum_{i=0}^{N_f-1} A_f(i) E_f \varepsilon_f(i) \cdot \left(\frac{h}{2} - d_f(i) \right)$$

(9.6)

Condition 4: neutral axis depth $c_{bal} \geq c \geq 0$. The cross-section failure is triggered by rupture of the FRP reinforcement when the neutral axis is located above the balanced location. When the neutral axis intersects the top fiber (c = 0), the entire cross section is in tension and the FRP bars are the only component engaged in resisting the load. The strain and stress distributions for the case of c = 0 are shown in Figure 9.8.

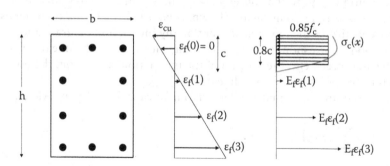

Figure 9.7 Neutral axis depth $h \geq c \geq c_{bal}$.

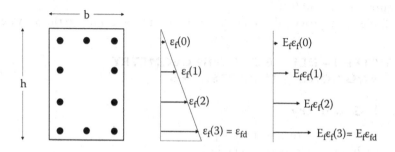

Figure 9.8 Neutral axis depth c = 0.

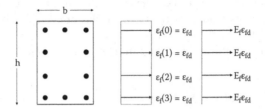

Figure 9.9 Pure tension.

The combined nominal axial and moment capacities, P_n and M_{nx}, are

$$P_n = \sum_{i=0}^{N_f-1} A_f(i)E_f\varepsilon_{fd}$$

$$M_{nx} = \sum_{i=0}^{N_f-1} A_f(i)E_f\varepsilon_f(i)\cdot\left(\frac{h}{2}-d_f(i)\right)$$

(9.7)

Condition 5: pure tension ($c = -\infty$). The strain and stress distributions for the case of pure tension are shown in Figure 9.9. The maximum tensile force constitutes the lowermost point of any interaction diagram and is the other point corresponding to zero eccentricity. Due to the linear behavior of the reinforcement, the portion of the interaction curve from the point of $c = 0$ to the point of $c = -\infty$ is linear.

The combined nominal axial and moment capacities, P_n and M_{nx}, are

$$P_n = \sum_{i=0}^{N_f-1} A_f(i)E_f\varepsilon_{fd}$$

$$M_{nx} = 0$$

(9.8)

FRP reinforcement. Bar layout and typical details are shown in Figure 9.10(a) and 9.10(b).

Table 9.4 is provided to convert the US customary units to the SI system.

9.3 STEP I—DEFINE COLUMN GEOMETRY AND CONCRETE PROPERTIES

9.3.1 Geometry

h := 20 in.   Cross-section depth
b := 20 in.   Cross-section width
c_c := 2.5 in.   Concrete cover to center of longitudinal bar

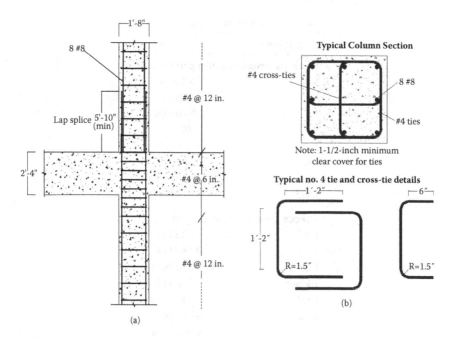

Figure 9.10 (a) Reinforcement layout; (b) typical detail.

9.3.2 Concrete properties

The following concrete properties are considered for the design:

$f_c := 5000 \text{ psi}$ — Compressive strength

$\varepsilon_{cu} := 0.003$ — Ultimate compressive strain

$\rho_c := 145 \dfrac{\text{lbf}}{\text{ft}^3}$ — Density

$E_c := 33 \text{ psi}^{0.5} \left(\dfrac{\rho_c}{\text{lbf} \cdot \text{ft}^{-3}} \right)^{1.5} \cdot \sqrt{f_c'}$ — Compressive modulus of elasticity

$E_c = 4074 \text{ ksi}$ — Computed as indicated in ACI 318-11

$f_{ct} := 7.5 \cdot \sqrt{f_c'} \cdot \text{psi}$ — Concrete tensile strength

$f_{ct} = 530.33 \text{ psi}$ — Computed as indicated in ACI 318-11

Table 9.4 Conversion table

US customary	SI units
Lengths, areas, section properties	
I in.	25.4 mm
	0.025 m
I ft	304.8 mm
	0.305 m
I in.^{2}	645 mm^2
I ft^2	0.093 m^2
I in.^{3}	16,387 mm^3
I in.^{4}	416,231 mm^4
Forces, pressures, strengths	
I lbf	4.448 N
I kip	4.448 kN
I lbf-ft	1.356 N.m
I kip-ft	1.356 kN·m
I psi	6.895 kPa
I psf	47.88 N/m^2
I ksi	6.895 MPa
I ksf	47.88 kN/m^2
I lbf/ft^3	157.1 N/m^3

9.4 STEP 2—COMPUTE ULTIMATE LOADS

The frame structure is considered a non-sway and any slenderness effect is neglected, as discussed in the previous section. The analysis results per each single load case limited to the first-story portion of column B3 are shown as follows:

- Dead load (D):
$P_D := 187$ kip
$M_{Dtop} := 52$ kip·ft
$M_{Dbottom} := -36$ kip·ft
$V_D := 6.6$ kip·ft

- Live load (L):
$P_L := 122$ kip
$M_{Ltop} := 0$ kip·ft
$M_{Lbottom} := -0$ kip·ft
$V_L := 0$ kip

- Wind (W$^+$):
$P_{W1} := -35$ kip
$M_{W1top} := -32$ kip·ft

- Wind (W$^-$):
$P_{W2} := -35$ kip
$M_{W2top} := 32$ kip·ft

$M_{W1bottom} := 98 \text{ kip·ft}$ $M_{W2bottom} := -98 \text{ kip·ft}$
$V_{W1} := -9.6 \text{ kip}$ $V_{W2} := 9.6 \text{ kip}$

The analysis results per load combination are shown as follows:

- Load combination 1: 1.4D
  $P_{u1} := 1.4P_D = 261.8 \text{·kip}$
  $M_{u1top} := 1.4M_{Dtop} = 72.8 \text{·kip·ft}$
  $M_{u1bottom} := 1.4M_{Dbottom} = -50.4 \text{·kip·ft}$
  $V_{u1} := 1.4V_D = 9.24 \text{·kip}$
- Load combination 2: 1.2D + 1.6L
  $P_{u2} := 1.2P_D + 1.6 \cdot P_L = 419.6 \text{·kip}$
  $M_{u2top} := 1.2M_{Dtop} + 1.6 \cdot M_{Ltop} = 62.4 \text{·kip·ft}$
  $M_{u2bottom} := 1.2M_{Dbottom} + 1.6 \cdot M_{Lbottom} = -43.2 \text{·kip·ft}$
  $V_{u2} := 1.2V_D + 1.6 \cdot V_L = 7.92 \text{·kip}$
- Load combination 3: 1.2D + 1.6W$^+$ + 1.0L
  $P_{u3} := 1.2P_D + 1.6 \cdot P_{W1} + 1.0P_L = 290.4 \text{·kip}$
  $M_{u3top} := 1.2M_{Dtop} + 1.6 \cdot M_{W1top} + 1.0M_{Ltop} = 11.2 \text{·kip·ft}$
  $M_{u3bottom} := 1.2M_{Dbottom} + 1.6 \cdot M_{W1bottom} + 1.0M_{Lbottom} = 113.6 \text{·kip·ft}$
  $V_{u3} := 1.2V_D + 1.6 \cdot V_{W1} + 1.0 \cdot V_L = -7.44 \text{·kip}$
- Load combination 4: 1.2D + 1.6W$^-$ + 1.0L
  $P_{u4} := 1.2P_D + 1.6 \cdot P_{W2} + 1.0P_L = 290.4 \text{·kip}$
  $M_{u4top} := 1.2M_{Dtop} + 1.6 \cdot M_{W2top} + 1.0M_{Ltop} = 113.6 \text{·kip·ft}$
  $M_{u4bottom} := 1.2M_{Dbottom} + 1.6 \cdot M_{W2bottom} + 1.0M_{Lbottom} = -200 \text{·kip·ft}$
  $V_{u4} := 1.2V_D + 1.6 \cdot V_{W2} + 1.0 \cdot V_L = 23.28 \text{·kip}$
- Load combination 5: 0.9D + 1.6W$^+$
  $P_{u5} := 0.9P_D + 1.6 \cdot P_{W1} = 112.3 \text{·kip}$
  $M_{u5top} := 0.9_{MDtop} + 1.6 \cdot M_{W1top} = -4.4 \text{·kip·ft}$
  $M_{u5bottom} := 09M_{Dbottom} + 1.6 \cdot M_{W1bottom} = 124.4 \text{·kip·ft}$
  $V_{u5} := 0.9V_D + 1.6 \cdot V_{W1} = -9.42 \text{·kip}$
- Load combination 6: 0.9D + 1.6W$^-$
  $P_{u6} := 0.9P_D + 1.6 \cdot P_{W2} = 112.3 \text{·kip}$
  $M_{u6top} := 0.9_{MDtop} + 1.6 \cdot M_{W2top} = 98 \text{·kip·ft}$
  $M_{u6bottom} := 09M_{Dbottom} + 1.6 \cdot M_{W2bottom} = -189.2 \text{·kip·ft}$
  $V_{u6} := 0.9V_D + 1.6 \cdot V_{W2} = 21.3 \text{·kip}$

9.5 STEP 3—DESIGN LONGITUDINAL FRP REINFORCEMENT

Number of FRP reinforcement levels $N_f := 3$
Number of FRP bars at top and bottom level $N_{bar0} := 3$

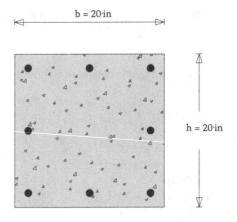

b = 20·in

h = 20·in

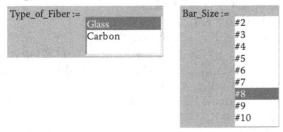

Properties of the selected FRP reinforcement

Type_of_Fiber :=
| Glass |
| Carbon |

Bar_Size :=
| #2 |
| #3 |
| #4 |
| #5 |
| #6 |
| #7 |
| #8 |
| #9 |
| #10 |

The manufacturer's guaranteed mechanical properties are the following:

$f_{fuu} = 80 \cdot ksi$ Ultimate guaranteed tensile strength of the FRP
$\varepsilon_{fuu} = 0.01404$ Ultimate guaranteed rupture strain of the FRP
$E_f = 5700 \cdot ksi$ Guaranteed tensile modulus of elasticity of the FRP

The geometrical properties of the selected bar are

$\phi_{f-bar} = 1 \cdot in.$ Bar diameter
$A_{f-bar} = 0.785 \cdot in.^2$ Bar area

FRP reduction factors: Table 7-1 of ACI 440.1R-06 is used to define the environmental reduction factor, C_E. The type of exposure has to be selected:

Type_of_Exposure :=
| Interior |
| Exterior |

$C_E = 0.8$ Environmental reduction factor for GFRP

Table 8-3 of ACI 440.1R-06 is used to define the reduction factor to take into account the FRP creep-rupture stress. Creep-rupture stress in the FRP has to be evaluated considering service conditions defined as the total unfactored dead loads and the sustained portion of the live load (20% of the total live load).

$k_{creep} = 0.2$ Creep-rupture stress limitation factor

FRP ultimate design properties: The design properties are calculated per Equations (5.2) and (5.3), defined in Chapter 5.

$f_{fu} := \min(C_E \cdot f_{fuu}, 0.01 \cdot E_f) = 57 \cdot ksi$ Design tensile strength
$\varepsilon_{fu} := \min(C_E \cdot \varepsilon_{fuu}, 0.01) = 0.01$ Design rupture strain

FRP creep-rupture limit stress: The FRP creep-rupture limit stress is calculated per Section 8.4 of ACI 440.1R-06.

$f_{f\_creep} := k_{creep} \cdot f_{fu} = 11.4 \cdot ksi$

P–M diagram. *Preliminary calculations:* The spacing between the FRP reinforcement levels can be calculated as follows.

$$s_f := \frac{h - 2 \cdot c_c - 2 \cdot \dfrac{\phi_{f\_bar}}{2}}{N_f - 1} = 7 \cdot in.$$

$$CheckSpacing := \begin{array}{|l} \text{"OK" if } s_f \geq \min(1 in, \phi_{f\_bar}) \\ \text{"Reduce number of reinforcement layers" otherwise} \end{array}$$

$CheckSpacing = \text{"OK"}$

The index indicates the reinforcement level is

$i := 0, 1 \ .. \ N_f - 1$

Depth of FRP reinforcement level is

$$d_f(i) := \left(c_c + \frac{\phi_{f\_bar}}{2} \right) + i \cdot s_f$$

Number of bars per level of reinforcement is

$$N_{bars}(i) := \begin{vmatrix} N_{bar0} & \text{if } i = 0 \\ 2 & \text{if } 1 \le i \le N_f - 2 \\ N_{bar0} & \text{if } i = N_f - 1 \end{vmatrix}$$

Total area of FRP reinforcement per level of reinforcement is

$$A_f(i) := A_{f-bar} \cdot N_{bars}(i)$$

The FRP reinforcement area shall not be smaller than 1% of the cross-section gross area:

$$A_{f-min} := 0.01 \cdot b \cdot h = 4 \cdot in.^2$$

$$\text{Check\_MinReinf} := \begin{vmatrix} \text{"OK" if } \sum_{i=0}^{N_f-1} A_f(i) \ge A_{f\_min} & \left(\sum_{i=0}^{N_f-1} A_f(i) = 6.28 \cdot in^2 \right) \\ \text{"Not satisfied" otherwise} \end{vmatrix}$$

Check_MinReinf = "OK"

9.5.1 Point I—Pure compression

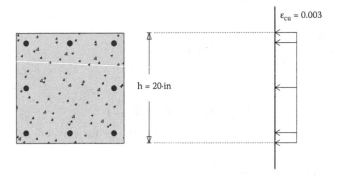

$\varepsilon_{cu} = 0.003$

$h = 20 \cdot in$

Nominal axial strength is

$$P_{n1} := 0.85 f'_c \cdot b \cdot h = 1700 \cdot kip$$

Nominal flexural strength is

$$M_{n1} := 0 \ kip \cdot ft$$

Maximum tensile stress in FRP bars is

$$\varepsilon_{f1} := 0$$

The strength reduction factor is

$$\phi_1 := \begin{vmatrix} 0.65 \text{ if } 1.15 - \dfrac{\varepsilon_{f1}}{2\varepsilon_{fu}} \leq 0.65 \\[3mm] 1.15 - \dfrac{\varepsilon_{f1}}{2\varepsilon_{fu}} \\[3mm] 0.75 \text{ if } 1.15 - \dfrac{\varepsilon_{f1}}{2\varepsilon_{fu}} \geq 0.75 \end{vmatrix}$$

$$\phi_1 = 0.75$$

The ACI compressive strength limitation is

$$\phi_1 \cdot 0.8 \cdot Pn1 = 1020 \cdot kip$$

9.5.2 Point 2—Neutral axis at the level of the lowest section fibers

Neutral axis depth is

$$c_2 := h = 20 \cdot in.$$

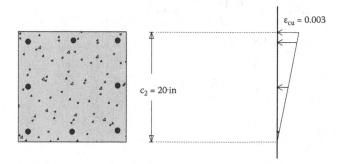

$$\varepsilon_{cu} = 0.003$$

$$c_2 = 20 \cdot in$$

Concrete compressive strain distribution is

$$\varepsilon_{c2}(x) := \frac{\varepsilon_{cu}}{c_2} x$$

The strain in FRP reinforcement is

$$\varepsilon_{f2}(i) := \begin{vmatrix} \dfrac{\varepsilon_{cu}}{c_2} \cdot \left(c_2 - d_f(i)\right) \text{ if } \dfrac{\varepsilon_{cu}}{c_2} \cdot \left(c_2 - d_f(i)\right) \geq -\varepsilon_{fu} \\[3mm] 0 \text{ if } \dfrac{\varepsilon_{cu}}{c_2} \cdot \left(c_2 - d_f(i)\right) \geq 0 \end{vmatrix}$$

Axial strength is

$$P_{n2} := 0.85 \cdot f_c' \cdot b \cdot (0.8 \cdot c_2) + \left[\sum_{i=0}^{N_f-1} \left(A_f(i) \cdot E_f \cdot \varepsilon_{f2}(i) \right) \right]$$

$P_{n2} = 1306 \cdot \text{kip}$

Flexural strength is

$$M_{n2} := 0.85 \cdot f_c' \cdot b \cdot (0.8 \cdot c_2) \cdot \left[\frac{h}{2} - (0.4 \cdot c_2) \right]$$

$$+ \left[\sum_{i=0}^{N_f-1} \left[A_f(i) \cdot E_f \cdot \varepsilon_{f2}(i) \cdot \left(\frac{h}{2} - d_f(i) \right) \right] \right]$$

$M_{n2} = 227 \cdot \text{kip} \cdot \text{ft}$

The strength reduction factor is

$$\phi_2 := \left| \begin{array}{l} 0.65 \text{ if } 1.15 - \dfrac{\varepsilon_{f2}(N_f - 1)}{2\varepsilon_{fu}} \le 0.65 \\[2em] 1.15 - \dfrac{\varepsilon_{f2}(N_f - 1)}{2\varepsilon_{fu}} \\[2em] 0.75 \text{ if } 1.15 - \dfrac{\varepsilon_{f2}(N_f - 1)}{2\varepsilon_{fu}} \ge 0.75 \end{array} \right.$$

$\phi_2 = 0.75$

9.5.3 Point 3—Neutral axis within the cross section

In this step, the neutral axis depth ranges between h and the balanced neutral axis depth c_{bal} (compression failure):

$$c_{bal} := \frac{\varepsilon_{cu}}{\varepsilon_{cu} + \varepsilon_{fu}} \cdot d_f (N_f - 1) = 3.9 \cdot \text{in.}$$

Intervals of variation of the neutral axis depth are

$$j := 0, \frac{h - c_{bal}}{50 \text{in}} \, .. \, \frac{h - c_{bal}}{\text{in}}$$

Neutral axis depth is

$$c_3(j) := h - j \cdot in.$$

Concrete compressive strain distribution is

$$\varepsilon_c(j, x) := \frac{\varepsilon_{cu}}{c_3(j)} \cdot x$$

Strain in FRP reinforcement is

$$\varepsilon_{f3}(j, i) := \begin{vmatrix} \frac{\varepsilon_{cu}}{c_3(j)} \cdot (c_3(j) - d_f(i)) \text{ if } \frac{\varepsilon_{cu}}{c_3(j)} \cdot (c_3(j) - d_f(i)) \geq -\varepsilon_{fu} \\ 0 \text{ if } \frac{\varepsilon_{cu}}{c_3(j)} \cdot (c_3(j) - d_f(i)) \geq 0 \end{vmatrix}$$

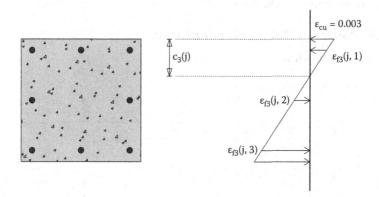

Axial strength is

$$P_n(j) := 0.85 \cdot f_c' \cdot b \cdot (0.8 \cdot c_3(j)) + \sum_{i=0}^{N_f-1} (A_f(i) \cdot E_f \cdot \varepsilon_{f3}(j, i))$$

Flexural strength is

$$M_n(j) := 0.85 \cdot f_c' \cdot b \cdot (0.8 \cdot c_3(j)) \cdot \left[\frac{h}{2} - (0.4 \cdot c_3(j)) \right]$$

$$+ \sum_{i=0}^{N_f-1} \left[A_f(i) \cdot E_f \cdot \varepsilon_{f3}(j, i) \cdot \left(\frac{h}{2} - d_f(i) \right) \right]$$

The strength reduction factor is

$$
\phi_3(j) := \begin{vmatrix} 0.65 \text{ if } 1.15 - \dfrac{\left|\varepsilon_{f3}(j, N_f - 1)\right|}{2\varepsilon_{fu}} \le 0.65 \\[2em] 1.15 - \dfrac{\left|\varepsilon_{f3}(j, N_f - 1)\right|}{2\varepsilon_{fu}} \text{ if } 1.15 - \dfrac{\left|\varepsilon_{f3}(j, N_f - 1)\right|}{2\varepsilon_{fu}} \ge 0.65 \wedge 1.15 \\[2em] \qquad\qquad\qquad\qquad - \dfrac{\left|\varepsilon_{f3}(j, N_f - 1)\right|}{2\varepsilon_{fu}} \le 0.75. \\[2em] 0.75 \text{ if } 1.15 - \dfrac{\left|\varepsilon_{f3}(j, N_f - 1)\right|}{2\varepsilon_{fu}} \ge 0.75 \end{vmatrix}
$$

The equations for computing the axial strength, the bending moment, and the phi-factor are shown above, and are functions of the neutral axis depth represented by the factor j.

9.5.4 Point 4—Balanced conditions

Neutral axis depth is

$$
c_4 := c_{bal} = 3.9 \cdot \text{in.}
$$

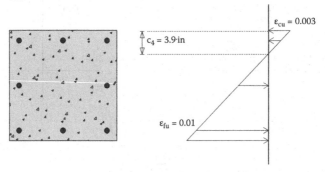

Concrete compressive strain distribution is

$$
\varepsilon_{c4}(x) := \frac{\varepsilon_{cu}}{c_4} x
$$

The strain in FRP reinforcement is

$$
\varepsilon_{f4}(i) := \begin{vmatrix} \dfrac{\varepsilon_{cu}}{c_4} \cdot \left(c_4 - d_f(i)\right) \text{ if } \dfrac{\varepsilon_{cu}}{c_4} \cdot \left(c_4 - d_f(i)\right) \ge -1.05\varepsilon_{fu} \\[2em] 0 \text{ if } \dfrac{\varepsilon_{cu}}{c_4} \cdot \left(c_4 - d_f(i)\right) \ge 0 \end{vmatrix}
$$

Axial strength is

$$P_{n4} := 0.85 \cdot f_c' \cdot b \cdot (0.8 \cdot c_4) + \left[\sum_{i=0}^{N_f-1} \left(A_f(i) \cdot E_f \cdot \varepsilon_{f4}(i) \right) \right]$$

$P_{n4} = 91 \cdot kip$

Flexural strength

$$M_{n4} := 0.85 \cdot f_c' \cdot b \cdot (0.8 \cdot c_4) \cdot \left[\frac{h}{2} - (0.4 \cdot c_4) \right] + \left[\sum_{i=0}^{N_f-1} \left[A_f(i) \cdot E_f \cdot \varepsilon_{f4}(i) \cdot \left(\frac{h}{2} - d_f(i) \right) \right] \right]$$

$M_{n4} = 266 \cdot kip \cdot ft$

Strength reduction factor

$$\phi_4 := \left| \begin{array}{ll} 0.65 & \text{if } 1.15 - \dfrac{\left| \varepsilon_{f4}(N_f-1) \right|}{2\varepsilon_{fu}} \le 0.65 \quad \phi_4 \qquad = 0.65 \\[3ex] 1.15 - \dfrac{\left| \varepsilon_{f4}(N_f-1) \right|}{2\varepsilon_{fu}} \\[3ex] 0.75 & \text{if } 1.15 - \dfrac{\left| \varepsilon_{f4}(N_f-1) \right|}{2\varepsilon_{fu}} \ge 0.75 \end{array} \right.$$

9.5.5 Point 5—Neutral axis at the level of the highest section fibers

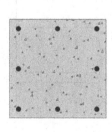 

$\varepsilon_{fu} = 0.01$

The strain in FRP reinforcement is

$$\varepsilon_{f5}(i) := \frac{-\varepsilon_{fu}}{d_f(N_f-1)} \cdot d_f(i)$$

Axial strength is

$$P_{n5} := \sum_{i=0}^{N_f-1}\left(A_f(i)\cdot E_f \cdot \varepsilon_{f5}(i)\right) = -211\cdot kip$$

Flexural strength is

$$M_{n5} := \sum_{i=0}^{N_f-1}\left[A_f(i)\cdot E_f \cdot \varepsilon_{f5}(i)\cdot\left(\frac{h}{2}-d_f(i)\right)\right] = 64.518\cdot kip\cdot ft$$

The strength reduction factor is

$$\phi_5 := \begin{vmatrix} 0.65 \text{ if } 1.15-\dfrac{\left|\varepsilon_{f5}\left(N_f-1\right)\right|}{2\varepsilon_{fu}} \le 0.65 \\[3ex] 1.15-\dfrac{\left|\varepsilon_{f5}\left(N_f-1\right)\right|}{2\varepsilon_{fu}} \\[3ex] 0.75 \text{ if } 1.15-\dfrac{\left|\varepsilon_{f5}\left(N_f-1\right)\right|}{2\varepsilon_{fu}} \ge 0.75 \end{vmatrix}$$

$$\phi_5 = 0.65$$

9.5.6 Point 6—Pure tension

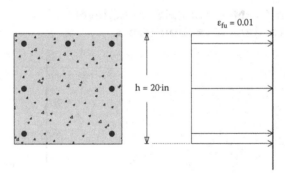

The strain in FRP reinforcement is

$$\varepsilon_{f6}(i) := -\varepsilon_{fu}$$

Axial strength is

$$P_{n6} := \left[\sum_{i=0}^{N_f-1}\left(A_f(i)\cdot E_f \cdot \varepsilon_{f6}(i)\right)\right] = -358\cdot kip$$

Flexural strength is

$M_{n6} := 0 \text{ kip·ft}$

The strength reduction factor is

$$\phi_6 := \left| \begin{array}{l} 0.65 \text{ if } 1.15 - \dfrac{\left|\varepsilon_{f6}\left(N_f - 1\right)\right|}{2\varepsilon_{fu}} \le 0.65 \\[3ex] 1.15 - \dfrac{\left|\varepsilon_{f6}\left(N_f - 1\right)\right|}{2\varepsilon_{fu}} \\[3ex] 0.75 \text{ if } 1.15 - \dfrac{\left|\varepsilon_{f6}\left(N_f - 1\right)\right|}{2\varepsilon_{fu}} \ge 0.75 \end{array} \right.$$

$\phi_6 = 0.65$

▶

P–M interaction diagram: Both nominal and design P–M curves are plotted next. The black *dots* correspond to the pairs of ultimate bending moment and axial force (M_u, P_u) relative to the load combinations defined in step 2. All eleven *dots* are enclosed within the design P–M curve and therefore the column section is adequate to carry the design loads.

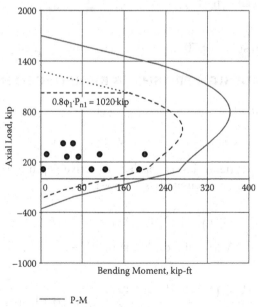

Development length and tension lap splice: The development length, *ld*, for straight bars can be calculated using Equation (11-3) of ACI 440.1R-06. The minimum between cover to bar center and half of the center-to-center bar spacing is

$$C_{b2} := \min\left(c_c + \frac{\phi_{f\_bar}}{2}, \frac{s_f}{2}\right) = 3 \cdot in$$

The bar location modification factor is:

$$\alpha_{Pos} := 1.5$$

The minimum development length is computed according to ACI 440.1R-06 Equation 11-6:

$$1_{d\_min} := \frac{\alpha_{Pos} \cdot \dfrac{f_{fu}}{\sqrt{f'_c \cdot psi}} - 340}{13.6 + \dfrac{C_{b2}}{\phi_{f\_bar}}} \phi_{f\_bar} = 52.4 \text{ in.}$$

The following development length is considered and adopted:

$$l_d := 53 \text{ in.}$$

The recommended development length of FRP tension lap splices is $1.3l_d$ (Section 11.4 of ACI 440.1R-06). The minimum recommended tension lap splice development length is

$$1.3l_d = 69 \text{ in.}$$

A tension lap splice of 70 in. is considered and adopted.

9.6 STEP 4—DESIGN FRP SHEAR REINFORCEMENT

Preliminary calculations: The maximum shear acting on the column is

$$V_u := \max(V_{u1}, V_{u2}, V_{u3}, V_{u4}, V_{u5}, V_{u6}) = 23 \cdot kip$$

The ratio of the neutral axis depth to effective depth of the longitudinal reinforcement can be calculated as follows:

$$A_{f\_e}(k, i) := \left| \begin{array}{ll} -A_f(i) & \text{if } d_f(i) \le k \cdot d_f(N_f - 1) \\ A_f(i) & \text{if } d_f(i) \ge k \cdot d_f(N_f - 1) \end{array} \right.$$

$$S_o(k) := b \cdot \frac{\left(k \cdot d_f(N_f - 1)\right)^2}{2} - \frac{E_f}{E_c} \cdot \sum_{i=0}^{N_f-1} \left[A_{f\_e}(k, i) \cdot \left(d_f(i) \cdot k \cdot d_f(N_f - 1)\right) \right]$$

First guess:

$$k_0 := 0.2$$

Given:

$$S_o(k_0) = 0$$

Solve for k:

$$k_{elastic} := root(S_o(k_0), k_0) = 0.15$$

The concrete shear capacity, V_c, can be calculated per Equation (9-1) of ACI 440.1R-06:

$$V_c := 5\sqrt{f'_c \cdot psib} \cdot \left(k_{elastic} \cdot d_f \left(N_f - 1\right)\right) = 18.1 \cdot kip$$

The shear reduction factor given by ACI 440.1R-06 is adopted:

$$\phi_v := 0.75$$

$$\frac{1}{2} \cdot \phi_v \cdot V_c = 7 \cdot kip$$

$$\text{Check\_ShearReinf} := \begin{vmatrix} \text{"OK. No shear reinforcement is needed."} & \text{if } \frac{1}{2} \cdot \phi_v \cdot V_c \geq V_u \\ \text{"Shear reinforcement is needed"} & \text{otherwise} \end{vmatrix}$$

Check_ShearReinf = "Shear reinforcement is needed"

Design shear reinforcement: The ACI 440.1R-06 minimum manufacturer's guaranteed mechanical properties of the selected bars are

Design shear reinforcement

Stirrup_Size :=

#2
#3
#4
#5
#6
#7
#8
#9
#10

$f_{fuuv} = 100 \cdot ksi$ Ultimate guaranteed tensile strength of the FRP
$\varepsilon_{fuuv} = 0.018$ Ultimate guaranteed rupture strain of the FRP
$E_f = 5700 \cdot ksi$ Guaranteed tensile modulus of elasticity of the FRP

The geometrical properties of the selected bars are

$\phi_{f-stirrup} = 0.5 \cdot in.$ Bar diameter
$A_{f-stirrup} = 0.196 \cdot in.^2$ Bar area
$r_{bsv} := 3\phi_{f-stirrup} = 1.5 \cdot in.$ Radius of the bend

The tail of the stirrup is to be at least 12 bar diameters:

$$1_{vtail} := 12 \cdot \phi_{f-stirrup} = 6 \cdot in.$$

The design tensile strength of the bend of the GFRP bar is Equation (7-3) of ACI 440.1R-06:
The design tensile strength is:

$$f_{fuv} := C_E \cdot f_{fuuv} = 80 \text{ ksi}$$

$$f_{fbsv} := \left(0.05 \cdot \frac{r_{bsv}}{\phi_{f\_stirrup}} + 0.3 \right) f_{fu} = 25.6 \cdot ksi$$

The stress level in the shear reinforcement is limited by Equation (9-3) of ACI 440.1R-06:

$$f_{fv} := \begin{vmatrix} (0.004 \cdot E_f) & \text{if } 0.004 \cdot E_f \leq f_{fbsv} \\ f_{fbsv} & \text{otherwise} \end{vmatrix}$$

$$f_{fv} = 22.8 \cdot ksi$$

The area of the two-leg tie and the cross-tie is

$$A_{fv} := 3 \cdot A_{f-stirrup} = 0.59 \cdot in.^2$$

The maximum stirrup spacing, s_{sv}, is Equation (9-4) of ACI 440.1R-06:

$$s_{sv\_max1} := \frac{A_{fv} \cdot \phi_v \cdot f_{fv} \cdot d_f (N_f - 1)}{V_u - \phi_v \cdot V_c} = 17.6 \cdot in.$$

The selected spacing is

$$s_{sv} := 12 \text{ in.}$$

The shear resistance provided by FRP stirrups and cross-tie, V_f, is Equation (9-2) of ACI 440.1R-06:

$$V_{f0} := \frac{A_{fv} \cdot f_{fv} \cdot d_f (N_f - 1)}{s_{sv}} = 19 \cdot kip$$

V_f cannot exceed $3V_c$, as discussed in Chapter 4:

$$V_f := \min(V_{f0}, 3V_c) = 19 \cdot kip \; (3V_c = 54.258 \cdot kip)$$

The nominal shear capacity, V_n, is

$$V_n := V_c + V_f = 37 \cdot kip$$

$$\phi_v \cdot V_n = 28 \cdot kip$$

$$\text{Check\_Shear} := \begin{vmatrix} \text{"OK"} & \text{if } \phi_v \cdot V_n \geq V_u \\ \text{"Not good"} & \text{otherwise} \end{vmatrix}$$

$$\text{Check\_Shear} = \text{"OK"}$$

The maximum stirrup spacing is the minimum of the least dimension of the column: 12 longitudinal bar diameters and 24 tie bar diameters:

$$s_{svmax} := \min\left[\min(b, h), 12 \cdot (\phi_{f\_bar}), 24 \cdot \phi_{f\_stirrup}\right] = 12 \cdot in.$$

$$\text{Check\_Spacing} := \begin{vmatrix} \text{"OK"} & \text{if } s_{sv\_max1} \geq s_{sv} \wedge s_{svmax} \geq s_{sv} \\ \text{"Not good"} & \text{otherwise} \end{vmatrix}$$

$$\text{Check\_Shear} = \text{"OK"}$$

9.7 STEP 5—CHECK CREEP-RUPTURE STRESS

Creep-rupture stress in the GFRP has to be evaluated considering the total unfactored dead loads and the sustained portion of the live load (20% of the total live load). The service maximum bending moment is

$$M_s := \left| M_{Dbottom} + M_{Lbottom} \right| = 36 \cdot kip \cdot ft$$

The bending moment due to dead load plus 20% of live load is

$$M_{creep} := M_s \cdot \frac{P_D + 0.20 P_L}{P_D + P_L} = 24.6 \cdot ft \cdot kip$$

The ratio of modulus of elasticity of bars to modulus of elasticity of concrete is

$$N_f := \frac{E_f}{E_c} = 1.399$$

The ratio of depth of neutral axis to reinforcement depth is

$$k_{elastic} = 0.15$$

The maximum tensile stress in the FRP is

$$f_{fmax\_creep} := -\frac{(P_D + 0.20 P_L)}{b \cdot h} + \frac{M_{creep}}{A_f (N_f - 1) \cdot d_f (N_f - 1) \cdot \left(1 - \frac{k_{elastic}}{3}\right)} = 7.2 \cdot ksi$$

$$Check\_Creep := \begin{vmatrix} \text{"OK"} & \text{if } f_{fmax\_creep} \leq f_{\_creep} & \left(f_{\_creep} = 11.4 \cdot ksi\right) \\ \text{"Not good"} & \text{otherwise} \end{vmatrix}$$

$$Check\_Shear = \text{"OK"}$$

Chapter 10

Design of square footing for a single column

10.1 INTRODUCTION

Shallow foundations must be designed to safely resist the effects of the applied factored axial and shear forces, and bending moments. The size (base area) of a shallow foundation is determined based on the permissible soil pressure, which is computed by principles of soil mechanics. For the specific case of an isolated footing, depth and required amount of flexural reinforcement are determined based on the design requirements governed by the provisions of Chapter 4 (design of flexural members). As a general consideration, a footing supporting a single column (isolated footing) must comply with all the provisions of flexure and shear for slabs. A combined footing may be designed as a beam in the longitudinal direction and as an isolated footing in the transverse direction over a defined width on each side of the supported columns. Similarly, a mat foundation normally behaves very much like a two-way slab without beams and could be designed accordingly.

This example illustrates the design of the isolated footing under column B3 (Figure 10.1) of the two-story medical facility building presented in Chapter 9.

The design of the isolated footing is presented as a sequence of seven steps as summarized here:

Step 1   Define concrete properties
Step 2   Compute service axial forces and bending moments
Step 3   Preliminary analysis (footing area and depth)
Step 4   Design FRP reinforcement
Step 5   Check creep-rupture stress
Step 6   Check crack width
Step 7   Recheck shear strength

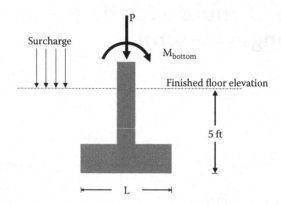

Figure 10.1 Schematic view of footing under column B3.

The results of the footing design are summarized next to facilitate the understanding of the seven sequential steps devoted to calculations.

10.2 DESIGN SUMMARY

The following soil properties are considered:

Soil density	130 psf
Permissible soil pressure	4000 psf

An additional surcharge of 100 psf due to the live load applied directly to the ground floor is also considered.

The axial forces and bending moments at the bottom of column B3 are to be considered for the design of the footing. The individual unfactored forces and moments resulting from dead, live, and wind loads are summarized in Table 10.1. The resulting service (unfactored) combinations are listed in Table 10.2. The shear force transferred by the column to the footing is disregarded in this example.

The soil pressure distribution resulting from the most demanding load combination is shown in Figure 10.2.

The critical sections for shear strength checks and other geometrical dimensions are defined in Figure 10.3. For lightly reinforced fiber-reinforced polymer (FRP) reinforced concrete members, shear strength (one way and two way) can govern the design.

The final reinforcement layout is shown in Figure 10.4 with the footing and column dimensions.

Table 10.3 is provided to convert US customary to the SI system.

Table 10.1 Individual unfactored forces and moments action on footing of column B3

Load case	Axial load: P (kip)	Moment: M (ft-kip)
Dead (D)	189.6	32.3
Live (L)	123.8	0.0
Wind (W)	−35.2	90.9

Table 10.2 Service combinations of column B3

Load combination	Axial load: P (kip)	Moment: M (ft-kip)
D	189.6	32.3
D + L	313.4	32.3
D + L + W	278.2	123.2

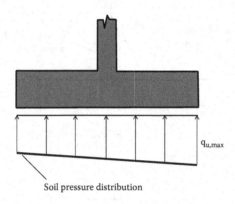

Soil pressure distribution

Figure 10.2 Soil pressure distribution.

10.3 STEP I—DEFINE CONCRETE PROPERTIES

10.3.1 Concrete properties

The following concrete properties are considered for the design:
Compressive strength is

$$f'_c := 5000 \text{ psi}$$

Ultimate compressive strain is

$$\varepsilon_{cu} := 0.003$$

Density is

$$\rho_c := 145 \frac{\text{lbf}}{\text{ft}^3}$$

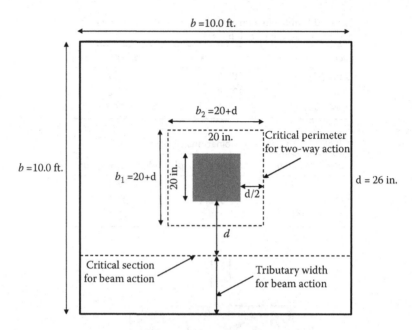

Figure 10.3 Critical cross sections for shear strength checks.

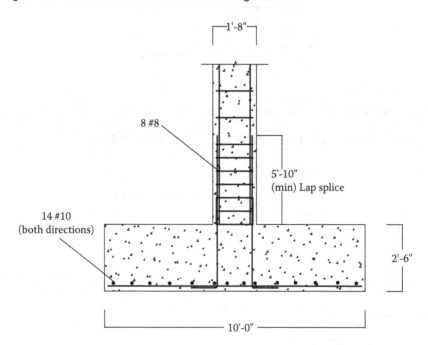

Figure 10.4 Final reinforcement layout.

Table 10.3 Conversion table

US customary	SI units
Lengths, areas, section properties	
1 in.	25.4 mm
	0.025 m
1 ft	304.8 mm
	0.305 m
1 in.^{2}	645 mm^2
1 ft^2	0.093 m^2
1 in.^{3}	16,387 mm^3
1 in.^{4}	416,231 mm^4
Forces, pressures, strengths	
1 lbf	4.448 N
1 kip	4.448 kN
1 lbf-ft	1.356 N.m
1 kip-ft	1.356 kN·m
1 psi	6.895 kPa
1 psf	47.88 N/m^2
1 ksi	6.895 MPa
1 ksf	47.88 kN/m^2
1 lbf/ft^3	157.1 N/m^3

Compressive modulus of elasticity is

$$E_c := 33 \, psi^{0.5} \left(\frac{\rho_c}{lbf.ft^{-3}} \right)^{1.5} \cdot \sqrt{f_c'}$$

The following is computed as indicated in Section 8.5.1 of ACI 318-11:

$$E_c = 4074 \cdot ksi$$

Concrete tensile strength is

$$f_{ct} := 7.5 \cdot psi^{0.5} \cdot \sqrt{f_c'}$$

This is computed as indicated in Section 9.5.3.2 of ACI 318-11:

$$f_{ct} = 530 \cdot psi$$

The stress-block factor, β_1, is computed as indicated in Section 10.2.7 of ACI 318-11:

$$\beta_1 := \begin{vmatrix} 0.85 \text{ if } f_c' = 4000\text{psi} & = 0.8 \\[2mm] 1.05 - 0.05 \cdot \dfrac{f_c'}{1000\text{psi}} \text{ if } 4000\text{psi} < f_c' < 8000\text{psi} \\[2mm] 0.65 \text{ otherwise} \end{vmatrix}$$

10.3.2 Analytical approximations of concrete compressive stress–strain curve—Todeschini's model

The compressive strain at peak is

$$\varepsilon_{c0} := \frac{1.71 \cdot f_c'}{E_c} = 0.0021$$

The compressive stress at peak is

$$\sigma''_c := 0.9 \cdot \frac{f_c'}{\text{psi}}$$

The stress–strain curve equation is

$$\sigma''_c(\varepsilon_c) := \frac{2 \cdot \sigma''_c \cdot \left(\dfrac{\varepsilon_c}{\varepsilon_{c0}}\right)}{1 + \left(\dfrac{\varepsilon_c}{\varepsilon_{c0}}\right)^2}$$

The concrete compressive stress–strain curve is

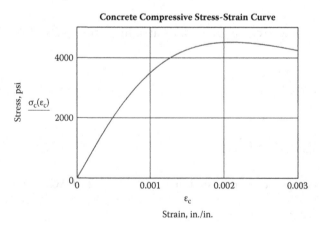

Concrete Compressive Stress-Strain Curve

10.4 STEP 2—COMPUTE SERVICE AXIAL LOADS AND BENDING MOMENTS

The axial forces and bottom moments of column B3 (see Chapter 9) are the forces to be considered for design of the footing. Analysis results per single load case are

Dead load (D):	Live load (L):	Wind (W):
$P_D := 190$ kip	$P_L := 124$ kip	$P_W := -35$ kip
$M_D := 32$ kip·ft	$M_L := 0$ kip·ft	$M_W := 91$ kip·ft

Analysis results per load combination are

- Load combination 1: D
  $P_{S1} := P_D = 190 \cdot \text{kip}$
  $M_{S1} := M_D = 32 \cdot \text{kip·fit}$
- Load combination 2: D + L
  $P_{S2} := P_D + P_L = 314 \cdot \text{kip}$
  $M_{S2} := M_D + M_L = 32 \cdot \text{kip·ft}$
- Load combination 3: D + L + W
  $P_{S3} := P_D + P_L + P_W = 279 \cdot \text{kip}$
  $M_{S3} := M_D + M_L + M_W = 123 \cdot \text{kip·ft}$

10.5 STEP 3—PRELIMINARY ANALYSIS

10.5.1 Design footing base area

To estimate the weight of soil and concrete above footing base, it was assumed that average density was

$\rho_{soil} := 130$ pcf

An additional surcharge due to the loads on the ground floor is considered:

$s_{additional} := 100$ psf

Foundation base depth is

$z := 5$ ft

The total pressure of surcharge at the foundation base depth is

$s_{total} := \rho_{soil} \cdot z + s_{additional} = 750 \cdot \text{psf}$

The permissible soil pressure is

$\sigma_{adm} := 4000$ psf

The net admissible soil pressure is

$$q_n := \sigma_{adm} - s_{total} = 3250 \cdot psf$$

The minimum required base area of the footing is

$$A_{min} := \frac{\max(P_{s1}, P_{s2}, P_{s3})}{q_n} = 96.6 \cdot ft^2$$

A square footing is selected of side b equal to

$$b = 10 \ ft$$

$$A_{footing} := b_2 = 100 \cdot ft^2$$

$$\text{CheckArea} := \begin{vmatrix} \text{"OK" if } A_{footing} \geq A_{min} \\ \text{"Not Good" otherwise} \end{vmatrix}$$

$$\text{CheckArea} = \text{"OK"}$$

10.5.2 Verify effects of eccentricity

The total weight of surcharge is

$$S_{total} := s_{total} \cdot A_{footing} = 75 \cdot kip$$

The eccentricity, e, relative to each load combination is
Load combination 1:

$$e_1 := \frac{M_{s1}}{P_{s1} + S_{total}} = 0.121 \cdot ft$$

Load combination 2:

$$e_2 := \frac{M_{s2}}{P_{s2} + S_{total}} = 0.082 \cdot ft$$

Load combination 3:

$$e_3 := \frac{M_{s3}}{P_{s3} + S_{total}} = 0.347 \cdot ft$$

The average soil pressure, q_{ave}, relative to each load combination can be computed as follows:

Load combination 1:

$$q_{ave1} := \frac{P_{s1}+S_{total}}{A_{footing}} = 2.65 \cdot ksf$$

Load combination 2:

$$q_{ave2} := \frac{P_{s2}+S_{total}}{A_{footing}} = 3.89 \cdot ksf$$

Load combination 3:

$$q_{ave3} := \frac{P_{s3}+S_{total}}{A_{footing}} = 3.54 \cdot ksf$$

The maximum soil pressure, q_{max}, relative to each load combination can be computed as follows:

Load combination 1:

$$q_{max1} := \left| \begin{array}{ll} \left[\left(1+\frac{6 \cdot e_1}{b}\right) \cdot q_{ave1}\right] & \text{if } \frac{e_1}{b} \le \frac{1}{6} \\[4mm] \dfrac{4}{\left[3 \cdot \left(1-\frac{2e_1}{b}\right)\right]} \cdot q_{ave1} & \text{if } \frac{e_1}{b} \ge \frac{1}{6} \wedge \frac{e_1}{b} \le \frac{1}{2} \end{array} \right. \qquad = 2.8 \cdot ksf$$

Load combination 2:

$$q_{max2} := \left| \begin{array}{ll} \left[\left(1+\frac{6 \cdot e_2}{b}\right) \cdot q_{ave2}\right] & \text{if } \frac{e_2}{b} \le \frac{1}{6} \\[4mm] \dfrac{4}{\left[3 \cdot \left(1-\frac{2e_2}{b}\right)\right]} \cdot q_{ave2} & \text{if } \frac{e_2}{b} \ge \frac{1}{6} \wedge \frac{e_2}{b} \le \frac{1}{2} \end{array} \right. \qquad = 4.1 \cdot ksf$$

Load combination 3:

$$q_{max3} := \begin{vmatrix} \left[\left(1+\dfrac{6\cdot e_3}{b}\right)\cdot q_{ave3}\right] \text{ if } \dfrac{e_3}{b} \le \dfrac{1}{6} & = 4.3\cdot ksf \\[3mm] \dfrac{4}{\left[3\cdot\left(1-\dfrac{2e_3}{b}\right)\right]}\cdot q_{ave3} \text{ if } \dfrac{e_3}{b} \ge \dfrac{1}{6} \wedge \dfrac{e_3}{b} \le \dfrac{1}{2} \end{vmatrix}$$

The maximum soil pressure shall not exceed the admissible soil pressure. Note that it is typically acceptable to increase the allowable soil pressure when the maximum pressure is investigated. Here, an increase of 20% is deemed appropriate. The ratio between the eccentricity and the side of the footing, e/b, is also normally limited to a maximum of 0.25. Furthermore, under seismic or wind loading the allowable soil pressure can be increased by 33%.

Load combination 1:

$$CheckSoilPressure1 := \begin{vmatrix} \text{"OK" if } q_{max1} \le 1.2\cdot\sigma_{adm} \\ \text{"Not good" otherwise} \end{vmatrix}$$

CheckSoilPressure1 = "OK"

Load combination 2:

$$CheckSoilPressure2 := \begin{vmatrix} \text{"OK" if } q_{max2} \le 1.2\cdot\sigma_{adm} \\ \text{"Not good" otherwise} \end{vmatrix}$$

CheckSoilPressure2 = "OK"

Load combination 3:

$$CheckSoilPressure3 := \begin{vmatrix} \text{"OK" if } q_{max3} \le 1.2\cdot 1.33\cdot\sigma_{adm} \\ \text{"Not good" otherwise} \end{vmatrix}$$

CheckSoilPressure3 = "OK"

10.5.3 Ultimate pressure under the footing

To proportion the footing for strength, factored loads are considered.
Load combination 1:

$P_{u1} := 1.4P_D = 266\cdot kip$

$M_{u1} := 1.4\cdot M_D = 44.8\cdot kip\cdot ft$

Load combination 2:

$$P_{u2} := 1.2P_D + 1.6P_L = 426.4 \cdot kip$$

$$M_{u2} := 1.2 \cdot M_D + 1.6M_L = 38.4 \cdot kip \cdot ft$$

Load combination 3:

$$P_{u3} := 1.2P_D + 1.0P_L + 1.6P_W = 296 \cdot kip$$

$$M_{u3} := 1.2 \cdot M_D + 1.0M_L + 1.6M_W = 184 \cdot kip \cdot ft$$

The eccentricity, e, relative to each factored load combination is
Load combination 1:

$$e_{u1} := \frac{M_{u1}}{P_{u1}} = 0.168 \cdot ft$$

Load combination 2:

$$e_{u2} := \frac{M_{u2}}{P_{u2}} = 0.09 \cdot ft$$

Load combination 3:

$$e_{u3} := \frac{M_{u3}}{P_{u3}} = 0.622 \cdot ft$$

The average soil pressure, q_{uave}, relative to each factored load combination can be computed as follows:
Load combination 1:

$$q_{uave1} := \frac{P_{u1}}{A_{footing}} = 2.7 \cdot ksf$$

Load combination 2:

$$q_{uave2} := \frac{P_{u2}}{A_{footing}} = 4.3 \cdot ksf$$

Load combination 3:

$$q_{uave3} := \frac{P_{u3}}{A_{footing}} = 3 \cdot ksf$$

The maximum soil pressure, q_u, relative to each factored load combination can be computed as follows:

Load combination 1:

$$q_{u1} := \left| \begin{array}{l} \left[\left(1 + \dfrac{6 \cdot e_{u1}}{b}\right) \cdot q_{uave1}\right] \text{if } \dfrac{e_{u1}}{b} \le \dfrac{1}{6} \\[2em] \dfrac{4}{\left[3 \cdot \left(1 - \dfrac{2e_{u1}}{b}\right)\right]} \cdot q_{uave1} \text{ if } \dfrac{e_{u1}}{b} \ge \dfrac{1}{6} \wedge \dfrac{e_{u1}}{b} \le \dfrac{1}{2} \end{array} \right. = 2.9 \cdot \text{ksf}$$

Load combination 2:

$$q_{u2} := \left| \begin{array}{l} \left[\left(1 + \dfrac{6 \cdot e_{u2}}{b}\right) \cdot q_{uave2}\right] \text{if } \dfrac{e_{u2}}{b} \le \dfrac{1}{6} \\[2em] \dfrac{4}{\left[3 \cdot \left(1 - \dfrac{2e_{u2}}{b}\right)\right]} \cdot q_{uave2} \text{ if } \dfrac{e_{u2}}{b} \ge \dfrac{1}{6} \wedge \dfrac{e_{u2}}{b} \le \dfrac{1}{2} \end{array} \right. = 4.5 \cdot \text{ksf}$$

Load combination 3:

$$q_{u3} := \left| \begin{array}{l} \left[\left(1 + \dfrac{6 \cdot e_{u3}}{b}\right) \cdot q_{uave3}\right] \text{if } \dfrac{e_{u3}}{b} \le \dfrac{1}{6} \\[2em] \dfrac{4}{\left[3 \cdot \left(1 - \dfrac{2e_{u3}}{b}\right)\right]} \cdot q_{uave3} \text{ if } \dfrac{e_{u3}}{b} \ge \dfrac{1}{6} \wedge \dfrac{e_{u3}}{b} \le \dfrac{1}{2} \end{array} \right. = 4.1 \cdot \text{ksf}$$

The assumed soil pressure distribution was shown in Figure 10.2. Based on the three load combinations, the following soil pressure distribution is used for design.

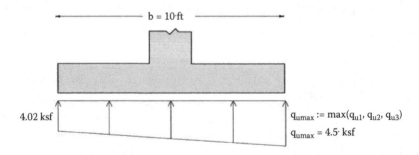

$q_{umax} := \max(q_{u1}, q_{u2}, q_{u3})$

$q_{umax} = 4.5 \cdot \text{ksf}$

10.5.4 Design for depth

A first assumption of reinforcement area based on that required for shrinkage and temperature ($\rho_{f,ts}$) is reasonable:
Assuming:

$$f_{fu'} := 0.7 \cdot 70 \text{ ksi} = 49 \cdot \text{ksi}$$

$$E_{f'} := 5700 \text{ ksi}$$

$$\rho_{f\_ts} := \min\left[0.0036, 0.0018 \cdot \left(\frac{60\text{ksi}}{f_{fu'}}\right) \cdot \frac{29000\text{ksi}}{E_{f'}}\right] = 0.0036$$

Note that $\rho_{f,ts}$ is calculated based on the gross area of the concrete cross section, whereas ρ_f is calculated based on the effective area as determined by the reinforcement depth. The FRP reinforcement ratio ρ_f is selected as $1.5 \rho_{f,ts}$:

$$\rho_f := 1.5 \cdot \rho_{f\_ts} = 0.0054$$

For lightly FRP reinforced sections, shear strength can be the factor that governs the thickness. The critical sections considered for the shear checks are indicated in Figure 10.3.
Beam (one-way) action: The column width is

$$b_{col} := 20 \text{ in}$$

As a starting point, an FRP reinforcement depth equal to the column width is assumed:

$$d_1 := b_{col} = 20 \cdot \text{in.}$$

The tributary width at a distance d_1 from the column can be calculated as follows:

$$w_{tributary} := \frac{b}{2} - \frac{b_{col}}{2} - d_1 = 30 \cdot \text{in.}$$

The tributary area is therefore equal to

$$A_{tributary} := b \cdot w_{tributary} = 25 \cdot \text{ft}^2$$

The ultimate shear force is equal to

$$V_u := \max(q_{u1}, q_{u2}, q_{u3}) \cdot A_{tributary} = 112.4 \cdot \text{kip}$$

Assuming an FRP modular ratio equal to

$$n_{f'} := 1.4$$

The neutral axis depth to FRP effective depth ratio of the cracked section is the larger of

$$k := \max\left[\sqrt{2 \cdot (\rho_f \cdot n_{f'}) + (\rho_f \cdot n_{f'})^2} - (\rho_f \cdot n_{f'}), 0.16 \right] = 0.16$$

The nominal shear strength is

$$V_{n1} := 5\sqrt{f'_c \cdot \text{psi}} \cdot b \cdot k \cdot d_1 = 136 \cdot \text{kip}$$

The shear strength reduction factor is

$$\phi_V := 0.75$$

$$\text{CheckShear1} := \left| \begin{array}{l} \text{"OK" if } V_u \leq \phi_v \cdot V_{n1} \\ \text{"Not good" otherwise} \end{array} \right.$$

$$\text{CheckShear1} = \text{"Not good"}$$

A new value of effective depth can be computed as follows:
First guess:

$$d_0 := b_{col} = 20 \cdot \text{in.}$$

Given:

$$f(d_2) := V_u - \phi_V \cdot 5 \cdot \sqrt{f'_c \cdot \text{psi}} \cdot k \cdot d_2 \cdot b$$

Solve for effective depth:

$$f(d_2) = 0$$

$$d_2 := \text{root}((f(d_0)), d_0) = 22.1 \cdot \text{in.}$$

A cover of 4.5 in. from the centroid of the reinforcement is considered. The following thickness is assumed:

$$h' := 27 \text{ in.}$$

The following effective depth is computed based on a 4.5 in. cover:

$$d' := 22.5 \text{ in.}$$

The new nominal shear strength is

$$V_{n'} := 5\sqrt{f_c' \cdot psi} \cdot b \cdot k \cdot d' = 153 \cdot kip$$

$$CheckShear2 := \left| \begin{array}{l} \text{"OK" if } V_u \leq \phi_V \cdot V_{n'} \\[2ex] \text{"Not good" otherwise} \end{array} \right. \qquad (V_u = 112 \cdot kip \text{ and } \\ \phi_V \cdot V_{n'} = 115 \cdot kip)$$

$$CheckShear2 = \text{"OK"}$$

A footing thickness of 27 in. is therefore acceptable for the one-way shear check.

Punching (two-way) action:
Load combination 1:

$$b_1' = b_{col} + d' = 42.5 \cdot in.$$

$$b_2' := b_1'$$

First, the shear forces and moments acting on the boundary of the critical section for punching shear must be calculated from the axial forces and moments at the bottom of the column, which are presented above. Since none of the three load combinations generates uplift, it can be concluded that shear on the boundary of the critical section, V_u, is proportional to the ratio of tributary area to total area. The tributary area is the part of the foundation outside the critical perimeter for the two-way action. V_u can be computed as follows:

$$V_{u\_ps1'} := P_{u1} \cdot \frac{\left(b^2 - b_{1'} \cdot b_{2'}\right)}{b^2} = 232.6 \cdot kip$$

Moment on the boundary of the critical section, M_u, is proportional to the ratio moment of inertia of tributary area to total area. M_u can be computed as follows:

$$M_{u\_ps1'} := M_{u1} \cdot \frac{\left(\dfrac{b^4}{12} - \dfrac{b_{1'}^3 \cdot b_{2'}}{4}\right)}{\dfrac{b^4}{12}} = 42.7 \cdot kip \cdot ft$$

The portion of this moment transferred by the eccentricity of the shear force, $\gamma_v M_u$, can be calculated as follows. This moment creates additional shear on the surface of the critical section for punching shear:

$$\gamma_{f'} := \frac{1}{1 + \frac{2}{3} \cdot \sqrt{\frac{b_{1'}}{b_{2'}}}} = 0.6$$

$$\gamma_{v'} := 1 - \gamma_{f'} = 0.4$$

$$\gamma_{v'} \cdot M_{u\_ps1'} = 17.1 \cdot \text{kip} \cdot \text{ft}$$

The same can be repeated for the other two load combinations:
Load combination 2:

$$V_{u\_ps2'} := P_{u2} \cdot \frac{\left(b^2 - b_{1'} \cdot b_{2'}\right)}{b^2} = 372.9 \cdot \text{kip}$$

$$M_{u\_ps2'} := M_{u2} \cdot \frac{\left(\dfrac{b^4}{12} - \dfrac{b_{1'}^3 \cdot b_{2'}}{4}\right)}{\dfrac{b^4}{12}} = 36.6 \cdot \text{kip} \cdot \text{ft}$$

$$\gamma_{v'} M_{u\_ps2'} = 14.6 \cdot \text{kip} \cdot \text{ft}$$

Load combination 3:

$$V_{u\_ps3'} := P_{u3} \cdot \frac{\left(b^2 - b_{1'} \cdot b_{2'}\right)}{b^2} = 258.9 \cdot \text{kip}$$

$$M_{u\_ps3'} := M_{u3} \cdot \frac{\left(\dfrac{b^4}{12} - \dfrac{b_{1'}^3 \cdot b_{2'}}{4}\right)}{\dfrac{b^4}{12}} = 175.3 \cdot \text{kip} \cdot \text{ft}$$

$$\gamma_{v'} M_{u\_ps3'} = 70.13 \cdot \text{kpi} \cdot \text{ft}$$

The maximum punching shear stress v_u is to be computed.
The depth of the concrete that resists shear is computed as

$$c_{v'} := \frac{5}{2} k \cdot d' = 9 \cdot \text{in.}$$

The total area of concrete resisting two-way shear is

$$A_c' := 2 \cdot (b_1' + b_2') c_v' = 1530 \cdot in.^2$$

The property of the assumed critical section analogous to the polar moment of inertia is

$$J_{c'} := \frac{c_{v'} \cdot b_{1'}^3}{6} + \frac{c_{v'}^3 \cdot b_{1'}}{6} + \frac{c_{v'} \cdot b_{1'}^2 \cdot b_{2'}}{2} = 465757 \cdot in^4$$

Load combination 1:

$$v_{u1'} := \frac{V_{u\_ps1'}}{A_{c'}} + \frac{\gamma_{v'} \cdot M_{u\_ps1'} \cdot b_{1'}}{2 J_{c'}} = 161.4 \cdot psi$$

Load combination 2:

$$v_{u2'} := \frac{V_{u\_ps2'}}{A_{c'}} + \frac{\gamma_{v'} \cdot M_{u\_ps2'} \cdot b_{1'}}{2 J_{c'}} = 251.7 \cdot psi$$

Load combination 3:

$$v_{u3'} := \frac{V_{u\_ps3'}}{A_{c'}} + \frac{\gamma_{v'} \cdot M_{u\_ps3'} \cdot b_{1'}}{2 J_{c'}} = 207.6 \cdot psi$$

$$\text{CheckPunchingShear'} := \left| \begin{array}{l} \text{"OK" if } \max(v_{u1'}, v_{u2'}, v_{u3'}) \leq \phi_v \cdot 4 \sqrt{f_c' \cdot psi} \\ \text{"Not good" otherwise} \end{array} \right.$$

CheckPunchingShear' = "Not good"

Because the check for punching shear is not verified, the depth of the footing is increased. The new depth, h, is considered:

h := 30 in.

The following reinforcement depth is computed based on a 4.5 in. concrete cover:

d := 25.5 in.

The preceding calculations are repeated.

Load combination 1:

$b_1 := b_{col} + d = 45.5 \cdot in.$

$b_2 := b_1$

$$V_{u\_ps1} := P_{u1} \cdot \frac{\left(b^2 - b_1 \cdot b_2\right)}{b^2} = 227.8 \cdot kip$$

$$M_{u\_ps1} := M_{u1} \cdot \frac{\left(\dfrac{b^4}{12} - \dfrac{b_1^3 \cdot b_2}{4}\right)}{\dfrac{b^4}{12}} = 42 \cdot kip \cdot ft$$

$$\gamma_f := \frac{1}{1 + \dfrac{2}{3} \cdot \sqrt{\dfrac{b_1}{b_2}}} = 0.6$$

$\gamma_v := 1 - \gamma_f = 0.4$

$\gamma_v \cdot M_{u-ps1} = 17.074 \cdot kip \cdot ft$

Load combination 2:

$$V_{u\_ps2} := P_{u2} \cdot \frac{\left(b^2 - b_1 \cdot b_2\right)}{b^2} = 365.1 \cdot kip$$

$$M_{u\_ps2} := M_{u2} \cdot \frac{\left(\dfrac{b^4}{12} - \dfrac{b_1^3 \cdot b_2}{4}\right)}{\dfrac{b^4}{12}} = 36 \cdot kip \cdot ft$$

$\gamma_v \cdot M_{u-ps2} := 14.408 \cdot kip \cdot ft$

Load combination 3:

$$V_{u\_ps3} := P_{u3} \cdot \frac{\left(b^2 - b_1 \cdot b_2\right)}{b^2} = 253.4 \cdot kip$$

$$M_{u\_ps3} := M_{u3} \cdot \frac{\left(\dfrac{b^4}{12} - \dfrac{b_1^3 \cdot b_2}{4}\right)}{\dfrac{b^4}{12}} = 172.6 \cdot kip \cdot ft$$

$$\gamma_v \cdot M_{u-ps3} = 69.036 \cdot kip \cdot ft$$

The depth of the concrete that resists shear is computed as

$$c_v := \frac{5}{2} k \cdot d = 10.2 \cdot in.$$

The total area of concrete resisting shear is

$$A_c := 2 \cdot (b_1 + b_2) c_v = 1856.4 \cdot in.^2$$

The property of the assumed critical section analogous to the polar moment of inertia is

$$J_c := \frac{c_v \cdot b_1^3}{6} + \frac{c_v^3 \cdot b_1}{6} + \frac{c_v \cdot b_1^2 \cdot b_2}{2} = 648583 \cdot in^4$$

Load combination 1:

$$v_{u1} := \frac{V_{u\_ps1}}{A_c} + \frac{\gamma_v \cdot M_{u\_ps1} \cdot b_1}{2J_c} = 130 \cdot psi$$

Load combination 2:

$$v_{u2} := \frac{V_{u\_ps2}}{A_c} + \frac{\gamma_v \cdot M_{u\_ps2} \cdot b_1}{2J_c} = 230 \cdot psi$$

Load combination 3:

$$v_{u3} := \frac{V_{u\_ps3}}{A_c} + \frac{\gamma_v \cdot M_{u\_ps3} \cdot b_1}{2J_c} = 166 \cdot psi$$

$$\text{CheckPunchingShear'} := \left|\begin{array}{l} \text{"OK" if } \max\left(v_{u1'}, v_{u2'}, v_{u3'}\right) \le \phi_v \cdot 4\sqrt{f_c' \cdot psi} \\ \text{"Not good" otherwise} \end{array}\right.$$

$$\text{CheckPunchingShear} = \text{"OK"}$$

10.6 STEP 4—DESIGN FRP REINFORCEMENT FOR BENDING MOMENT CAPACITY

Properties of the selected FRP reinforcement: The manufacturer's guaranteed mechanical properties are

Properties of the selected FRP reinforcement

Type_of_Fiber :=
| Glass |
| Carbon |

Bar_Size ;=
| #3 |
| #4 |
| #5 |
| #6 |
| #7 |
| #8 |
| #9 |
| #10 |

$f_{fuu} = 70 \cdot ksi$ Ultimate guaranteed tensile strength of the FRP
$\varepsilon_{fuu} = 0.01228$ Ultimate guaranteed rupture strain of the FRP
$E_f = 5700 \cdot ksi$ Guaranteed tensile modulus of elasticity of the FRP

The geometrical properties of the selected bar are

$\phi_f\_bar = 1.27 \cdot in.$ Bar diameter
$A_f\_bar = 1.267 \cdot in.^2$ Bar area

FRP reduction factors: Table 7-1 of ACI 440.1R-06 is used to define the environmental reduction factor, C_E. The type of exposure has to be selected:

Type_of_Exposure :=
| Interior |
| Exterior |

$C_E = 0.7$ Environmental reduction factor for GFRP

Table 8-3 in ACI 440.1R-06 is used to define the reduction factor to take into account the FRP creep-rupture stress. Creep stress in the FRP has to be evaluated considering the total unfactored dead loads and the sustained portion of the live load (20% of the total live load):

$k_{creep} = 0.2$ Creep-rupture stress limitation factor

Crack width is checked using Equation (8-9) of ACI 440.1R-06. A crack width limit, w_{lim}, of 0.020 in. is used for exterior exposure:

$w_{lim} = 0.02 \cdot in.$ Crack width limit

FRP ultimate design properties: The ultimate design properties are calculated per Section 7.2 of ACI 440.1R-06:

$f_{fu} := C_E \cdot f_{fuu} = 49 \cdot ksi$ Design tensile strength
$\varepsilon_{fu} := C_E \cdot \varepsilon_{fuu} = 0.0086$ Design rupture strain

FRP creep-rupture limit stress: The FRP creep-rupture limit stress is calculated per Section 8.4 of ACI 440.1R-06:

$f_{f-creep} := k_{creep} \cdot f_{fu} = 9.8 \cdot ksi$

Minimum FRP reinforcement ratio: The minimum FRP reinforcement ratio, $\rho_{f,smin}$, to limit cracks due to shrinkage and temperature is given by Equation (10-1) of ACI 440.1R-06 (computed in step 3):

$\rho_{fsmin} := \rho_{f-ts} = 0.0036$

The following minimum clear concrete cover is selected:

$c_c := 2.0$ in.

The effective reinforcement depth is

$d_f := h - c_c - \dfrac{3 \cdot \phi_{f\_bar}}{2} = 26.1 \cdot in.$

As a starting point for the design, the FRP reinforcement area can be taken equal to 1.5 times the minimum FRP reinforcement area:

$Af_{-min} := (b \cdot d_f) \cdot 1.5 \rho_{fsmin} = 17.085 \cdot in.^2$

The number of FRP bars can be estimated as follows:

$N_{barEstimated} := \dfrac{A_{f\_min}}{A_{f\_bar}} = 13.487$

The number of bars is therefore:

$N_{bar} := 14$

$A_f := N_{bar} \cdot A_{f\_bar} = 17.735 \cdot in.^2$

The FRP reinforcement ratio is

$$\rho_{f\_des} := \frac{A_f}{b \cdot d_f} = 0.00561 \qquad \text{Equation (8-2) of ACI 440.1R-06}$$

$$\text{Check\_MinReinf} := \begin{vmatrix} \text{"OK" if} \quad \rho_{f\_des} \geq \rho_{fsmin} \\ \text{"Not satisfied" otherwise} \end{vmatrix}$$

Check_MinReinf = "OK"

FRP bar spacing: The bar clear spacing is

$$s_{f0} := \begin{vmatrix} b - \phi_{f\_bar} \text{ if } N_{bar} = 1 \\ \dfrac{b - 2c_c - N_{bar} \cdot \phi_{f\_bar}}{N_{bar}^{-1}} \text{ otherwise} \end{vmatrix}$$

$$s_{f0} = 7.402 \cdot \text{in.}$$

The minimum required bar spacing is

$$s_{f-min} := \max(1 \cdot \text{in}, \phi_{f-bar}) = 1.27 \cdot \text{in.}$$

$$\text{Check\_BarSpacing} := \begin{vmatrix} \text{"OK" if } s_{f0} \geq s_{f\_min} \\ \text{"Too many bars" otherwise} \end{vmatrix}$$

Check_BarSpacing = "OK"

The bar spacing is

$$s_{f1} := s_{f0} + \phi_{f-bar} = 8.7 \cdot \text{in.}$$

Design flexural strength: The balanced reinforcement ratio, ρ_{fb}, is computed per Equation (8-3) of ACI 440.1R-06:

$$\rho_{fb} := 0.85\beta_1 \cdot \frac{f'_c}{f_{fu}} \cdot \frac{E_f \cdot \varepsilon_{cu}}{E_f \cdot \varepsilon_{cu} + f_{fu}} = 0.01795$$

Based on cross-section compatibility, the effective concrete compressive strain at failure can be computed as a function of the neutral axis depth, x:

$$\varepsilon_c(x) := \left| \begin{array}{l} \varepsilon_{cu} \text{ if } \quad \rho_f \ge \rho_{fb} \\ \\ \dfrac{\varepsilon_u}{d_f - x} \cdot x \text{ if } \rho_f < \rho_{fb} \end{array} \right.$$

Based on cross-section compatibility, the effective tensile strain in the FRP reinforcement can be computed as a function of the neutral axis depth, x:

$$\varepsilon_f(x) := \left| \begin{array}{l} \varepsilon_{fu} \text{ if } \rho_f < \rho_{fb} \\ \\ \min\left[\dfrac{\varepsilon_{cu}}{x} \cdot (d_f - x), \varepsilon_{fu}\right] \quad \text{ if } \rho_f \ge \rho_{fb} \end{array} \right.$$

The compressive force in the concrete as a function of the neutral axis depth, x, is

$$C_c(x) := b \cdot \int_{0\text{in}}^x \frac{2 \cdot \sigma''_c \cdot \left(\dfrac{\varepsilon_c(x)}{\varepsilon_{c0}}\right)}{1 + \left(\dfrac{\varepsilon_c(x)}{\varepsilon_{c0}}\right)^2} \text{psi } dx \qquad \text{Compressive force in the concrete}$$

The tensile force in the FRP reinforcement as a function of the neutral axis depth, x, is

$$T_f(x) := A_f \, E_f \cdot \varepsilon_f(x)$$

The neutral axis depth, c_u, can be computed by solving the equation of equilibrium $C_c - T_f = 0$:
First guess:

$$x_{01} := 0.1 d_f$$

Given:

$$f_0(x) := C_c(x) - T_f(x)$$

$$c_u := \text{root}(f_0(x_{01}), x_{01})$$

The neutral axis depth is

$$c_u = 3.314 \cdot \text{in.}$$

The maximum concrete strain is: $\varepsilon_c(c_u) = 0.00124$.
The maximum FRP strain is: $\varepsilon_f(c_u) = 0.0086$.

The nominal bending moment capacity can be computed as follows:

$$M_n := b \cdot \int_0^{c_u} x \cdot \frac{\left(2 \cdot \sigma''_c \cdot \dfrac{\varepsilon_c(x)}{\varepsilon_{c0}}\right)}{\left[1 + \left(\dfrac{\varepsilon_c(x)}{\varepsilon_{c0}}\right)^2\right]} \text{psi } dx + T_f(c_u) \cdot (d_f - c_u) = 1826 \cdot \text{ft} \cdot \text{kip}$$

The concrete crushing failure mode is less brittle than the one due to FRP rupture. The ϕ-factor is calculated according to Jawahery and Nanni [1]:

$$\phi_b := \left|\begin{array}{l} 0.65 \text{ if } 1.15 - \dfrac{\varepsilon_f(c_u)}{2\varepsilon_{fu}} \leq 0.65 \\[4mm] 0.75 \text{ if } 1.15 - \dfrac{\varepsilon_f(c_u)}{2\varepsilon_{fu}} \geq 0.75 \\[4mm] 1.15 - \dfrac{\varepsilon_f(c_u)}{2\varepsilon_{fu}} \text{ otherwise} \end{array}\right.$$

$\phi_b = 0.65$

The design flexural strength equation is computed per Equation (8-1) of ACI 440.1R-06:

$\phi_b \cdot M_n = 1187 \cdot \text{kip} \cdot \text{ft}$

$$M_u := \frac{1}{2} \max(q_{u1}, q_{u2}, q_{u3}) \cdot b \cdot \left(\frac{b - b_{col}}{2}\right)^2 = 390.139 \cdot \text{kip} \cdot \text{ft}$$

$$\text{Check\_Flexure} := \left|\begin{array}{l} \text{``OK'' if } \phi_b \cdot M_n \geq M_u \\[2mm] \text{``Not good'' otherwise} \end{array}\right.$$

$\text{Check\_Flexure} = \text{``OK''}$

Development length: The development length, l_d, for straight bars can be calculated using Equation (11-3) of ACI 440.1R-06:

The minimum between cover to bar center and half of the center-to-center bar spacing is

$$C_{b2} := \min\left(c_c + \frac{\phi_{f\_bar}}{2}, \frac{s_{f0}}{2}\right) = 3.635 \cdot \text{in.}$$

The bar location modification factor for bottom reinforcement is

$$\alpha_{Pos} := 1$$

The minimum development length is computed according to ACI 440.1R-06 Equation (11-6):

$$l_{d\_min} := \frac{\alpha_{Pos} \cdot \dfrac{f_{fu}}{\sqrt{f'_c \cdot \text{psi}}} - 340}{13.6 + \dfrac{C_{b2}}{\phi_{f\_bar}}} \phi_{f\_bar} = 27.23 \cdot \text{in}$$

The following development length is considered and is available to develop the required moment capacity:

$$l_d = 28 \text{ in.}$$

10.7 STEP 5—CHECK CREEP-RUPTURE STRESS

Creep-rupture stress in the FRP has to be evaluated considering the total unfactored dead loads and the sustained portion of the live load (20% of the total live load). The maximum value of the service bending moment is the following:

$$M_{sMax} := \frac{1}{2} q_{ave2} \cdot b \cdot \left(\frac{b - b_{col}}{2}\right)^2 = 337.7 \cdot \text{kip} \cdot \text{ft}$$

The blending moment due to total load plus 20% of live load is

$$M_{creep} := M_{sMax} \cdot \frac{P_D + 0.20 P_L}{P_D + P_L} = 231 \cdot \text{kip} \cdot \text{ft}$$

The ratio of modulus of elasticity of bars to modulus of elasticity of concrete is

$$n_f := \frac{E_f}{E_c} = 1.399$$

The ratio of depth of neutral axis to reinforcement depth, calculated per Equation (8-12), is

$$k_f := \sqrt{2\rho_{f\_des} \cdot n_f + (\rho_{f\_des} \cdot n_f)^2} - \rho_f \cdot n_f = 0.119$$

with $\rho_{fdes} = 0.00561$

The tensile stress in the FRP is

$$f_{fcreep} := \frac{M_{creep}}{A_f \cdot d_f \cdot \left(1 - \dfrac{k_f}{3}\right)} = 6.2 \cdot ksi$$

$$\text{Check\_Creep} := \begin{vmatrix} \text{"OK" if } f_{fcreep} \leq f_{f\_creep} & (f_{f\_creep} = 9.8 \cdot ksi) \\ \text{"Not good" otherwise} \end{vmatrix}$$

$$\text{Check\_Creep} = \text{"OK"}$$

10.8 STEP 6—CHECK CRACK WIDTH

Crack width is checked using Equation (8-9) of ACI 440.1R-06. A crack width limit, w_{lim}, of 0.020 in. is used for exterior exposure. The ratio of modulus of elasticity of bars to modulus of elasticity of concrete is

$$n_f = 1.399$$

The ratio of depth of neutral axis to reinforcement depth is

$$k_f = 0.119$$

Tensile stress in GFRP under service loads is

$$f_{fs} := \frac{M_{sMax}}{A_f \cdot d_f \cdot \left(1 - \dfrac{k_f}{3}\right)} = 9.1 \cdot ksi$$

with $M_{sMax} = 337.7 \cdot kip \cdot ft.$

The ratio of distance from neutral axis to extreme tension fiber to distance from neutral axis to center of tensile reinforcement is

$$\beta_{11} := \frac{h - k_f \cdot d_f}{d_f \cdot (1 - k_f)} = 1.17$$

The thickness of concrete cover measured from extreme tension fiber to center of bar is

$$d_c := h - d_f = 3.9 \cdot in.$$

The bond factor (provided by the manufacturer) is

$$k_b := 0.9$$

The crack width under service loads is

$$w := 2 \frac{f_{fs}}{E_f} \beta_{11} \cdot kb \cdot \sqrt{d_c^2 + \left(\frac{s_{f1}}{2}\right)^2} = 0.02 \cdot in \quad \text{Equation (8-9) of ACI 440.1R-06}$$

The crack width limit for the selected exposure is

$$w_{lim} = 0.02 \cdot in.$$

$$\text{Check\_Crack1} := \begin{array}{|l} \text{"OK" if } w \leq w_{lim} \\ \text{"Not good" otherwise} \end{array}$$

$$\text{Check\_Crack1} = \text{"OK"}$$

10.9 STEP 7—RECHECK SHEAR STRENGTH

Beam (one-way) shear: The ultimate shear force is

$$V_u = 112.36 \cdot kip$$

The reinforcement effective depth is

$$d_f = 26.1 \cdot in.$$

The new nominal shear strength is

$$V_n := \left[5 \sqrt{f_c' \cdot psi} \cdot b \cdot (k_f) \cdot d_f \right] = 132 \cdot kip$$

but not less than

$$V_{n\_min} := \left(0.8\sqrt{f_c' \cdot psi} \cdot b \cdot d_f\right) = 179 \cdot kip$$

The ϕ-factor for shear is

$$\phi_v = 0.75$$

$$\text{Check\_Oneway\_Shear} := \begin{array}{|l} \text{``OK'' if } V_u \le \phi_V \times V_{n\_min} \\ \\ \text{``Not good'' otherwise} \end{array} \begin{pmatrix} V_u = 112 \cdot kip \text{ and} \\ \phi_v \times V_{n\_min} = 134 \cdot kip \end{pmatrix}$$

Check_Oneway_Shear = "OK"

Punching (two-way) shear: The depth of the concrete that resists shear is computed as

$$c_{vf} := \frac{5}{2}(k) \cdot d_f = 10.5 \cdot in.$$

The total area of concrete resisting shear is

$$A_{cf} := 2 \cdot (b_1 + b_2)c_{vf} = 1919.4 \cdot in^2$$

The property of the assumed critical section analogous to the polar moment of inertia is

$$J_{cf} := \frac{c_{vf} \cdot b_1^3}{6} + \frac{c_{vf}^3 \cdot b_1}{6} + \frac{c_{vf} \cdot b_1^2 \cdot b_2}{2} = 671158 \cdot in^4$$

Load combinations 1, 2, and 3 are

$$v_{u1} = 130 \text{ psi}, \quad v_{u2} = 203 \text{ psi}, \quad v_{u3} = 166 \text{ psi}$$

$$\text{CheckPunchingShear} := \begin{array}{|l} \text{``OK'' if } \max(v_{u1}, v_{u2}, v_{u3}) \le \phi_v \cdot 4\sqrt{f_c'} \cdot psi \\ \\ \text{``Not good'' otherwise} \end{array}$$

CheckPunchingShear = "OK"

REFERENCE

1. H. Jawahery Zadeh and A. Nanni. Reliability analysis of concrete beams internally reinforced with FRP bars. *ACI Structural Journal* 110 (6): 1023–1032 (2013).

Index

Printed in the United States
by Baker & Taylor Publisher Services